T0337857

Essential Methods of Instrumental Analysis

Essential Methods of Instrumental Analysis

Frank M. Dunnivant
Jake W. Ginsbach

Published by John Wiley & Sons, Inc., Hoboken, New Jersey.
Published simultaneously in Canada.

Library of Congress Cataloging-in-Publication Data

Names: Dunnivant, Frank M., author. | Ginsbach, Jake W., author.
Title: Essential methods of instrumental analysis / Frank M. Dunnivant, Jake W. Ginsbach.
Description: Hoboken, New Jersey : Wiley, [2024] | Includes index.
Identifiers: LCCN 2024023825 (print) | LCCN 2024023826 (ebook) | ISBN 9781394226719 (hardback) | ISBN 9781394226726 (adobe pdf) | ISBN 9781394226733 (epub)
Subjects: LCSH: Instrumental analysis.
Classification: LCC QD79.I5 D86 2024 (print) | LCC QD79.I5 (ebook) | DDC 542/.3–dc23/eng/20240617
LC record available at https://lccn.loc.gov/2024023825
LC ebook record available at https://lccn.loc.gov/2024023826

Cover Design: Wiley
Cover Image: Courtesy of Frank M. Dunnivant and Jake W. Ginsbach

Set in 9.5/12.5pt STIXTwoText by Straive, Chennai, India

SKY10080622_072724

Contents

About the Authors

Frank M. Dunnivant is Professor of Chemistry at Whitman College. He has worked in the environmental/analytical field for 44 years as a consulting environmental chemist, as a postdoc researcher with Oak Ridge National Laboratory (DOE) and the University of Zurich-Swiss Federal Institute for Water and Wastewater Pollution, as a senior scientist at the Idaho National Engineering Laboratory (DOE), and has taught chemistry courses for 28 years. This is his fifth book on environmental and chemistry issues.

Jake W. Ginsbach was a chemistry major at Whitman College and an undergraduate student in Frank M. Dunnivant's lab. At Whitman, he contributed to the *Essential Methods of Instrumental Analysis* and researched the kinetics of chlorinated organic pollutants. He then completed his PhD at Stanford University studying the activation of dioxygen at copper active sites in metalloenzymes. His graduate work investigating the molecular basis of human disease led him to a career in the private sector, working at a number of healthcare companies including ZS (a management consulting company), Vim (a digital health startup), and Kaiser Permanente (an integrated provider of health care and insurance coverage).

Periodic Table of the Elements

Carbon 6 C 12.01	Element Element # Symbol Atomic Mass

	IA	IIA	IIIB	IVB	VB	VIB	VIIB	VIII	VIII	VIIII	IB	IIB	IIIA	IVA	VA	VIA	VIIA	VIIIA
	Hydrogen 1 H 1.01																	Helium 2 He 4.00
2	Lithium 3 Li 6.94	Beryllium 4 Be 9.01											Boron 5 B 10.81	Carbon 6 C 12.01	Nitrogen 7 N 14.01	Oxygen 8 O 16.00	Fluorine 9 F 19.00	Neon 10 Ne 20.18
3	Sodium 11 Na 22.99	Magnesium 12 Mg 24.31											Aluminum 13 Al 26.98	Silicon 14 Si 28.09	Phosphorous 15 P 30.97	Sulfur 16 S 32.07	Chlorine 17 Cl 35.45	Argon 18 Ar 39.95
4	Potassium 19 K 39.10	Calcium 20 Ca 40.08	Scandium 21 Sc 44.96	Titanium 22 Ti 47.88	Vanadium 23 V 50.94	Chromium 24 Cr 52.00	Manganese 25 Mn 54.94	Iron 26 Fe 55.85	Cobalt 27 Co 58.93	Nickel 28 Ni 58.69	Copper 29 Cu 63.55	Zinc 30 Zn 65.39	Gallium 31 Ga 69.72	Germanium 32 Ge 72.61	Arsenic 33 As 74.92	Selenium 34 Se 78.96	Bromine 35 Br 79.90	Krypton 36 Kr 83.80
5	Rubidium 37 Rb 85.47	Strontium 38 Sr 87.62	Yttrium 39 Y 88.91	Zirconium 40 Zr 91.22	Niobium 41 Nb 92.91	Molybdenum 42 Mo 95.94	Technetium 43 Tc 99	Ruthenium 44 Ru 101.07	Rhodium 45 Rh 102.91	Palladium 46 Pd 106.42	Silver 47 Ag 107.87	Cadmium 48 Cd 112.41	Indium 49 In 114.82	Tin 50 Sn 118.71	Antimony 51 Sb 121.75	Tellurium 52 Te 127.60	Iodine 53 I 126.90	Xenon 54 Xe 131.29
6	Cesium 55 Cs 132.91	Barium 56 Ba 137.33	Lanthanum 57 La 138.91	Hafnium 72 Hf 178.49	Tantalum 73 Ta 180.95	Tungsten 74 W 183.85	Rhenium 75 Re 186.21	Osmium 76 Os 190.2	Iridium 77 Ir 192.22	Platinum 78 Pt 195.08	Gold 79 Au 196.97	Mercury 80 Hg 200.59	Thallium 81 Tl 204.38	Lead 82 Pb 207.2	Bismuth 83 Bi 208.98	Polonium 84 Po 209	Astatine 85 At 210	Radon 86 Rn 222
7	Francium 87 Fr 223	Radium 88 Ra 226	Actinium 89 Ac 227	Rutherfordium 104 Rf 261	Dubnium 105 Db 262	Seaborgium 106 Sg 263	Bohrium 107 Bh 262	Hassium 108 Hs 265	Meitnerium 109 Mt 266	Ununnilium 110 Uun 269	Unununium 111 Uuu 272	Ununbium 112 Uub 277						

Lanthanide Series	Cerium 58 Ce 140.12	Praseodymium 59 Pr 140.91	Neodymium 60 Nd 144.24	Promethium 61 Pm 147	Samarium 62 Sm 150.36	Europium 63 Eu 151.97	Gadolinium 64 Gd 157.25	Terbium 65 Tb 158.93	Dysprosium 66 Dy 162.50	Holmium 67 Ho 164.93	Erbium 68 Er 167.26	Thulium 69 Tm 168.93	Ytterbium 70 Yb 173.04	Lutetium 71 Lu 174.97
Actinide Series	Thorium 90 Th 232.04	Protactinium 91 Pa 231	Uranium 92 U 238.03	Neptunium 93 Np 237	Plutonium 94 Pu 244	Americium 95 Am 243	Curium 96 Cm 247	Berkelium 97 Bk 247	Californium 98 Cf 251	Einsteinium 99 Es 252	Fermium 100 Fm 257	Mendelevium 101 Md 258	Nobelium 102 No 259	Lawrencium 103 Lr 260

Preface

There are several excellent textbooks on instrumental analysis dedicated to a broad, yet terse, multi-semester coverage of instruments. Our text offers a unique focus on common instruments used in the undergraduate laboratory for analytical chemistry, where analytes are only present in the parts-per-trillion to parts-per-million concentration ranges. The content of this book evolved since 2008 based on experiments developed in a undergraduate instrumental analysis course and includes numerous insightful figures, graphs, and animations to explain complex chemical and physical concepts and covers instrumentation for the analysis of inorganic and organic compounds. We have also incorporated a variety of applications involving real-world samples, analytical techniques, case studies at the end of most chapters, and other practical information for future analytical chemists. This text has been reviewed by students and faculty and revised over the years in a one-semester, junior–senior level class. The extensive collection of group-based and multi-laboratory projects for real world samples and includes student results for additional problem solving make this text a particularly useful resource. We hope you find this book student-accessible and a helpful companion to your coursework.

Frank M. Dunnivant
Jake W. Ginsbach
July 2024

About the Companion Website

This book is accompanied by the companion website:

www.wiley.com/go/essentialmethodsofinstrumentalanalysis1e

The website includes supplementary materials

- Videos
- PPTs

1

Proper Laboratory Protocols for Analytical Instrumentation

The proper operation of analytical instruments requires hands-on experience, even though studying the various components of instrument and understanding the theory is important. Experiential learning is a must for most students, especially when dealing with analytical instruments. This chapter is divided into two sections: Section 1.1 contains preliminary information that is not included in most textbooks but is necessary to truly understand the difficulty in measuring analytes in the parts-per-billion (ppb) and parts-per-trillion (ppt) concentrations; Section 1.2 deals with trace analysis and concerns the proper setup and testing of instruments to ensure that accurate and reproducible numbers are obtained and a brief discussion on quality assurance/quality control is provided.

1.1 Laboratory Preliminaries

The analysis of metals in the ppb and ppt concentrations is a difficult and tedious process that requires optimum laboratory conditions and analytical skills. Many laboratories even have their sample handling and preparation sections of the lab physically separated from their instrument rooms in order to avoid contaminating instruments and cross contamination of samples. Error associated with laboratory work can be divided into three general categories: sampling, sample preparation, and analysis. Entire books and statistics courses have been dedicated to proper sampling design and implementation in order to avoid errors (referred to as bias) in the collection of samples, such as in the proper sampling of lake water. Laboratory error has been significantly reduced by standardizing sample extraction procedures for a particular sample, element, compound, industry, and even a specific technician.

This beginning section of this chapter is dedicated to a few of the more important factors associated with the instrumental analysis of samples. Before actual experimental procedures and results are presented, it is necessary to comment on sample preparation, special chemicals, and the calibration of an instrument. These topics are not normally covered in textbooks but are important for first-time instrument users. The types of samples that are commonly analyzed by analytical instruments will be discussed, including proper procedures that are critical to avoid common laboratory errors.

1.1.1 Types of Samples and Sample Preparation

There are various sample types that need to be analyzed for their analyte concentration. For example, a geochemist may be interested in determining the concentration of a particular

Essential Methods of Instrumental Analysis, First Edition. Frank M. Dunnivant and Jake W. Ginsbach.

Companion Website: www.wiley.com/go/essentialmethodsofinstrumentalanalysis1e

metal, such as lead (Pb), in a terrestrial or oceanic rock, or a fisheries biologist may be interested in determining the concentration of mercury in a particular fish species. In addition, a wide range of instrumentation is available for the analysis of organic compounds such as hydrocarbons and pesticides. However, the various instruments that could perform this task at a trace level (ppm or lower) are unable to accommodate the introduction of a rock, fish, or soil matrix. As a result, sample preparation, specifically sample digestion, is needed before an instrument can be utilized. The chemical process to transform the solid sample into an aqueous sample must be undertaken carefully with the proper reagents and glassware. After the concentration of the digested sample is determined, simple calculations can be performed to accurately determine the concentrations of Pb in a rock, mercury in a fish, or pesticide in a soil or sediment sample. If sample preparation is not performed properly, no degree or expense of instrumental components or laboratory technique can correct for a sampling or digestion error. Despite the obviousness of some of these techniques, this component of analytical chemistry is still a considerable portion of error associated with the final analyte concentration.

1.1.1.1 Samples

Before a procedure of sample preparation can be created (commonly referred to as a standard operating procedure, or SOP), the analyte(s) of interest must be identified as well as the matrix. The most common type of sample matrix for metals and organic compounds is aqueous. The first step when analyzing water samples is to determine whether the total or dissolved metal concentration is of interest. All natural water samples contain particles; some are extremely small, while others are visible to the naked eye. Frequently, particle size is used to differentiate between dissolved and suspended solids. For example, in environmental chemistry and other disciplines, most scientists accept that constituents of a sample are *dissolved* if they pass through a 0.45-μm glass-fiber filter; anything not passing through the filter is considered *particulate*. This will be the distinction between the dissolved and particulate phases used in this text. If the total analyte concentration needs to be determined, the water sample may require strong acidification and heating for metals or solvent extraction for organic compounds prior to analysis depending on the concentration in the solid phase. If the dissolved solid concentration is desired, the water is first filtered and then acidified or extracted, usually with high-purity nitric acid (HNO_3) (1–3% acid) in a plastic container or with high-purity (chromatography-grade) organic solvents. The purpose of the extraction is to permanently dissolve any analyte that is adsorbed to colloidal (very small) particles or the container walls. Some procedures require that samples be stored at 4 °C for less than one month prior to analysis. Each industry, governmental agency, or company will have an SOP that must be rigidly followed.

Other sample matrices, such as gaseous, solid, and biological tissues, are usually chemically converted to aqueous samples through digestions or extractions. Metals and organic analytes contained in the atmosphere from geological processes, dust from wind, and high-temperature industrial processes are frequently measured to model the fate and transport of pollutants. Analytes may be in the atomic gaseous state, such as mercury vapor, or associated with suspended particles, such as cadmium (Cd) or Pb ions adsorbed to clay particles. One common sampling method is to utilize a vacuum to pull large volumes of air (tens to hundreds of cubic meters) through a filter. The filter is then removed and the metals or organics are extracted into an aqueous or organic solution that is then analyzed. This type of sampling and analysis is commonly performed in urban areas to monitor atmospheric metal emissions and organic pollution. Standardized methods for these types of measurements have been developed by the US Environmental Protection Agency (EPA) and other governmental agencies.

The metal concentration in a solid sample or geological material is also frequently analyzed by flame atomic absorption spectrometry (FAAS) and inductively coupled plasma-mass spectrometry (ICP-MS). Organic analytes are commonly measured by gas chromatography (GC) with flame ionization detection (FID) or MS detection, or high-pressure liquid chromatography-mass spectrometry (HPLC-MS). Sometimes analytes are analyzed directly by special attachments to FAAS and ICP units (as described in Chapters 4 and 5) but are more commonly digested or extracted with acids followed by analysis. Similarly, organic compounds in soil, sediment, and biological tissue are extracted with organic solvents and analyzed by GC or HPLC. The resulting solution is then analyzed as described in Section 1.1.3. However, the specific type of digestion/extraction depends on the ultimate goal of the analysis. If adsorbed metal concentrations are of interest, soil/sediment samples are simply placed in acid (usually 1–5% HNO_3), heated for a specified time, diluted, filtered, and then the filtrate is analyzed as if it were a liquid sample. If the total metal concentration is desired, as in many geological applications, the sample is completely digested with a combination of hydrofluoric (HF), HNO_3, and $HClO_4$; again, the samples are then diluted, filtered, and analyzed as an aqueous sample. The extraction of organic analytes can be far more complicated and may involve liquid–liquid extraction for water samples, digestion of biological tissues followed by extraction, or extensive Soxhlet extraction for soil and sediment samples. Procedures are available from the US EPA and the US Geological Survey (USGS).

Biological tissues present considerable problems as do soil and sediment samples. In order to create an aqueous sample, the tissue sample is first digested in acid to oxidize all biological matter and "free" up any present metal or organic analyte. This can be accomplished with a combination of sulfuric acid (H_2SO_4) and HNO_3 (1–5% each) and 30% hydrogen peroxide (H_2O_2), after digesting for 24 hours, and when necessary, heat. After the tissue has been dissolved, the sample is diluted, filtered, and analyzed as an aqueous sample. Each of the digestion procedures described above dilutes the original metal concentration in the sample. Thus, a careful accounting of all dilutions of a sample is required. After a concentration is obtained from an instrument, it must be adjusted for the dilution(s) to determine the analyte concentration in the original sample. Calculations concerning these dilutions will be discussed in the final paragraphs of this section.

1.1.1.2 Chemical Reagents

A variety of chemicals are used in the trace analysis. Mostly, these include reagents used in the digestion of solid samples or tissues to release metals into an acidic aqueous solution. The majority of these chemicals fall into two categories: oxidants and acids. Common digestion reagents include oxidants such as H_2O_2, potassium permanganate ($KMnO_4$), and potassium perchlorate ($KClO_4$) for oxidizing organic matter or tissue. Common acids include hydrochloric acid (HCl), HF, H_2SO_4, HNO_3, and perchloric acids ($HClO_4$). Organic solvents used in extraction of organic analytes must be extremely pure. For example of 20-L barrel of standard-grade acetone may cost $10–20 (US), but an analytical grade 4-L bottle of acetone could be as high as $50–100 (US). As manufacturers continue to produce commercially available instruments with lower and lower (better) detection limits, the purity of the reagents that are utilized to prepare samples must also improve in order to take advantage of these instrumental advances. Many of the reagents commonly used to prepare samples are commercially available in ultrapure grades. However, purchasing these ultrapure reagents is often expensive because of the relatively large volume of the acids or the oxidants needed to oxidize and remove the matrix prior to analysis. In addition, some ultrapure acids can still contain some metals at the ppt-to-ppb concentrations, especially for analytes such as Pb and mercury. Some laboratories that analyze large sample volumes choose to prepare their own acids by redistilling even the ultrapure acids. While this may seem like a major inconvenience, the cost savings are obvious when a 2-L bottle of ultratrace metal HNO_3 (double distilled and packaged under clean room conditions) currently lists for several hundreds of dollars while a 2.5-L bottle of ACS-grade HNO_3 costs approximately only tens of dollars. A Teflon® sublimation distillation apparatus used to produce purer versions of these acids in-house only costs approximately $5000–10 000 (US).

While the purity of the chemical reagents is important, the purity of dilution water is also important. Due to the presence of trace metals and salts, the use of tap water is out of the question. House resin-deionized (DI) water may be sufficiently pure for some flame-based methods but may be inadequate for emission and mass spectrometry techniques. For these techniques, house DI is usually passed through a commercially available secondary resin filter system (such as the Nanopure® water filtration system from Millipore and Bronstead manufacturers) that removes metal concentrations down to the parts-per-trillion concentration or less, and organic concentrations down to the ppt levels. If concentrations in samples are to be measured below this level, cleaner water must be obtained. Commonly available resin-based systems cost approximately $5000–10 000 (US).

Despite the extensive preparations and costs associated with acquiring pure reagents, there is usually some detectable metal concentration in the reagents. In order to correct for the presence of trace metals, their concentration in a reagent blank must be determined. A reagent blank is a solution that contains every reagent in the digestion and analysis procedure but lacks the actual sample. This blank allows for any metal concentrations present in the reagents (including dilution water) to be accounted for and subtracted from the metal concentrations found in the sample. This process ensures that the final concentration is representative of the analyte's true presence in the original sample and not from contamination in the added reagents.

1.1.1.3 Glassware

FAAS, flame atomic emission spectroscopy (FAES), and ICP instruments yield concentration data with 3–4 significant figures, while chromatography instruments yield fewer significant figures. Thus, quantitative glassware with a similar number of significant figures must be used in order for the instrumentally significant figures to accurately represent the concentration of the sample. The most accurate dilutions are achieved with Class A pipettes and volumetric flasks. These pipettes and flasks are almost always made out of glass which can present a problem for trace metal analysis in techniques such as ICP-MS. Trace levels of some metals in unacidified water commonly "stick" (adsorb) to glass and some glass containers actually contain metals in their matrix and can release measurable concentrations when acidic solutions are added to them. In order to avoid this source of contamination, plastic or Teflon materials may be required for delivery, digestion, and storage of these samples. The use of manual pipetting systems with plastic tips, however, results in slightly less precise data (approximately three significant figures), but these devices are rapidly being adopted in the laboratory. Similar Class A glassware used in the analysis of organic compounds can be adequately cleaned with high-purity acetone or methanol.

It should be noted that for metal analysis, contact with metal surfaces, such as Al weigh boats should be avoided. Plastic and glass are preferred. For most organic analyses, all contact with plastics should be avoided.

1.1.2 Calibration Techniques

Once a sample has been prepared properly, it is necessary to calibrate the instrument to determine the concentration in the sample. Uncalibrated scientific instruments are random number generators. These number generators are transformed into valuable analytical tools with the use of calibration techniques. There are various ways to determine the concentration of a sample through the use of external standards, standard addition, or semiquantitative methods.

1.1.2.1 External Standards

The most common technique to determine the concentration of a sample is through the use of external standards. For example, before a sample of water is analyzed for magnesium (Mg), a series of known Mg concentrations in acidified water ranging from 0.010 to 50.0 ppm would be analyzed. The respective absorbance or emission values for each standard are tabulated and a plot of concentration versus detector signal is made. Next, a plot of detector response (in this case, absorbance or counts per second) is made against the respective metal concentration on the *y*- and *x*-axes, respectively. Such a plot is shown in Figure 1.1 where a FAAS was used to measure the concentration of Mg in various external standards.

Note the linearity of the data plot; most calibration "curves" are lines. This linear detector response is preferred since it allows for easier statistical calculations (discussed in Chapter 2). Once an instrument is calibrated with an external standard, it can be used to estimate the concentration of an analyte in an unknown sample. This is accomplished by performing a linear least-squares (LLS) regression on the dataset (see Chapter 2) and obtaining an equation for the calibration line. For Figure 1.1, this equation is $y = 0.012x + 0.000$ where 0.012 is the slope of the line and 0.000 is the *y*-intercept. Next, the equation is rearranged to solve for x (the reader should do this now). When the actual sample is analyzed on the instrument, the readout will give the absorbance of the sample that corresponds to the value of y in the linear regression. For example, if an unknown sample yields an absorbance reading of 0.062, the Mg concentration would be estimated at 5.03 ppm with the rearranged linear regression equation (the reader should perform this calculation now). Another important statistical calculation is determined by r^2 (Figure 1.1) and is a measure of the "goodness of fit" between the linear model and the experimental data. The closer to 1.000 the r^2 value, the more the linear model for the data approaches a perfect fit. Most modern instruments have the built-in ability to perform a LLS and r^2 calculation. But an even more important statistical parameter, s_c, the standard deviation of replicate samples will be discussed in Chapter 2.

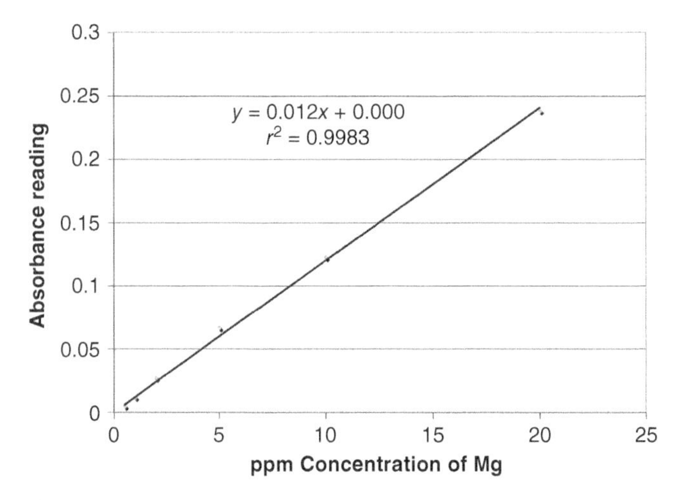

Figure 1.1 Instrument response (absorbance) versus Mg concentration (ppm) by FAAS. (Source: Dunnivant and Ginsbach).

Some calibration lines do not automatically intersect at the origin ($x = 0$, $y = 0$) because of the presence of small concentrations of analyte in the blank sample or because of "unbalanced" electronics (the detector not reading zero absorbance for the blank due to noise). This is especially true as the analyst approaches the detection limit of the instrument. In the past, the blank absorbance reading was simply manually subtracted from other readings prior to plotting the data. Today, modern instruments that are connected to computers perform many of these tasks automatically. Frequently, calibration lines, subtraction of blanks, and estimating unknown sample concentrations and even dilutions can all be calculated or accounted for automatically by the computer. This eliminates the need to re-enter data into another computer and decreases typographical and transposing errors. However, these automatic procedures should be monitored to ensure that accurate data are being generated.

1.1.2.2 Standard Addition

Another form of calibration, standard addition, can be used to account for matrix effects (such as surface tension or viscosity) or other problems (such as chemical interferences). For example, in FAAS and FAES analysis, the sample is drawn into the nebulizer by a constant vacuum source; and if the viscosity of the solutions being analyzed is not all the same, then different flow rates of sample or standard solution will reach the flame and therefore different masses of metal will be measured for each. (This is not a problem in ICP since a peristaltic pump is used to place sample in the nebulizer.) or in chromatography instruments. A sample containing significant amounts of sugar, commonly found in food products, will have a higher viscosity and will move more slowly through the inlet tube in FAAS and FAES systems than standards that are made up of relatively pure water; hence, the resultant signal obtained from this external standard-based analysis will be inaccurately low for the metal of interest. If the standard addition method described below is used, more accurate metal concentrations for the samples will be obtained.

In standard addition analysis, several equal volume aliquots of a sample are added to a volumetric flask (i.e. 15 mL of sample to each 25 mL flask). Increasing reference concentrations of the analyte of interest are added to each flask beginning at zero and increasing to the end of the linear range of the instrument (i.e. one flask will have 0.00 ppm, the next will have 1.00 ppm, the next will have 5.00 ppm, etc.). Acid is added to each flask in the analysis of metals to reach a specific and consistent percent (usually 1.00%). Finally, each flask is filled to the 25-mL volumetric line with high-purity water. Thus, each flask has equal volumes of sample, and a linear increase of known analyte is added starting at 0.00 ppm in the "blank" flask and ending at the highest analyte concentration. After the samples are analyzed, the concentration is plotted as a function of analyte added. An LLS regression ($y = mx + b$) is performed and the linear equation can be rearranged to solve for x, the sample concentration, as a function of y, the detector response. The concentration of the blank sample (the diluted sample containing no reference standard) is determined by computing the x intercept of the line (where $y = 0.00$).

An example of a standard addition analysis of a complex matrix (Ca in beer) is shown in Figure 1.2.

1.1.3 Figures of Merit

The individual calibration lines that instruments generate are dependent upon numerous variables such as laboratory technique, instrument components, and operating parameters. In order to compare different instruments to one another, three *figures of merit* have been developed to quantitatively compare different analytical techniques. These are sensitivity, detection limit, and signal to noise ratio (S/N).

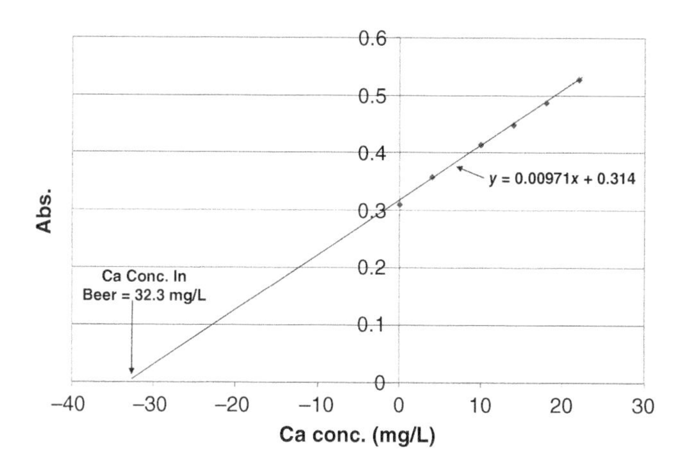

Figure 1.2 Standard addition analysis of Ca in beer. (Source: Dunnivant and Ginsbach).

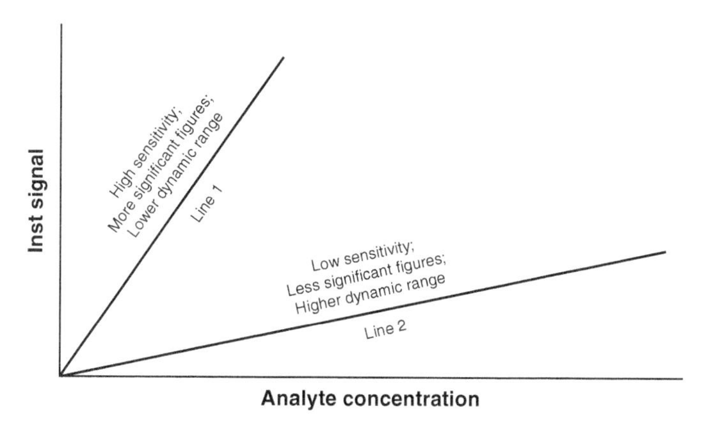

Figure 1.3 Calibration lines with differing sensitivity. (Source: Dunnivant and Ginsbach).

Sensitivity refers to the slope of the calibration line. Figure 1.3 shows two calibration lines, one with a steep slope and another with a shallower slope. The determination of which level of sensitivity is best depends on the situation. Calibration "line 1" is referred to as being more sensitive since it will allow the analyst to distinguish between smaller differences in concentration (i.e. allows the determination of 30.1 from 30.2 as opposed to 30 versus 40). This type of calibration line would be of interest when high degrees of accuracy are needed. If the screening of samples for gross differences in concentration is required, "line 1" would be of little use since it has only a limited dynamic range (small range in analyte concentrations) and its use could require the dilution of samples outside the calibration range. In this case, calibration "line 2" would be useful since it covers a larger dynamic range, but again, yields less accuracy (fewer significant figures).

Another related parameter, calibration sensitivity, is mathematically defined as

$$m = \frac{\Delta \text{Signal}}{\Delta \text{Concentration}} \tag{1.1}$$

where m is the slope of the calibration line from the linear regression. Another form of sensitivity is analytical sensitivity, defined as

$$\frac{m}{\text{std dev}} \tag{1.2}$$

where std dev is the standard deviation of the slope estimate. A range of sensitivities for the identical analyte are usually possible on the same instrument. High sensitivity is accomplished by "tuning" an instrument to its maximum sensitivity before testing the external standards. A lower sensitivity can be accomplished by "detuning" the instrument to obtain a lower slope and a broader range of useful analyte concentrations. The term sensitivity is commonly, but incorrectly, interchanged with detection limit.

Another figure of merit is the detection limit; this is one of the most important ways to compare two analytical techniques and brands of instruments, and is a major competitive selling/advertising metric used by instrument manufacturers. The detection limit determines the limitations of the instrument or technique and is commonly and incorrectly referred to as "zero" concentration of the analyte. For example, it is often stated that "no" pollutant was found in a sample. What does "no" mean? Absolute zero? Probably not. Zero concentration does not typically mean that no analyte exists since there are almost always a few atoms or molecules of any substance in any sample. Today, with modern instrumentation, there is no "zero concentration" for more environmental analytes. For example, the EPA-recommended limit (maximum contaminant level or MCL) for Pb in drinking water is 15 ppb, but the detection limit by ICP-MS is 0.1 ppb. No Pb concentration is healthy for consumption, but if we set the MCL to "zero," virtually no water would be safe to drink. Additionally, most people think the air they breathe is "free" of hazardous chemicals; few know it, but the air that they just took into their lungs contains measurable concentrations of PCBs and Hg. However, these concentrations are only measurable with rigorous analytical procedures. Instead, it would be more accurate to report the detection limit of the technique and indicate that the sample was below that limit (for example, the sample has less than 26.1 ppt of pollutant). So, how does an analyst quantitatively determine the detection limit? To illustrate this problem, Figure 1.4 shows three signal responses (these can be measures of absorbance or counts per second). Which of these signals can we confidently say is attributed to the analyte of interest instead of random background noise?

Most of everyone would agree that the right-hand sample signal would be attributed to the analyte. Some would also agree with the middle figure, but fewer analysts, if any, would call the left-hand spike the analyte. A quantitative way of distinguishing noise from a sample signal has been developed using the results from the linear regression process described earlier. In order to complete this process, a range of calibration standards is analyzed on an instrument with blank samples being run before, during, and after the standards. The responses for 5–10 blanks are averaged and a standard deviation is calculated. Next, a linear regression is completed for the results of the external calibration data. This data is then utilized to determine the detection limit by using the following equations.

There are two ways of determining the detection limit. First, from a purely statistical viewpoint, the reported detection limit should be the lowest concentration of the reference standard. But some analytical chemists find this definition too restrictive.

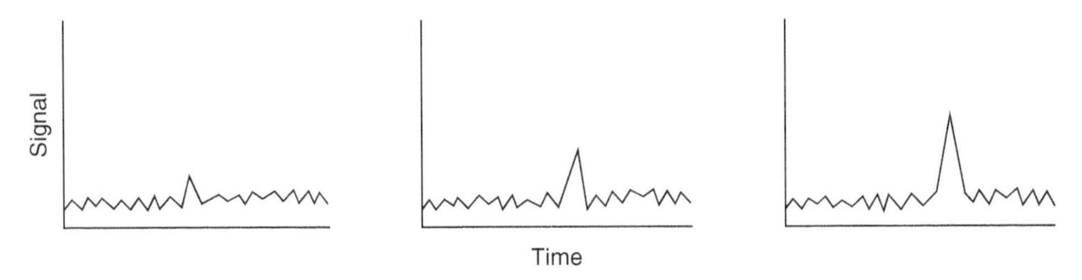

Figure 1.4 Illustration of signal to noise and instrument responses for a sample. (Source: Dunnivant and Ginbach).

From a more analytical standpoint, the minimum signal that can be distinguished from the background is determined by defining a constant k, usually equal to three (3.00..., a defined number), in the equation

$$S_m = S_{blank} + k(s.\,d._{blank}) \tag{1.3}$$

where S_m is the minimum signal discernable from the noise, S_{blank} is the average blank signal, and $s.\,d._{blank}$ is the standard deviation of the blank measurements. The k value of 3 is common across most disciplines and basically means that a signal can be attributed to the analyte if it has a value that is larger than the noise (blank concentration) plus three times the standard deviation.

Next, recall the calibration line equation

$$S = mC + b \tag{1.4}$$

Redefining S to S_m and b to S_{blank} (to be consistent with most statistics textbooks) yields

$$S_m = S_{blank} + k(s.\,d._{blank}) = mC + C_{blank}$$

And upon rearranging yields $C = \frac{k(s.\,d._{blank})}{m}$.

Recalling that $S_m = S_{blank} + k(s.\,d._{blank})$ and upon rearranging yields: $S_m - S_{blank} = k(s.\,d._{blank})$.

Substitution into the above equation yields

$$C = \frac{S_m - S_{blank}}{m}$$

where C is the minimum detection limit.

This method gives a quantitative and consistent way of determining the detection limit once a value of k is determined (again, usually 3.00).

The final figure of merit quantifies the noise in a system by calculating the S/N ratio. Sources of noise are commonly divided into environmental, chemical, and instrumental sources that were discussed in the detector section of the FAAS chapter (Chapter 4) and are relevant to almost all detector systems.

The S/N is determined by:

$$\frac{S}{N} = \sqrt{n}\frac{S_x}{N_x} \tag{1.5}$$

where S_x and N_x are the signal and noise readings for a specific setting and n is the number of replicate measurements. One of the most common noise reduction techniques is to take as many readings as are reasonably possible. The more replicate readings a procedure utilizes, the greater the decrease of S/N by the square root of the number of measurements. By taking two measurements, one can increase the S/N by a factor of 1.4; or by taking four measurements, the analyst can cut the noise in half.

For the topics covered in this book, FAAS and ICP-AES instruments generally allow for multiple measurements or for an average measurement, for example, over 5–10 seconds, to be taken. For ICP-MS, again, multiple measurements are usually taken, usually 3–5 per mass/charge value in a few milliseconds. And, of course, the analyst can analyze a sample multiple times given the common presence of automatic samplers in the modern laboratory. For chromatography techniques, multiple sample runs require considerable time and are rarely conducted.

These statistical calculations, since they are dependent upon the linear calibration curve, can only be accurately applied over the range of measured external standards. This is because of the fact that detectors do not always give linear responses at relatively low or relatively high concentrations. As a result, all sample signals must be within the range of signals contained in the calibration line. This is referred to as "bracketing." If sample signals (and therefore concentrations) are too high, the samples are diluted and reanalyzed. If the samples are too low, they are concentrated by a variety of techniques or they are reported to be below the minimum detection limit. In some cases, the limit of quantification (LOQ) is at the lowest reference standard. The limit of linearity (LOL) is where the calibration line becomes nonlinear, at the upper and lower ends of the line. In a case where the sample concentration is below the lowest standard but above the calculated detection limit, it is acceptable to record that sample as containing "trace concentrations" for the analyte but usually no actual number is quoted.

Technically, a fourth figure of merit, selectivity, exists but this is more obvious. Selectivity refers to the ability of a technique or instrument to distinguish between two different analytes (i.e. calcium (Ca) versus Mg or ^{206}Pb versus ^{207}Pb or 2-2′-dichlorobiphenyl versus 2,4-dichlorobiphenyl). For example, AAS and inductively coupled plasma-optical emission spectrometry (ICP-AES) can distinguish (are selective) between Ca and Mg but not different isotopes of the same element. ICP-MS can be selective for isotopes and elements depending on the mass resolution of the instrument.

1.1.4 Calculating Analyte Concentrations in the Original Sample

Dilutions and digestions are routine practices in the analysis of metal compounds. Once the concentration of the aqueous sample is prepared, an interesting problem arises, "How does the analysis relate the concentration in a diluted sample to the concentration in the original sample?" In general, this is accomplished by carefully performing and recording all steps of the sample preparation. A more detailed explanation of the calculation is offered in the following example problem:

Problem Statement: An analysis takes 0.500 g of wet sediment and prepares it for Cd analysis by FAAS. The purpose of the analysis is to determine the concentration of Cd adsorbed to the sediment on a dry weight basis. The water content in the wet sample was gravimetrically determined to be 24.8%. The wet sample is then measured out and digested in 25 mL of DI water and 3.00 mL of ultrapure HNO_3 on a hot plate at 80 °C for one hour. The sample is then cooled and filtered through a 0.45-μm filter into a 100-mL volumetric flask that is then filled to the mark with DI. Previous analysis indicated that the sample needed to be diluted by a factor of 10 to be within the linear concentration range of the instrument. The diluted sample was analyzed on an instrument and a concentration of 1.58 mg/L was measured for the aqueous sample. What is the concentration of Cd in the original, dry-weight sample?

First, the data are tabulated.

Next, the analyte concentration measured by the instrument is back-calculated to the concentration in the original sample. Note the accounting of each of the measurements in this calculation.

$$\text{Conc. in sample in}\frac{\text{mg}}{\text{kg}} = \frac{\left(\frac{1.58\text{ mg}}{\text{L}}\right)\left(\frac{1\text{ L}}{1000\text{ mL}}\right)\left(\frac{100.0\text{ mL}}{10.00\text{ mL}}\right)(100\text{ mL})}{(0.500\text{ g wet sediment})\left(\frac{1\text{ kg}}{1000\text{ g}}\right)\left(\frac{100-28\text{ g dry}}{100\text{ g wet}}\right)}$$

$$\text{Conc. in sample in}\frac{\text{mg}}{\text{kg}} = 4390$$

1.2 Standard Practices

1.2.1 Ensuring Adequate Rinsing Between Samples and Standards

The cleanliness of all instruments at the beginning of a run or between every sample and reference standard is insured by running a blank solvent sample through the system. For FAAS and ICP-MS, this is relatively easy and does not require much time. For chromatography systems, this can be very time-intensive. With the advent of computer-controlled instruments, this is automatically performed by most modern AAS systems by the automatic sampler. Additional rinsing after a series of samples will usually prevent any buildup of analytes in the inlet of the instruments.

The part-per-trillion detection limits of ICP-MS and in the GC-ECD (electron capture detector) systems create the need for more thorough rinsing between samples and standards. Usually, the rinsing part of the sample sequence is the most time-consuming.

1.2.2 Quality Assurance/Quality Control

Most laboratories have a quality assurance/quality control (QA/QC) program, and usually, one person is in control of all QA/QC matters. QA/QC starts with a laboratory technician being certified on a particular instrument and for a particular analysis. Certification is completed with extensive training, a required analysis time on the instrument, and by passing a test examining known concentrations of analyte in reference check samples (described later). Laboratories and personnel have to be certified to conduct most governmental, consulting, and industry analyses.

QA/QC begins with a sampling plan, which is a document detailing how, when, and where samples will be taken, handling of the samples, preservation of the samples, and storage of the samples. Next, an SOP will be created outlining exactly how any sample will be analyzed. The SOP can contain, at a minimum, what quality of chemicals will be used, detailed procedures of sample digestion, detailed procedures for analyte extraction for undigested and digested samples, instruments to be used in the measurement of analytes, data analysis, statistical analysis, and reporting and long-term storage of the data. Many of these exact details of the SOP are government- and industry-specific.

A rigorous SOP will include field blanks, reagent blanks, a daily calibration curve, duplicate (or more) sample analysis, and analysis of blanks between samples, spiked samples, and reference check samples. Field blanks are samples containing any preservative present in the actual samples that are taken to the field to serve as a blank for the sample container and any conditions that the samples may be exposed to. Reagent blanks are samples that are analyzed on the instrument that contain any chemicals used in preservation, digestion, and extraction of the samples. Calibration of the instruments was discussed earlier in this chapter, but is usually run every day on the instrument of interest. Many sampling plans require more than one sample to be taken at a specific location or time. It may also be required to analyze a particular sample multiple times to determine the reproducibility of a technique. It is always a good analytical practice to run reagent blanks between each sample and standard to ensure that no carryover of analyte occurs in the introduction of sample to an instrument. Autosamplers have made this large increase in the number of samples possible in a timely and economical manner. It is also common practice to take a replicate sample and spike it with a known quantity of analyte. This allows an analyte recovery check to be conducted on the digestion and extraction procedures. Finally, to ensure that your SOP, instrument, and overall analysis are accurate, reference check samples of many sample matrices (i.e. water, sediment, fish, etc.), containing known concentrations of analytes, can be purchased and analyzed.

A typical instrument run will include:

(1) several blank analyses at the beginning of the run sequence to ensure that all parts of the instrument are clean;
(2) a set of calibration standards ranging from the lowest possible detection limit to the end of the dynamic range of the instrument, from low to high concentrations, and with reagent blanks between each standard to ensure that no carryover of analyte occurs;
(3) the samples to be analyzed, including replicates, blanks, field blanks, reagent blanks, and spiked samples; and
(4) in some cases, the calibration standards are analyzed again at the end to ensure that instrument signal drift has not occurred.

Finally, the results from the instrument must be analyzed in a way to relate the instrument response to sample concentration, as discussed earlier in this chapter. Calibration data and all sample results can be stored on the instrument computer or a printed-out hard copy placed in laboratory notebooks. Most results must be archived for at least three years.

The true test of an adequate QA/QC program is documentation in the form of a legally defensible lab notebook (with procedures given in the laboratory experiments chapter at the end of this book). The term "legally defensible" is used since almost all data today in government, industry, and academia can be introduced as evidence in court and there are agency- and discipline-specific rules on documentation. Students start their education in lab notebook documentation in college, but this will continue for the rest of their professional lives as the system, and legal requirements continue to evolve. In academia, we mostly still use hardcopy, paper notebooks. Many industry groups have shifted to electronic notebooks that start with a barcode being attached to each sample entering the laboratory that is used to track the sample through the complete experimental analysis, data analysis, data reporting, and billing. This approach deters data fraud and aids in long-term storage of data, a requirement in some sectors of chemical analysis.

1.2.3 Computer-Controlled Instruments

There have been many notable advances in the history of instrumentation and these will be discussed in the following chapters, instrument by instrument. But over the entire history of instrumentation, the addition of a computer to each instrument has been the most important. Prior to the 1990s, an instrumental methods lab would consist of bored students manually injecting standards into the instrument, and maybe one or two samples before the lab period ended. Today, the student focuses more on sample preparation and programming of the computer and instrument that runs for hours or days collecting the results of their experiment. The computer-controlled system consists of an autosampler, the instrument, and a computer to manage the instrument and collect the data. The computer control is divided into two programs, the method and the sequence.

The method controls the physical parts of the instrument. In FAAS, FAES, and ICP, it controls the gas flows, turns on the flame or plasma, selects the intensity and wavelength(s) to measure, and monitors the instrument for safety. In GC systems, the computer controls the incoming gas pressures, the various gas flows via mass flow regulators, and the injector, oven, and detector temperatures. In LC, the computer controls the high-pressure pumps, the proportioning valve, the UV–vis

Table 1.1 Selected costs of analysis by chemical parameter.

Metal analysis (water and wastewater) (US dollars)	
ICP/FAAS digestion	\$10.00
ICP-OES (per metal)	\$11.00
ICP-MS (per metal)	\$18.00
Pb and Cu	\$35.00
Hg	\$25.00
Heavy metal suite	\$150.00
Cr^{+6}	\$70.00
Organic analysis (water and wastewater (US dollars)	
Aromatics (GC-FID)	\$100.00
Chlorinated HCs (GC-ECD)	\$185.00
Volatiles (GC purge and trap)	\$225.00
Gasoline components	\$65.00

(Source: Basis 2022 data from various laboratories)

wavelengths of interest, and parameters of the mass spectrometer. In capillary electrophoresis (CE), the computer controls the potential across the column and temperature zones.

The sequence program is a data management tool. It controls and correlates the position of the sample in the autosampler and the method file to the data storage file. By combining these two computer assets, an instrument user can program the system in a few minutes, press the start key, and the instrument can operate for hours or days depending on the duration of each sample run.

The net result of a computer-controlled instrument is multifold. First, a student can spend time learning the programming of the instrument instead of waiting to inject the next sample. The cost and time savings of a computer-controlled system are enormous. Today, a technician can essentially program several days' worth of work in an hour or so. To illustrate this, we have compared the cost of a single chemical parameter analysis in the 1970s to today, and the costs are essentially the same. Examples are given in Table 1.1 for specific parameters and the costs have not significantly increased in these decades due to computer-controlled systems and mass throughput (the ability to process large numbers of samples in a shorter time period, while the cost of a chemical technician has doubled or tripled.

1.3 Questions

1. List three types of sample matrices.
2. Why does analytical chemistry not use standard reagent-grade chemicals such as acids and solvents?
3. Which types of weigh boats are best for use with (i) metals and (ii) organic chemicals?
4. Explain how external calibration works.

5 Explain how internal calibration works.

6 What is the most important "figure of merit" in the sale of an instrument?

7 What is meant by the statement "There is no zero concentration anymore"?

8 What was the most important advancement in instrumentation?

9 Describe the difference between sensitivity and selectivity.

10 Describe the results of a linear least-squares analysis.

1.4 Problems

1 Benzene is commonly analyzed using GC or HPLC. The following dataset was obtained by using the external standard calibration technique and an unknown was analyzed five times. The dataset follows:

Standard conc. ppm benzene	Peak area resulting from analysis
100	653
250	1312
600	2848
800	3726
1200	5482
Unknown	1507, 1525, 1501, 1516, 1490

Analyze the data using the linear least-squares technique on your calculator. Prepare a summary (one-page maximum) of your results and discuss (at a minimum): the equation governing the calibration curve, the errors associated with the y-intercept and slope, some measure of how linear the line is, the minimum detection limit that you would report (and why you selected the number that you did), and the estimated value for your unknown.

2 Most quantitative measurements require several steps in a given procedure, including weighing, dilution, and various quantification approaches. Each of these processes has an error associated with it. Suppose that you are analyzing a liver sample for a given toxin X. You weigh out 1.05 g of liver, dry it, extract it, dilute it, and analyze your dilution. These following steps, and the error associated with each step, are summarized in the following data summary:

Operation	Value of operation	Error associated with each operation (as $\pm$ s)
Weight (of wet liver)	1.05 g	0.01 g
Determination of dry weight (g dry liver/g wet liver)	0.40	0.05
Total volume that toxin is extracted into	100 mL	0.05 (mL)

Operation	Value of operation	Error associated with each operation (as ± s)
Extraction efficiency	0.90	0.05
Dilution 1 (1–10)	0.10	0.001
Dilution 2 (1–100)	0.01	0.0001
Volume of solvents (in most dilute solutions) analyzed	1.00 µL	0.05 µL
Error from least squares	5.62 pg	0.08 pg
Analysis and calibration curve (the amount detected in 1.00 µL of injected solvent)		

Calculate the concentration of toxin *X* in your original sample (on a dry liver basis) and the total error associated with the measurement (propagation of error; as from your quantitative analysis course). Report concentrations in µg of toxin per gram of dry liver. Show all calculations for credit.

2

Statistical Analysis

Linear Least-Squares Analysis and Student's *t*-test

2.1 Introduction

Prior to the coupling of computers to instruments, a good understanding of statistical analysis was essential. Older instruments, prior to the 1990s had poor, sometimes completely absent, forms of automatic calibration. In chromatography, prior to the 1970s, it even involved using a strip-chart recorder that captured a chromatographic peak on paper, the paper peak was cut out, then weighed, and a calibration line created correlating peak weight to analyte concentration. Later, primitive electronic integrators were introduced that roughly estimated the area under the chromatographic peaks. The introduction of computerized instrumentation in the 1990s significantly eased the workload of the chemical technician and today allows for direct generation of linear least-squares (LLS) analysis and calculation of analyte concentrations in a sample.

One of the first lessons that you need to learn in instrumental analysis is that few, if any, instruments report direct measurements of concentration or activity without calibration. Even analytical balances need periodic calibration. More complicated instruments need even more involved calibration. Instruments respond to calibration standards in either a linear or exponential manner, and exponential responses can easily be converted to a linear plot by the log or natural log transformation. **The goals** of this first computer exercise are to create an LLS spreadsheet for analyzing calibration data and learn to interpret the results of your spreadsheet. Although most instruments do this automatically, students need to learn spreadsheet skills and understand the statistical equations for LLS analysis. **The goal** of the second computer exercise is to create a spreadsheet for conducting the student's *t*-test for (1) comparing your analysis results to a sample of known concentration and (2) comparing the results of two different groups' results to each other. The student's *t*-test allows you to tell whether the results are acceptable. An excellent description of the origin of the student's *t*-test can be found at https://en.wikipedia.org/wiki/Student\stquotes_t-test.

Essential Methods of Instrumental Analysis, First Edition. Frank M. Dunnivant and Jake W. Ginsbach.

Companion Website: www.wiley.com/go/essentialmethodsofinstrumentalanalysis1e

Today, most calculators and instruments can perform an LLS analysis but the output from these calculators is limited. The spreadsheet you will create in this exercise will give error estimates for every parameter that you estimate. Error estimates are very important in telling "how good" a number is. For example, if your estimate of the slope of a line is 2.34 and the standard deviation is plus or minus 4.23, the estimate is not very good. In addition, one of the *most important* parameters we will estimate with your spreadsheet is the standard deviation for your replicate sample concentrations (s_c, the standard deviation of the unknown sample concentration discussed later). First, we will conduct an LLS analysis for a calibration curve. Next we will use your unknown sample area, millivolts, or peak height to estimate the unknown sample concentration. Finally, we will calculate the standard deviation of your concentration estimate (again, s_c). This s_c parameter that calculators, nor instruments, calculate and incorporates all factors in a propagation of uncertainty. And as we will learn in the final section of this chapter, the largest uncertainty sometimes originates from very expensive instruments.

2.2 Linear Least-squares Analysis

2.2.1 Equipment Needed

- Access to a computer lab or a laptop computer
- A basic knowledge of spreadsheets
- A calculator for checking your work

2.2.2 Programming Hints

First, here are a few hints on using Microsoft Excel®:

- calculations must start with a "="
- the "$" locks a cell address; you can lock rows, columns, or both
- mathematical symbols are as you expect except "^" is used to raise a number to a power or with "e" notation
- text is normally entered as text, but sometimes you may have to start the line with a single quote symbol, ′

2.2.3 Linear Least-squares Equations

The first step in analyzing unknown samples is to have a measurement (mV, peak area, peak height, absorbance, etc.) in reference to the instrument signal (instruments do not directly read concentration). To relate the signal to concentration, we create a standard curve (line).

All of our calibration curves will be some form of linear relationship (line) of the form $y = mx + b$. For example, we can relate signal to concentration with the equation

$$S = mc + S_{bl} \tag{2.1}$$

where S is the signal response (abs, peak area, etc.), m is the slope of the straight line, c is the concentration of the analyte, and S_{bl} is the instrumental signal (abs, etc.) for the blank.

This is the calibration equation for a plot of S on the y-axis and c on the x-axis. The slope is m and the y-intercept is S_{bl}. The detection limit (S_m) will be $S_m = S_{bl} + ks_{bl}$ (where the right side of the equation is the signal of the blank plus three times (k) the standard deviation of the blank).

We will usually collect a set of data correlating S to analyte concentration. Examples of S include

1) light absorbance in spectroscopy,
2) peak height in chromatography, or
3) peak area in chromatography.

Next, plot the dataset in a spreadsheet and develop an LLS equation for the line connecting the data points. We will define the difference between the point on the line and the measured data point as the residual (in the x- and y-directions).

For calculation purposes, we will use the following equations (S_{xx}, S_{yy}, S_{xy} are the sum of squared error or residuals)

$$S_{xx} = \Sigma(x_i - \bar{x})^2 = \Sigma\left(x_i^2\right) - \frac{(\Sigma x_i)^2}{N} \tag{2.2}$$

$$S_{yy} = \Sigma(y_i - \bar{y})^2 = \Sigma\left(y_i^2\right) - \frac{(\Sigma y_i)^2}{N} \tag{2.3}$$

$$S_{xy} = \Sigma(x_i - \bar{x})(y_i - \bar{y}) = \Sigma(x_i y_i) - \frac{\Sigma x_i \Sigma y_i}{N} \tag{2.4}$$

where x_i and y_i are individual observations, N is the number of data pairs, and x-bar and y-bar are the average values of the observations. Six useful quantities can be computed from these.

The slope of the line (m) is

$$m = S_{xy}/S_{xx} \tag{2.5}$$

The y-intercept (b) is

$$b = \bar{y} - m\bar{x} \tag{2.6}$$

The standard deviation s_y of the residuals, which is given by

$$s_y = \sqrt{\frac{S_{yy} - m^2 S_{xx}}{N - 2}} \tag{2.7}$$

The standard deviation of the slope s_m:

$$s_m = \frac{s_y}{\sqrt{S_{xx}}} \tag{2.8}$$

The standard deviation s_b of the intercept:

$$s_b = s_y\sqrt{\frac{\Sigma\left(x_i^2\right)}{N\Sigma\left(x_i^2\right) - (\Sigma x_i)^2}} = s_y\sqrt{\frac{1}{N - \frac{(\Sigma x_i)^2}{\Sigma(x_i^2)}}} \tag{2.9}$$

The standard deviation s_c for analytical results obtained with the calibration curve that includes all weighting, pipetting, and instrument propagation of uncertainty is

$$s_c = \frac{s_y}{m}\sqrt{\frac{1}{L} + \frac{1}{N} + \frac{(\bar{y}_c - \bar{y})^2}{m^2 S_{xx}}} \tag{2.10}$$

where y_c-bar is the mean signal value for the unknown sample, L is the number of times the sample is analyzed, N is the number of standards in your calibration curve, and y-bar is the mean signal value of the y calibration observations (from standards). So, you will have a reported value of plus or minus a value.

Again, it is important to note what s_c refers to—it is the error of your sample concentration resulting from the LLS analysis, and by default including all analytical uncertainties (weighing of reference standards, pipetting, dilutions, and LLS errors).

Most of your calculators have an r or r^2 key, and you probably know that the closer this value is to 1.00, the better. This number comes from this equation:

$$r = \frac{\Sigma x_i y_i}{\sqrt{\Sigma \left(x_i^2\right) \, \Sigma \left(y_i^2\right)}} \tag{2.11}$$

r is called the coefficient of regression.

Table 2.1 is the printout of a spreadsheet using the equations described earlier. Note that only the numbers in bold are entry numbers and all other cells contain equations for calculating the given parameters. This spreadsheet can be used in all of the exercises in this manual for analyzing your instrument calibration data. The data in Table 2.1 were obtained from students measuring magnesium (Mg) on a flame atomic absorption spectrometer.

2.3 Student's *t*-test Equations

After you obtain average value for a sample, you will want to know whether this is in an acceptable range of the *true or known value*, or you may want to compare mean values obtained from two different techniques. We can do the first, comparison of the measured value to the true or known value, with a statistical technique called the student's *t*-test Equation (2.12). To perform this test, we simply rearrange the equation for the confidence limits to

$$\bar{x} - \mu = \pm \frac{t_{s.d.}}{\sqrt{N}} \tag{2.12}$$

where *x*-bar is the mean of your measurements, μ is the known or true value of the sample, t is the value from the t table, *s.d.* is the standard deviation, and N is the number of replicates that you analyzed.

Basically, we are looking at the acceptable difference between the measured value and the true value. The basis for comparison is dependent on a t value, the standard deviation, and the number of observations. The "t" values are taken from tables such as the one contained in your quantitative analysis or on the Internet, and you must pick a confidence interval and the degrees of freedom (this will be $N - 1$ for this test). If the experimental value of $(x - \mu)$ is larger than the value calculated from the right side of the equation earlier, the presence of bias in the method is suggested. If, on the other hand, the value calculated by the right side of the equation is larger, no bias has been demonstrated.

A more useful but difficult procedure can be performed to compare the mean results between two experiments or techniques. This uses the following equation:

$$\bar{x}_1 - \bar{x}_2 = \pm \frac{t s.d._{\text{pooled}}}{\sqrt{\frac{n_1 n_2}{n_1 + n_2}}} \tag{2.13}$$

where $\overline{x_1}$ and $\overline{x_2}$ are the respective standard deviations of each mean, and n_1 and n_2 are the number of observations in each mean.

$$s.d._{\text{pooled}} = \sqrt{\frac{s_1^2(n_1 - 1) + s_2^2(n_2 - 1)}{n_1 + n_2 - 2}} \tag{2.14}$$

In this case, the degrees of freedom in the t-table will be $(N - 2)$ (2 because you are using two s-squared values). As in the procedure above, if the experimental (observed) value of $(\bar{x}_1 - \bar{x}_2)$ is

Table 2.1 Spreadsheet for conducting a linear-least squares regression analysis.

Replicates for Sample Unknowns = 5

Units of Standard = ppm

Number	*x*-Value Conc.	*y*-Value Signal	*xy*	*x*-squared	*y*-squared
1	**0.5**	**0.005**	0.0025	0.25	0.000025
2	**1**	**0.012**	0.012	1	0.000144
3	**2**	**0.027**	0.054	4	0.000729
4	**5**	**0.067**	0.335	25	0.004489
5	**10**	**0.122**	1.22	100	0.014884
6	**20**	**0.238**	4.76	400	0.056644
7			0	0	0
8			0	0	0
9			0	0	0
10			0	0	0
11			0	0	0
Sums	38.5	0.471	6.3835	530.25	0.076915
Means	6.416667	0.0785			

Calc. of Minimum Detection Limit

Blanks Signal	5 replicates $(x\text{-xbar})^2$
−0.001	4E-08
−0.002	6.4E-07
−0.001	4E-08
0.000	1.44E-06
−0.002	6.4E-07

mean of blanks = −0.001

std. dev. of blks = 0.000837

S(m) = mean(S(blk)) + 3 s(blk)

S(m) = 0.00131

c(m) = 0.211 Minimum detection limit

$S_{xx} = 283$	Sum of squared error for the mean of *x*
$S_{yy} = 0.0399$	Sum of squared error for the mean of *y*
$S_{xy} = 3.36$	Sum of squared error for *x* and *y*
$m = 0.01187$	Slope
$b = 0.00234$	y-Intercept
$S_y = 0.00349$	Standard deviation of the residuals
$S_m = 0.000207$	Standard deviation of the slope
$S_b = 0.00195$	Standard deviation of the intercept
$r = 1.000$	
$r^2 = 0.999$	

Unknowns	**Signal** Rep 1	Conc 1	**Signal** Rep 2	Conc 2	**Signal** Rep 3	Conc 3	**Signal** Rep 4	Conc 4	**Signal** Rep 5	Conc 5	Mean abs.	Mean ppm	s_c
Cold water	**0.062**	5.026	**0.063**	5.111	**0.06**	4.858	**0.062**	5.026	**0.06**	4.858	0.0614	4.98	0.180
Hot water	**0.063**	5.111	**0.062**	5.026	**0.059**	4.774	**0.061**	4.942	**0.062**	5.026	0.0614	4.98	0.180
City tap	**0.019**	1.403	**0.020**	1.488	**0.018**	1.319	**0.018**	1.319	**0.018**	1.319	0.0186	1.37	0.198
Pond water	**0.025**	1.909	**0.024**	1.825	**0.023**	1.740	**0.023**	1.740	**0.024**	1.825	0.0238	1.81	0.195
Lab sample	**0.064**	5.195	**0.066**	5.363	**0.064**	5.195	**0.064**	5.195	**0.063**	5.111	0.0642	5.21	0.179

s_c = std. dev. of an unknown

Bold-face numbers are the only data one needs to add to the spreadsheet; all other numbers are fixed or are equations.

Table 2.2 Student's t-test Equations.

Statistics for Replicate Analyses

DATA SET 1

Units for observation =	**Abs**	
Number of replicates =	**5**	
DF	**4**	
alpha	**0.05**	
***t*–Value from table =**	**2.776451**	= TINV(alpha, DF)

Replicate	Observation **ppm Mg**	(x-mean)	(x-mean)^2
1	**5.195**	−0.0168	0.0002822
2	**5.363**	0.1512	0.0228614
3	**5.195**	−0.0168	0.0002822
4	**5.195**	−0.0168	0.0002822
5	**5.111**	−0.1008	0.0101616
6			
7			
8			
9			
10			
Sum	26.059		0.033868

Mean	=	5.21	
s	=	0.0920	Standard deviation of the mean
CV (%)	=	1.77	Coefficient of variation
CL (+ or −)		0.114	
Upper CL		5.33	Upper confidence limits on the mean
Lower CL		5.10	Lower confidence limits on the mean

DATA SET 2

Units for observation =	**Abs**	
Number of replicates =	**5**	
DF	**4**	
alpha	**0.05**	
***t*–Value from table =**	**2.7764**	= TINV(alpha, DF)

Replicate	Observation **ppm Mg**	(x-mean)	(x-mean)^2
1	**9.774**	0.0928	0.008612
2	**9.658**	−0.0232	0.000548
3	**9.542**	−0.1392	0.019387
4	**9.658**	−0.0232	0.000548
5	**9.774**	0.0928	0.008612
6			
7			
8			
9			
10			
Sum	48.406		0.03768

Mean	=	9.6812	
s	=	0.0971	Standard deviation of the mean
CV (%)	=	1.002	Coefficient of variation
CL (+ or −)		0.1205	
upper CL		9.802	Upper confidence limits on the mean
lower CL		9.561	Lower confidence limits on the mean

Say that the true value is = 5 ppm

Exp mean – true value = 0.2118
Calc. difference from Eqn = 0.1142547 Eqn
If (Exp mean – true mean) is greater than (calc. difference from eqn) than the presence of bias is suggested
If (calc. difference from eqn) is greater than (Exp mean – true mean) than no bias has been demonstrated
Is bias present? YES = IF(ABS(+D32) > D33, "YES", "NO")

Say that the true value is = **5 ppm**

Exp mean – true value = 4.6812
Calc. difference from eqn = 0.120507 Eqn
If (Exp mean – true mean) is greater than (calc. difference from eqn) than the presence of bias is suggested
If (calc. difference from eqn) is greater than (Exp mean – true mean) than no bias has been demonstrated
Is bias present? YES = IF (ABS(+D32) > D33, "YES", "NO")

POOLED DATA: COMPARING MEANS

DF	**8**	
alpha	**0.05**	DF = $N(1) + N(2) - 2$
***t*–Value from table =**	**2.306**	= TINV(alpha, DF)
s(pooled) =	0.0945685	
Mean(1) – Mean(2) =	–4.4694	Observed
Eqn $x(1) - x(2)$ = (+ or –)	0.137923	Difference in means from equation $(t \times s)/n$'s

If (Obs mean(1) – mean(2)) is greater than (calc. difference mean from eqn) than the presence of bias is suggested
If (calc. difference mean from eqn) is greater than (Obs mean(1) – mean(2)) than no bias has been demonstrated
Is bias present? YES IF(ABS(+D45)>D46, "YES", "NO")

Again, the bold-face cells are the only data you need to edit; all other cells are fixed numbers or equations.

larger than the value calculated from the right of Equation (2.13), then there is a basis for saying that the two techniques are different. If, on the other hand, the value calculated by the equation is larger, no basis is present for saying that the two techniques are different (i.e. the value from the equation gives your tolerance or level of acceptable error). Also, note that by using the 95% CI (confidence interval; $\alpha = 0.05$), you will be right 95 times out of 100 and wrong 5 times out of 100.

Table 2.2 conducts both of the *t*-tests mentioned earlier and will serve as your template for creating your own spreadsheet. Again numbers in bold are the only numbers that you will change when using this spreadsheet. The other cells contain equations for calculating each parameter estimate.

2.4 Assignments

1) Your task is to create a spreadsheet that looks identical to the ones attached (your professor may choose to give you this in an effort to save time and so you can spend more time developing your analytical technique in the laboratory). During the first laboratory period, or as homework, you will create an LLS analysis sheet. Then you will create a spreadsheet for conducting a student's *t*-test. The cells containing bold numbers are the only numbers that should be entered when you actually use the spreadsheet for calibrating an instrument. All other cells should contain equations that will not be changed (and can be locked to ensure that these cells do not change).
2) Compare your LLS results to those obtained in Chapter 1 calculated from your calculator. Does your calculator obtain the correct results?

2.4.1 What Do You Turn In?

A one-page printout (print to fit on one page) of each spreadsheet.

In constructing your spreadsheets, make sure the cell locations exactly make those in Tables 2.1 and 2.2.

Before you turn in your spreadsheets, change the format of all column data so that they only show 3 or 4 significant figures (whichever is more correct).

Explain your LLS analysis and student's *t*-test results (approximately 1 page each, typed).

Here are some things to include in your write-up. Basically, you should give an intelligent, statistically sound discussion of your data.

Give:

The equation of the line, and the minimum detection limit.

- Was bias indicated in your analysis of the unknown (the 5 ppm sample) and the true value?
- Were the results from the two groups comparable?
- How do the numbers compare to the results from your calculator?
- What shortcomings does your calculator have (if any)?

Further Reading

https://en.wikipedia.org/wiki/Student\stquotes_t-test (accessed September 25, 2023).

3

A Review of Optical Physics

3.1 Introduction

The concentration of metal and organic compound species in a variety of sample matrices has frequently been measured by observing an analyte's interaction with electromagnetic radiation (e.g. light). Since atomic species produce a unique line spectrum, the observation of these transitions can be used to identify a particular metal (spectroscopy) and its concentration (spectrometry). Molecular analytes produce a broader band of absorbed and emitted wavelengths. While these terms are sometimes used interchangeably in common speech, spectroscopy is the *study of the interaction* of radiation with matter and the term spectrometry is applied to *the measurement of concentrations* of a compound (analyte). Likewise, an instrument is a device that measures something (in our case, concentration of an analyte), while a machine is a device that completes work for you (the computer). This distinction has become more blurred since the addition of computers to instruments. Before spectroscopic techniques or instruments are presented in the following chapters, it is important to review a general understanding of the ways electromagnetic radiation interacts with sample atoms and molecules.

An analyte's interaction with electromagnetic radiation is critical in detecting its presence and concentration. It is necessary to manipulate the signal radiation to measure small concentrations of analyte. This is performed by placing optical components, for example, a mirror or prism, in the path of the radiation to alter the property or path of the radiation, then some form of a detector. Detectors will be discussed in the second half of this chapter, given their importance to various types of molecular spectrometry, atomic absorbance spectrometry (AAS), inductively coupled plasma—atomic emission spectrometry (ICP–AES), and later in high-pressure/performance liquid chromatography (HPLC). Once a general understanding of optical functions has been illustrated, the following chapters will illustrate how these components are utilized in series to achieve different purposes.

Before this discussion is initiated, it is necessary to become familiar with abbreviations that are commonly used in analytical laboratories. Given the length of many instrument's names, it has become common practice to abbreviate them. LC and nuclear magnetic resonance (NMR) are likely two of these acronyms that students should be familiar with from organic chemistry. Table 3.1 contains the common acronyms of instruments and techniques that are used to measure analyte concentrations in this and following chapters.

Essential Methods of Instrumental Analysis, First Edition. Frank M. Dunnivant and Jake W. Ginsbach.

Companion Website: www.wiley.com/go/essentialmethodsofinstrumentalanalysis1e

Table 3.1 Common types of spectrometric analyses and their acronyms.

Acronym	Technique/Instrument
AAS	Atomic absorption spectrometry
AES	Atomic emission spectrometry
OES	Optical emission spectrometry
FAAS	Flame atomic absorption spectrometry
FAES	Flame atomic emission spectrometry
ICP	Inductively coupled plasma
ICP–AES or OES	Inductively coupled plasma–atomic emission spectrometry (Optical emission spectroscopy)
ICP–MS	Inductively coupled plasma–mass spectrometry
HPLC	High-pressure liquid chromatography
CE	Capillary electrophoresis

3.2 Interaction of Electromagnetic Radiation with Sample Molecules

Various types of interactions (both absorption and emission spectrometry) are utilized to measure the concentration of analytes. Despite these different techniques, they both rely upon the excitation of atomic or molecular electrons. This is caused by the absorption of energy from a collision or energy from a photon. The absorption of a photon by a chemical species occurs when electromagnetic energy is transferred to the chemical species. In all cases, these transitions are wavelength-specific, meaning that the energy of the photon must very closely match the energy of the electronic transition. For a given molecular species, the energy is the sum of the electronic, vibrational, and rotational energies. However, an atomic species cannot undergo vibrational or rotational excitations since there are no chemical bonds; thus only electronic energy levels are shown in the figures later. After a transition occurs, a number of relaxations can be measured for a wide variety of analytical purposes. For the purposes of this chapter, only electronic transitions are discussed in determining the presence and concentration of a metal analyte, but in HPLC and CE, molecular transitions occur. Figure 3.1 summarizes the possible ways an energy excitation could occur in an atomic (or molecular) species. Recall that the difference in two adjacent energy levels is not constant but is instead a function of the type of transition. For example, with organic analytes, moving up an electronic energy level requires more energy than moving up a single vibration or rotational level. If a photon is not absorbed by a chemical species, it is transmitted through a sample and is left unaltered by the sample media.

The energy of electromagnetic radiation utilized in the analytical technique must match the energy of the transition of interest. It is important to recall the energy/frequency/wavelength associated with different types of radiation described in Figure 3.2. Table 3.2 summarizes the usual energy transitions that result from each type of radiation shown in Figure 3.2. The analytical techniques discussed in this text usually only utilize ultraviolet (UV) or visible radiation since these photons have sufficient energy to excite valence electrons in atomic species.

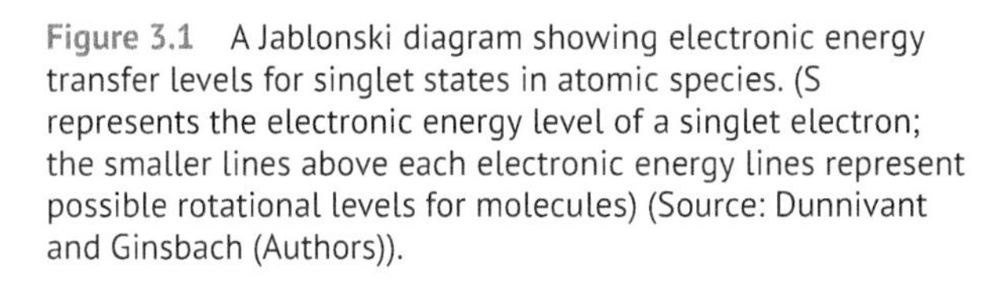

Figure 3.1 A Jablonski diagram showing electronic energy transfer levels for singlet states in atomic species. (S represents the electronic energy level of a singlet electron; the smaller lines above each electronic energy lines represent possible rotational levels for molecules) (Source: Dunnivant and Ginsbach (Authors)).

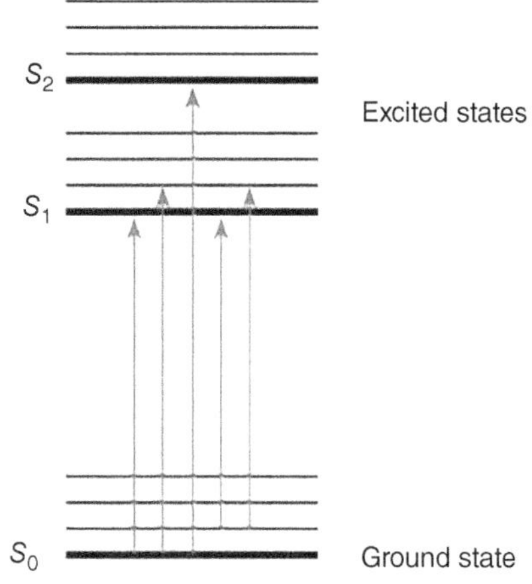

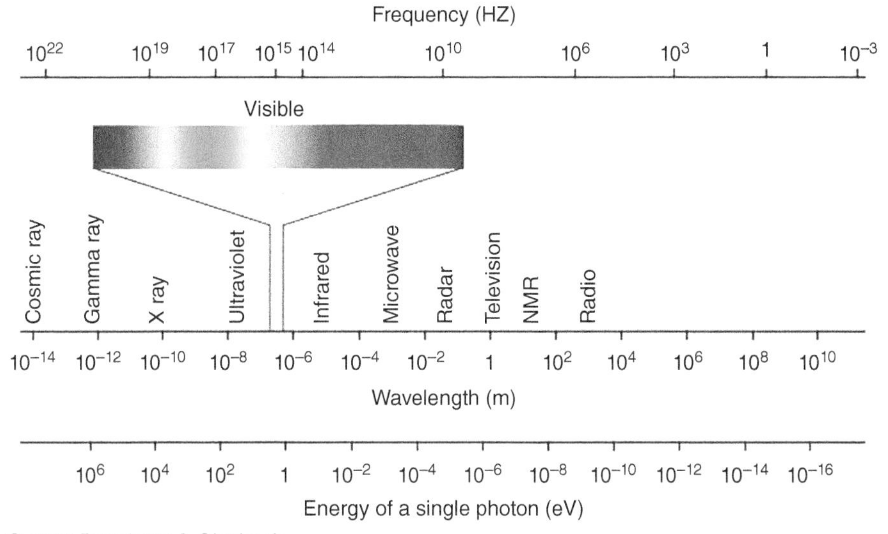

Figure 3.2 The electromagnetic spectrum. (Source: Dunnivant and Ginsbach (Authors)).

Table 3.2 Typical atomic and molecular energy transitions associated with each type of photon and photon wavelength.

X-rays	Angstroms to 0.1 nm	Can promote or remove inner (core) electrons
Vacuum UV	10–190 nm	Can break molecular bonds. Results in the removal or promotion of electrons to excited states
UV	190–300 nm	Can break molecular bonds. Results in the removal or promotion of electrons to excited states
Visible	350–800 nm	Results in the removal or promotion of electrons to excited states
IR	0.8–300 μm	Increases the amplitude of vibrations
Microwaves	~1 to 4 mm	Increases the rate of molecular rotation

The other factor that determines the wavelength of a given absorption is the type of atomic or molecular species. For example, Figure 3.3 shows the possible electronic transitions of a 3s electron of atomic sodium. The most likely transitions are indicated by the thicker yellow lines at 589.0 and 589.6 nm. These spectral lines are utilized for analytical measurements and are so intense that they can be observed by the human eye. Yellow light is emitted when sodium-containing compounds are placed in a common flame. These sodium atoms are excited by the flame and subsequently relax by releasing a photon that is then observed by our eye. These types of electronic transitions and energy relations should be familiar to the reader from basic first-year general chemistry principles where fundamental equations such as

$$c = \lambda \upsilon \tag{3.1}$$

and

$$E = \upsilon h \tag{3.2}$$

were used in combination with Bohr's equation (for hydrogen)

$$E = -Rhc\left(\frac{1}{n_f^2} - \frac{1}{n_i^2}\right) \tag{3.3}$$

where E is energy, c is the speed of light in a vacuum, λ is a photon's wavelength, υ is a photon's frequency, h is Planck's constant, R is Rydberg's constant, n_i is the initial electronic energy level,

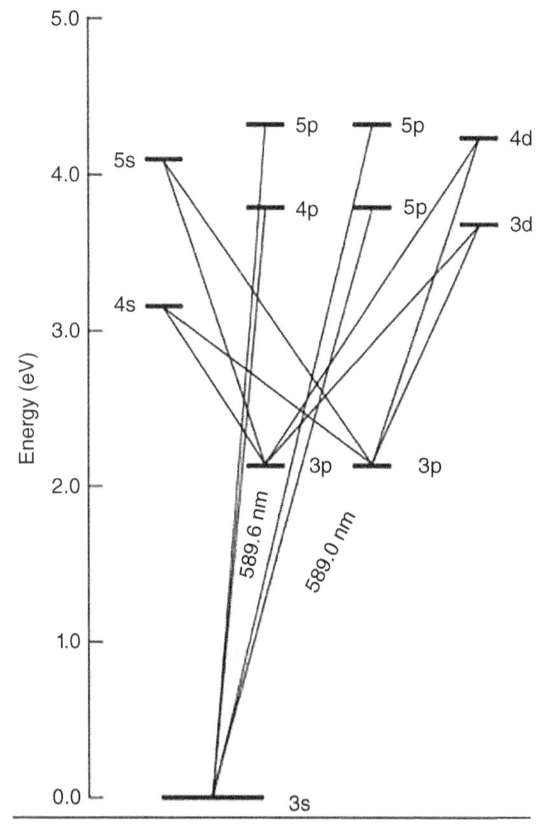

Figure 3.3 Possible electronic transitions for a 3s electron in a sodium atom. (Note the two different 3p energy levels, one with a higher energy for $j = 3/2$ and a lower energy with $j = 1/2$) (Source: Dunnivant and Ginsbach (Authors)).

and n_f is the final electronic energy level. Thus, electronic transitions such as n_1 to n_2, can be related to energy, frequency, and wavelength by the equations given above.

The possible transitions shown in Figure 3.3 are specific to a particular element and such a diagram can be constructed for every metal. When energy (or a photon) corresponding to a specific transition is absorbed, the electronic state is changed (i.e. from 3s to 3p). Some metals have only a few likely electronic transitions, while others will have more. Usually, the most prominent wavelength produced by this transition is used in instrumental absorption and emission measurements. Atomic absorption spectrometry (AAS), AES, and ICP–atomic emission spectroscopy (ICP–AES) are only concerned with the absorption of a photon by a ground state atomic species, and the emission of photon from a singlet electronic level.

This absorption of a photon is one potential way that FAAS and HPLC will produce signal data. Sometimes, the absorbance of a sample is converted to transmission by the following equation:

$$\text{Absorbance} = -\log(\text{Transmission}) \tag{3.4}$$

$$\text{Transmission} = \frac{I}{I_0} \tag{3.5}$$

where I is the intensity of light passing through a sample and I_0 is the intensity of light passing through a blank sample (reference sample cell).

In addition to absorption and transmission of photons, scattering of UV/visible radiation is another common interaction of electromagnetic radiation with matter. The scattering of radiation by large molecules or small particles can be a very useful phenomenon, and is the basis for other analytical techniques such as Raman spectroscopy or photocoupled spectroscopy, respectively. However, for the trace analytical instruments discussed here, photon scattering is a detriment to analytical measurements since it reduces the intensity of radiation that interacts with a sample or radiation that reaches the detector. Scattering is caused when the electric field of a photon interacts nonresonantly with the energy level structure of an atom or molecule. Scattering is a problem when stray photons occasionally reach the detector creating a false analyte signal. In optical spectrometry, scattered light is prevented from reaching the detector by nonreflective black surfaces inside the instrument, and by enclosing the detector in a separate compartment.

3.2.1 Correlating Absorbance to Concentration

For the majority of metal analysis performed by AAS and molecular analysis performed by HPLC and CE, the absorbance A of a beam of photons is measured by a detector. In order for this signal to be useful, this physical process must be correlated to the concentration of the analyte of interest. This can be accomplished by Beer's law

$$A = \varepsilon bC \tag{3.6}$$

where ε is the molar extinction coefficient (a measure of the absorbance of a particular chemical or chromophore), b is the cell path length (usually 1.0 cm), and C is the molar concentration of the analyte. Beer's law, as shown earlier, is the result of some mathematical simplifications that are only applicable over absorbance ranges from 0.000 (100% transmission) to approximately 2 (1% transmission). This limitation is usually enforced by sample preparation techniques that concentrate or dilute the sample to values within the acceptable range. Also note that Beer's law only governs absorption. Emission measurements are not limited from 0.0 to 2.0 "absorbance" units. In emission spectrometry, the intensity of the emitted light is typically measured in units of counts per time. Emission measurements often have superior detection limits due to the elimination of noise from the light source lamp.

The difficult task of measuring low concentrations of an analyte requires the reduction of instrumental noise to a minimum level. Ideally, instruments would only measure the exact wavelength that the given analyte absorbs. However, the light that reaches the detector is not always spectrally pure due to various imperfections in instrumental components and interactions with instrument components. The slit width of the monochromator (a common instrument component used to select a specific wavelength of electromagnetic radiation) limiting the bandwidth must be wide enough to allow sufficient radiation into the detector to produce a sufficient signal (over the noise), but be narrow enough to provide sufficient wavelength resolution. If the entrance slit is too wide, other wavelengths can create background noise in the signal, creating a nonlinearity in the calibration. These two conflicting needs are balanced by selecting a proper exit slit width in the monochromator.

3.2.2 Relaxation Processes

The absorption of a photon by an atom or molecule results in the creation of an excited energy state. Given that this atom or molecule would prefer to exist in the ground state, the energy that was absorbed is quickly dissipated by various pathways of relaxation, or (rarely) by breaking a bond. Nonradiative relaxation, emissions, fluorescence, and phosphorescence are all types of relaxations that occur without breaking a bond. For purposes of clarity, it should be noted that while some of these relaxations are not analytically useful for the techniques described here, they are of importance in other spectroscopic techniques. Some of these are only relevant to molecular spectrometry, while others are relevant to both atomic and molecular spectrometry. For completeness, all three will be described in this text.

Nonradiative relaxations are not completely understood but occur through the transfer of very small amounts of energy through molecular or atomic collisions. These result in the generation or transfer of heat energy but do not cause the emission of a UV or visible photon. Emission of a photon that has the same amount of energy as the absorbed photon is also a very common form of relaxation.

Fluorescence is a relatively rare form of relaxation in polyatomic compounds. Fluorescence occurs when an analyte is excited to a higher energy state by the absorption of a photon. The molecule then undergoes a vibrational relaxation (again, in polyatomic compounds) to a lower vibrational state within the same electronic state. Fluorescence emission occurs when the electron falls from one excited electronic state back to the ground electronic state (a singlet-to-singlet transition). This type of transition occurs since most vibrational transitions occur faster (10^{-15} to 10^{-12} seconds) than electronic transitions (around 10^{-8} seconds). Since the vibrational relaxation emits energy in the form of heat, the energy of the fluorescence emission is lower (and consequentially the wavelength is longer) than the absorbed energy. This is illustrated in Figure 3.4. Measurements of fluorescence usually yield detection limits that are from one to three orders of magnitude more sensitive than typical absorption or emission detection limits. The lower detection limits are due to lower background noise since the exciting radiation is not detected in the analytical measurement.

Phosphorescence occurs when a radiation-less vibrational transition occurs between two different spin states (for example, from a singlet (S_1) to a triplet (T_1) excited state). After this spin-forbidden transition, another spin-forbidden transition from the excited state to the singlet state must occur (from a triplet to a singlet state in Figure 3.5). Phosphorescence transitions are easy to distinguish from fluorescence since they occur much more slowly; fluorescence occurs over time scales of 10^{-5} seconds, while phosphorescence occurs over time scales of minutes to hours since spin-forbidden transitions have a low probability of occurring (a low transition rate).

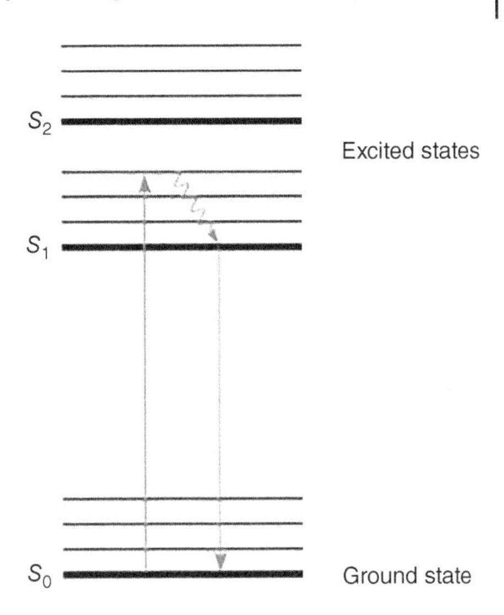

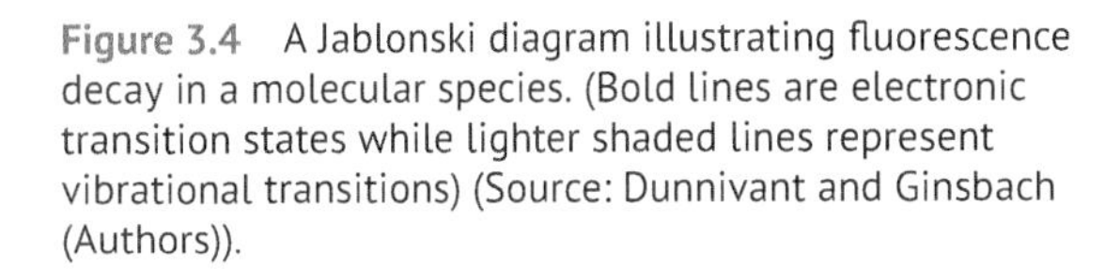
Figure 3.4 A Jablonski diagram illustrating fluorescence decay in a molecular species. (Bold lines are electronic transition states while lighter shaded lines represent vibrational transitions) (Source: Dunnivant and Ginsbach (Authors)).

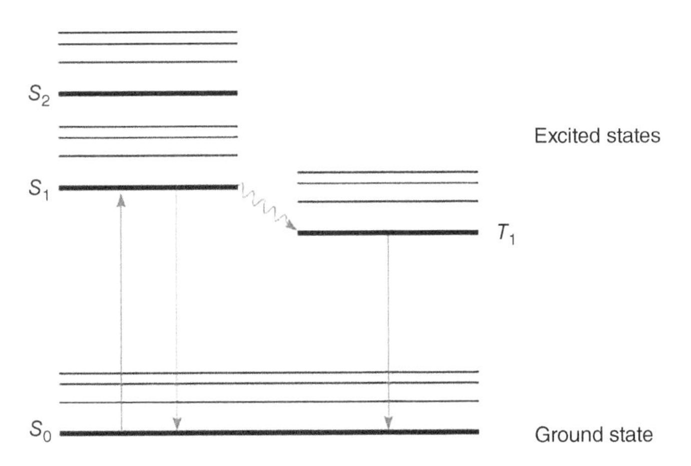

Figure 3.5 A Jablonski diagram illustrating phosphorescence decay in a molecular species (Source: Dunnivant and Ginsbach (Authors)).

In molecular spectroscopy fluorescence and phosphorescence instruments, the incident (source) radiation never reaches the detector. Instead, the detector aperture is orthogonal (90° offset) to the source lamp's path in order to reduce background noise from the incident source light. Source light is focused through a sample and generally passes straight through, with the exception of scattered radiation. Fluorescence and phosphorescence are emitted from the sample at all angles, so orientation of the detector at 90° minimizes the source light scattered to the detector. A very useful technique is possible with phosphorescence since emission can still be detected after the source lamp is turned off or blocked from the detector. It is possible to place a shutter between the excited sample and the detector. The shutter is opened after the source lamp is turned off and phosphorescence is occurring. This allows the detector to measure only phosphorescence. Phosphorescence has no use in atomic spectrometry but can be used for the analysis of metals in polyatomic states, such as those occurring for uranium oxide species. Phosphorescence has multiple uses in organic chemistry.

For quantitation of fluorescence (F) and phosphorescence (P) of polyatomic species, Beer's law is slightly modified to

$$F \text{ or } P = I\Phi\varepsilon bC \tag{3.7}$$

where I is the intensity of radiation passing through the sample, Φ is the yield of fluorescence or phosphorescence (or how much of the source radiation is converted to fluorescence or phosphorescence), ε is molar extinction coefficient, b is the cell path width, and C is the molar concentration.

Other forms of fluorescence are possible, but they involve molecular vibrational relaxations that are not relevant to atomic species. One common type of fluorescence in atomic species is referred to as "resonance fluorescence." This occurs when an electron in the ground state is excited to a higher energy state by a photon whose energy is nearly resonant with the transition energy. This is followed by the emission of a photon of the same energy as the electron relaxes (basically a normal absorption/emission process; this is the same type of emission that was discussed earlier). Emission of a photon from resonance fluorescence almost always occurs in absorbance spectroscopy but is usually not relevant because the fluorescence intensity is relatively weak in comparison to the intensity of the light source.

3.3 Interaction of Electromagnetic Radiation with Surfaces

In addition to the interactions between electromagnetic radiation and samples, the analyst must also be concerned with interactions between electromagnetic radiation and the components of the instrument. Most of these interactions are a result of the instrument design and help ensure proper function. At the most basic level, these components also act like sample molecules; they can potentially absorb electromagnetic radiation. The absorptive properties of black surfaces are utilized to absorb scattered light to prevent it from generating noise. However, some of these interactions will act in a detrimental way to interfere with the proper operation of the instrument. The absorption of light by any component in the path of the radiation would detrimentally affect the accuracy of the measurement.

3.3.1 Reflection

Mirrors are used to direct the beam of electromagnetic radiation in specific directions. Curved mirrors can be used to ensure that the beam is focused on the components of the instruments such as monochromator slits and the detector. In addition to mirror reflections, reflection also occurs any time electromagnetic radiation passes from a medium with one index of refraction into a medium with a different index of refraction. The amount of reflection tends to increase as the difference in refractive indexes between two adjacent media increases. The refractive index is defined as

$$\eta_1 = \frac{c}{c_i} \tag{3.8}$$

where c is the speed of light in a vacuum and c_i is the speed of light in a specific medium; thus, refractive index values in normal materials are always greater than or equal to 1.00. The importance of the refractive index will be discussed in the following sections on reflection and dispersion.

For dielectric (nonconducting) materials, as the difference in refractive indexes between two media increases, more reflection occurs. For a beam of light entering an interface at a right angle, the fraction of radiation reflected is calculated by

$$\frac{I_r}{I_o} = \frac{(n_2 - n_1)^2}{(n_2 + n_1)^2} \tag{3.9}$$

where I_r is the reflected intensity, I_o is the intensity of the incoming beam, and η_1 and η_2 are the refractive indexes of the two media. The mathematical formula for metal surfaces is more complicated, but the basic principle of reflection is the same.

Reflection at metal surfaces is a quantum phenomenon. Mirrors are typically made by vacuum deposition of an elemental metal on a rigid glass surface. Metals in their elemental state are held together in a "sea" of delocalized valance electrons; in the case of transition metals, these are d orbital electrons. These electrons have a charge or dipole and vibrate at a frequency similar to the incoming, visible radiation, and set up an oscillation of absorption and emission at the same incoming and outgoing frequency and wavelength. Given the range of visible radiation in sunlight and uneven surface of the metal, the emitted frequency for transition metals yields a gray color in natural light, with the exception of copper and gold. Similar quantum phenomena also occur in the near- and far-infrared.

Reflective surfaces are also present in grating systems. Common mirrors create a phenomenon that is referred to as "ordinary reflection" where an object is seen as it reflects off a mirror (Figure 3.6). Grating systems utilize higher-order diffraction arising from wave interference, which will be discussed in more detail in Section 3.3.2. Ordinary reflection can be thought of effectively as zero-order diffraction.

Reflection can also be a detriment to optimal instrumental function at the surfaces of sample containers and lenses. As the radiant light passes through the glass, the difference in refractive index of glass and air can cause reflection, which subsequently lowers the amount of incident radiation reaching the sample. Changes in the refractive indexes between these media can be overcome with the use of sample blanks and dual beam instruments (a reference light beam and a sample beam). The problem is more serious with reflective surfaces in the instrument. The composition of reflective surfaces and coatings is selected to minimize reflective losses. Reflection can decrease the source intensity since some of the light is lost. In these cases, reflection decreases sensitivity and detection limits, due to a loss of radiant intensity of the wavelength of interest. Generally, 2% of incoming incident radiation is lost at each reflective surface.

3.3.2 Diffraction

Diffraction is a process in which a collimated beam of radiation spreads out as it passes (1) through a narrow opening or (2) through a sharp barrier. An interesting distraction effect was observed in a laboratory setting by Thomas Young in 1801 when he performed the two-slit experiment (discussed later). Because diffraction depends on the wavelength of light, it can be used to separate

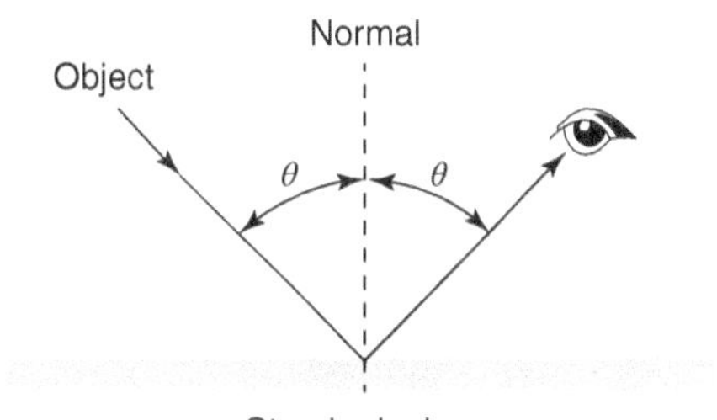

Figure 3.6 Ordinary reflection (zero-order diffraction) (Source: Dunnivant and Ginsbach (Authors)).

polychromatic light (white light) into its constituent optical frequencies; in the case of white light, this would result in a rainbow of colors. The manipulation of diffraction forms the basis of selecting a narrow range of wavelengths of light in a monochromator.

In order to more easily understand how diffraction can be used to disperse wavelengths, picture the passage of monochromatic radiation through a narrow barrier. Figure 3.7 illustrates such a barrier. As the slit width narrows waves of radiation spread out more strongly. The Heisenberg Uncertainty Principle indicates that as the uncertainty in the position of a photon decreases (in a direction transverse to its propagation direction), the uncertainty in its transverse momentum must increase; this uncertainty in transverse momentum is what causes the light to spread out. When the slit is sufficiently wide, the diffraction is small because there is still great uncertainty in position (Figure 3.7 top figure). As the slit becomes more narrow (Figure 3.7 bottom figure), the uncertainty in position becomes smaller, which results in greater diffraction.

Diffraction is intimately related to the constructive and destructive interference of waves. Start by considering monochromatic electromagnetic radiation passing through two slits (Figure 3.8). The monochromatic light is passing through one narrow slit before being passed through two more slits. As radiation passes through the second two slits, diffraction occurs at both slits creating two beams that spread out and overlap. Constructive and deconstructive interference between these beams can be observed on a flat plate (a detector in this case). Regions on the detector in which the path length difference between these two waves is an integer multiple of the wavelengths constructively interfere and produce a large intensity. Positions on the detection plate that correspond to odd integer multiples of half a wavelength create deconstructive interference, with no resulting intensity. Overall the intensity distribution on the detector varies sinusoidally, with high intensity corresponding to constructive interference and zero intensity corresponding to destructive interference. Each of these situations is illustrated in Figure 3.8.

If the two slits are replaced by many narrow, equally spaced slits (i.e. a diffraction grating), the regions of constructive interference become very narrow. Instead of a sinusoidal variation in intensity, a series of narrow bright bands appears on the detector.

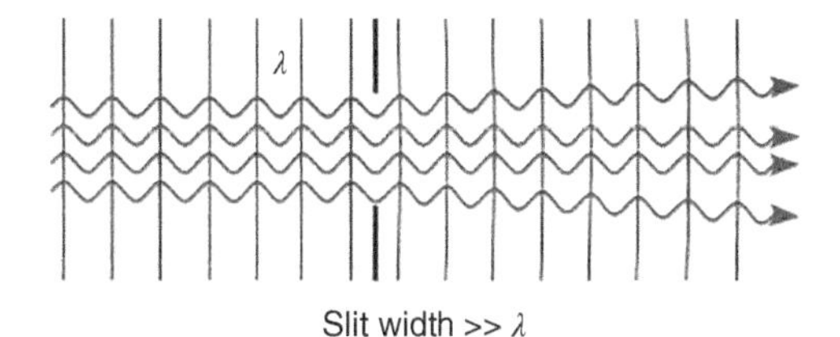

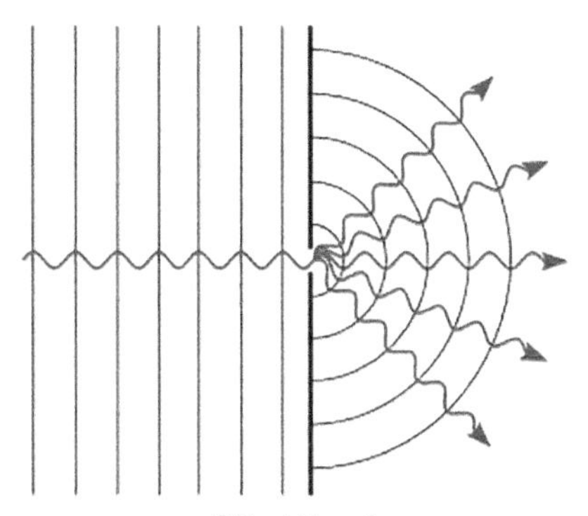

Figure 3.7 Variation in slit width openings (Source: Dunnivant and Ginsbach (Authors)).

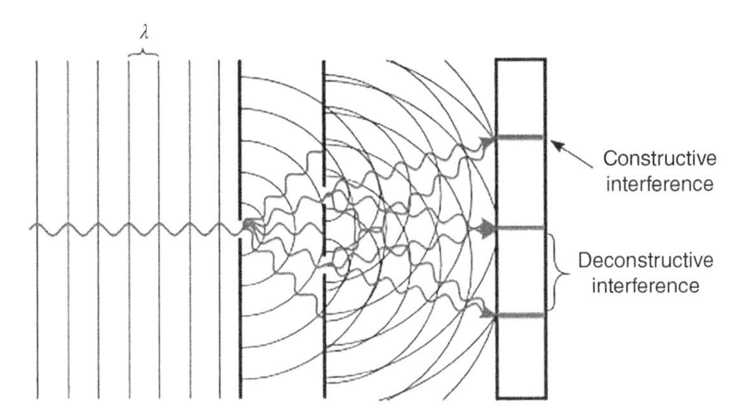

Figure 3.8 An idealized illustration of the two-slit experiment illustrating the constructive and deconstructive nature of electromagnetic radiation. Distinct bands are show here for illustrative purposes only; the bands are actually broader areas of light (constructive interference) that gray to areas with no light (deconstructive interference). The blue lines above correspond to the alignment of maximum constructive interference (Source: Dunnivant and Ginsbach (Authors)).

The earlier discussion was about monochromatic radiation. In theory, if the concepts in Figure 3.8 are extended one step further and polychromatic light is used, the different colors of radiation are vaguely separated. All of the different wavelengths will constructively interfere in the center of the detector. However, since each color has a specific wavelength, the locations of the other interference maxima depend on the wavelength. As a result, two vague "rainbow" spectra of light would be centered around the top and bottom constructive interference lines shown in Figure 3.8. A useful and instructive cartoon animation of the two-slit experiment can be found on the Internet by searching for Dr. Quantum (currently in English at https://www.youtube.com/watch?v=Q1YqgPAtzho).

While the diffractive properties of multiple transmissive slits could be utilized in analytical instruments, this would result in the creation of very narrow but long instruments. As a result, mirrors that also create diffraction are utilized to make instruments more compact. This is accomplished by bending light at a sharp angle, as observed later in the case of reflection grating monochromators. The principle is the same as transmission grating; each of the sawtooth-shaped steps in Figure 3.9 (referred to as a blazed surface) acts as a "slit." Waves are diffracted off of each of these slits, and wavelength-dependent constructive and destructive interference creates a rainbow spectrum in reflection. Monochromatic light is used for illustration purposes here to lessen the complexity of the figure. However, the principles discussed later for a single wavelength apply to a mixture of wavelengths.

Typically radiant waves constructively interfere in two directions (+1 and −1) depending on the incident angle of incoming radiation and the grating spacing d. These two directions are analogous to the upper and lower constructive interference bands in Figure 3.8. There is one more

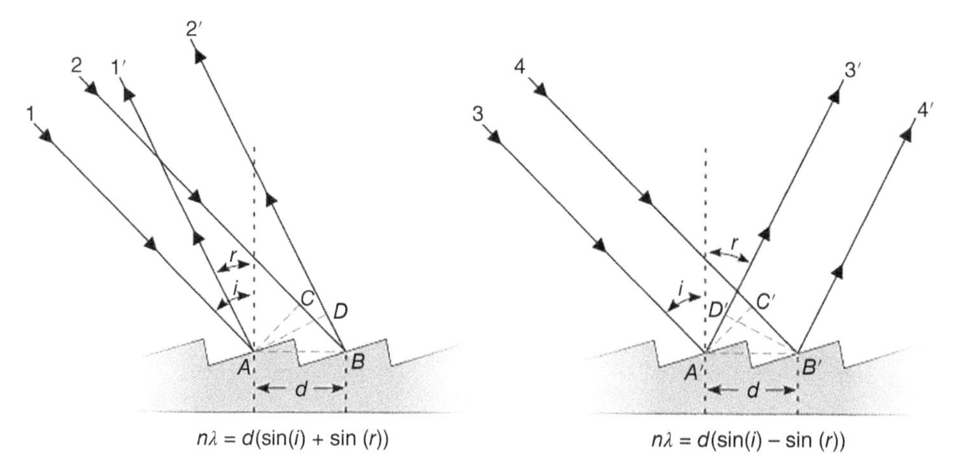

Figure 3.9 Diffraction resulting from a echelette (conventional) grating (Source: Dunnivant and Ginsbach (Authors)).

complication that has been added to Figure 3.9. In Figure 3.8, incident radiation was perpendicular to the slits (parallel to the grating normal), whereas in Figure 3.9, incident radiation makes an angle i with respect to the grating normal (the dashed vertical line in Figure 3.9). Radiation will be diffracted to the left and to the right of the grating normal; these will be mathematically characterized below separately for two types of grating surfaces. The left-of-normal diffraction will be discussed first (the left-hand figure). The lower right-hand diagonal lines 1 and 2 are incoming monochromatic radiation and the upper right-hand diagonal lines 1′ and 2′ are the diffracted monochromatic radiation.

The basis of the following derivation is that in order for constructive interference to occur, the path length differences between waves from adjacent slits must be an integer number of wavelengths. Before beginning the derivation, it is first necessary to define the problem and parameters. The line "d" is the distance between corresponding points on each grating line (a grating surface contains numerous blazed parallel lines of reflective material, in this case points A and B). Line segments AC and AD are the distances at 90° angles from the point of diffraction to the diffracted wave on the next grating surface (parallel ray). The maximum constructive interference will occur at the diffracted angle "r" where the path length difference between rays 1 and 2 is an integer number of wavelengths. As shown in Figure 3.9, ray 2 travels a greater distance than ray 1 and the difference in the paths is equal to the addition of line segment distances CB and BD. Again, in order for constructive interference to occur, the path difference (CB + BD) must be equal to the wavelength times an integer (n, not to be confused with the index of refraction), thus

$$n\lambda = (\mathrm{CB} + \mathrm{BD}) \tag{3.10}$$

The integer n is referred to as the diffraction order and λ is the wavelength of interest. Observe that the angle of CAB is equal to angle "i" and the angle DAB is equal to "r." From basic trigonometry principles, we obtain

$$\overline{\mathrm{CB}} = d\sin(i) \tag{3.11}$$

and

$$\overline{\mathrm{BD}} = d\sin(r) \tag{3.12}$$

where d is the distance between points A and B on two adjacent blazed surfaces. By substituting these two-line segment equations into the previous equation, we obtain the governing equation correlating the angle of reflection to wavelength

$$n\lambda = d(\sin(i) + \sin(r)) \tag{3.13}$$

Equation (3.13) is called the grating equation.

A similar diffraction pattern can occur in the opposite direction (to the right in Figure 3.9). As drawn in this figure, the low blazing angles will result in a higher intensity of reflected light to the right as compared to the light reflected in the left-hand side of Figure 3.9; this occurs due to the ordinary angle of reflection off of each surface being equal to the diffraction angle, and the intensity maximized. In this case, illustrated in the right-hand side of Figure 3.9, rays 3 and 4 entering from the left are reflected at the surface angles i plus r. Using line segments A', B', C', and D', a similar mathematic derivation can be made for the constructive interference shown by lines 3′ and 4′ as was made earlier for the left-hand side of Figure 3.9. This can also be derived by noting that the integer n can have plus or minus values of 1.

$$d\sin(i) = \overline{\mathrm{CD}} \tag{3.14}$$

$$d\sin(r) = \overline{\mathrm{AD}} \tag{3.15}$$

$$n\lambda = (\overline{\mathrm{CB}} - \overline{\mathrm{AD}}) \tag{3.16}$$

Thus

$$n\lambda d(\sin(i) - \sin(r)) \tag{3.17}$$

Thus, constructive interference can occur at a negative and a positive angle, relative to the normal plane, and is indicated by a plus and minus sign in the governing equation, respectively.

$$n\lambda = d(\sin(i) \pm \sin(r)) \tag{3.18}$$

A positive sign results when the incident and diffracted beams are on the same side of the grating normal line (the vertical line in the middle of each grating), with the converse for a negative sign. It should be noted again that the angle of diffraction is highly dependent on the angle of the incoming incident radiation and the wavelength of radiation. This will be important when different types of grating monochromators are discussed in Section 3.4.

The diffraction pattern illustrated in Figure 3.9 for monochromatic light is the same concept as that illustrated for the two-slit experiment described in Figure 3.8. Both the slit and each individual blaze behave as point sources of light. Both systems cause constructive interference when radiation from each blaze or slit travels the same distance or a distance that is an integer multiple of the wavelength ($n\lambda$). When polychromatic light is reflected by a grated surface as shown in Figure 3.10, a "rainbow" dispersal of wavelengths occurs on the surface of the detector. Subsequently, each wavelength of the rainbows is collimated with a collimating mirror (or lens) and focused onto a sample container or instrumental component, such as the entrance slit to a detector.

Further analysis of the governing equations given earlier indicates that several wavelength solutions exist for each diffraction angle r, depending on the integer n, referred to as the order of diffraction. The first wavelength, referred to as first-order diffraction ($n = 1$), is always the longer wavelength (i.e. 400 nm), with second-order diffraction being half the wavelength (i.e. 200 nm), and so on. Fortunately, for our purposes of obtaining as pure radiation as possible (with respect to wavelength), usually more than 90% of the incident intensity is first-order diffraction. When second-order and third-order diffraction is a problem, these wavelengths can be removed with a secondary filter or second monochromator (as presented in Chapters 4 and 5). In some cases, such as in the presence of high concentrations of analyte, second- and third-order diffraction is used for quantification since the corresponding wavelength is less intense and will not damage the detector.

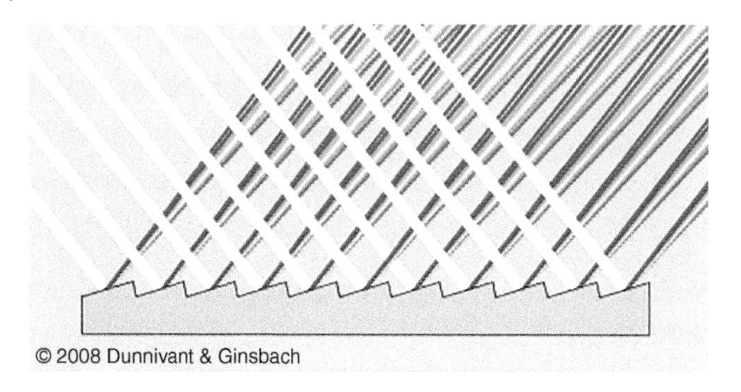

Figure 3.10 Separation of white light by a grating monochromator (Source: Dunnivant and Ginsbach (Authors)).

For an Echelette-type grating, high dispersion (separation of wavelengths, which is desired in the right-hand side of Figure 3.9) is obtained by making the groove width d as small as possible (having a high number of grooves or blazes) and by making the focal length large (the focal length is the distance between the monochromator or the focusing lens and the exit slit). However, achieving high dispersion is difficult since a large focal length results in a reduction in light intensity and dense gratings are relatively expensive to produce. Improvements in resolution result if the blaze angle is made steeper (called an Echelle-style grating), and a correspondingly large value for the diffraction order ($n \sim 100$) is used. In the typical Echelle grating configuration, the angles r and i (in the left-hand side of Figure 3.9) are now approximately equal, and r or i is represented by β. For this situation, the governing equation reduces to

$$n\lambda = 2d\sin(\beta) \tag{3.19}$$

The Echelle grating (left-hand image in Figure 3.9) achieves higher dispersion while also maintaining high diffraction intensity by making the angle β and the order of diffraction (n) large (i.e. by increasing the blaze angle and diffraction order). Echelle gratings have fewer blazes (grooves) than Echelette gratings. Again, while the previous normal grating resulted in first-, second-, and third-order diffraction ($n = 1, 2, 3$), n values for an Echelle are much larger (75th to over 100th). For example, as shown in Table 3.3, a diffraction order of 1 for a normal grating correlates to $n = 75$

Table 3.3 Comparisons of figure of merit for echelette and echelle grating monochromators.

Feature	Echellette (Conventional) Grating	Echelle Grating
Focal length (m)	0.5	0.5
Groove density (grooves/mm)	1200	79
Angle of diffraction	10°22′	63°26′
Width of grating (mm)	52	128
Order of 300 nm	1st	75th
Resolution at 300 nm	62 400	763 000
Linear dispersion at 300 nm (mm/nm)	0.61	6.65
Reciprocal linear dispersion at 300 nm (nm/mm)	1.6	0.15
f-Number	f/9.8	f/8.8

(Source: Keliher and Wohlers, 1976)

for an Echelle grating. This higher order of diffraction is selected by increasing the blaze angle to maximize the reflection of radiation at the 75th order of diffraction. This results in superior resolution of spectral lines over conventional grating systems. For example, in a conventional grating, the emission lines of Zn (at 202.55 nm) and Mg (at 202.58) overlap and cannot be resolved. In the Echelle grating, considerable baseline resolution is achieved for these two spectral lines.

Echelle-grating monochromators can produce highly pure spectral separations of radiation; these will be used in applications in the next two chapters where a combination of dispersive devices (prisms, Echelle, and Echellette gratings) are used to separate wavelengths of light. The use of an Echelle grating will allow more powerful imaging and detectors to be used in emission spectrometry where spectral overlap can be a problem. The wavelength separations in combined gratings systems are nearly as pure as the emitted radiation from a hollow cathode lamp (the light source in FAAS instruments), which will minimize most spectral interferences from the flame emissions or from the absorption and emission of other elements in the sample. Grating monochromators are the heart of modern absorption (AAS) and emission (AES) instruments, as well as in UV–vis detectors in HPLC.

3.3.3 Dispersion

Dispersion has been alluded to several times in the previous discussions, but a more formal description will be provided here. First, dispersion is the separation of polychromatic radiation by a material that alters the velocity of a wave based upon its frequency. Dispersion can be both an important tool and a nuisance. For example, some mirrors are coated with polymers to protect them from corrosive laboratory environments. If this coating disperses different wavelengths of light, it can deteriorate the integrity of the light by acting like a prism (and separate wavelengths even further).

The positive aspects of dispersion occur in prisms, where separation of wavelengths of light is desirable. When a change in refractive index occurs at an interface between substances in the path of a beam of radiation, the net effect is a bending of the light. This bending results from the fact that the waves travel at different speeds in the two substances. Because the amount of bending depends on the speed, and different wavelengths have different speeds, different wavelengths of radiation (light) bend at different angles at the interface. The net result is a separation of wavelengths.

Devices that separate light, such as monochromators and prisms, are "rated" by their ability to separate wavelengths, with a number of parameters to quantify their resolution. If the equation, derived earlier, that describes the constructive interference of radiation

$$n\lambda = d(\sin(i) \pm \sin(r)) \tag{3.20}$$

is differentiated while holding i constant, the angular dispersion of a grating can be obtained

$$\frac{\Delta r}{\Delta \lambda} = \frac{n}{d\cos(r)} \tag{3.21}$$

This equation relates the angle of diffraction to the wavelength, referred to as angular dispersion. This can be taken one step further by relating the angular dispersion to linear dispersion (D), a measure of the quality of the monochromator, by

$$D = \frac{\Delta y}{\Delta \lambda} = \frac{F\Delta r}{\Delta \lambda} \tag{3.22}$$

where F is the focal length of the monochromator (the distance between the monochromator grating to the exit slit). In this equation, the analyst is concerned with the variation of wavelength as a function of the distance along the line AB in the monochromator figure (Figure 3.8). A more useful measure of dispersion (and the quality of the monochromator) is the reciprocal linear dispersion (D^{-1}) that is commonly used by the manufacturing industry

$$D^{-1} = \frac{\Delta\lambda}{\Delta y} = \frac{1}{F}\frac{\Delta\lambda}{\Delta r} \tag{3.23}$$

where D^{-1} typically has values expressed in nm/mm or Angstroms/mm, and small numbers represent superior instruments.

Furthermore, substitution of the angular dispersion equation into the above equation yields the reciprocal linear dispersion, another figure of merit, for a grating monochromator

$$D^{-1} = \frac{\Delta\lambda}{\Delta y} = \frac{d\cos(r)}{nF} \tag{3.24}$$

From this equation, it is evident that angular dispersion increases as the distance, d, between the parallel grating lines decreases (as the number of grating lines per millimeter increases). Typical numbers of parallel grating lines in monochromators range from tens to hundreds of lines per millimeter for IR wavelengths, to several thousands of lines per millimeter for UV/visible wavelengths.

Finally, the resolving ability (where R is the resolution) of a monochromator refers to its power to separate adjacent wavelengths, represented mathematically by

$$R = \frac{\lambda}{\Delta\lambda} \tag{3.25}$$

Values of R for UV–visible monochromators of interest here range from 1000 to 10 000. It can also be shown that R is related to the total number of illuminated grating lines by

$$R = \frac{\lambda}{\Delta\lambda} = nN \tag{3.26}$$

where n is the diffraction order, and N is the number of grating lines illuminated by the radiation of interest. The number of illuminated lines is usually determined by the width of the entrance slit to the monochromator and the "*f*" number (*f*#) of the monochromator.

Decades ago, an analyst could request that a monochromator in an instrument be made to specific specifications based on reciprocal dispersion. However, today the earlier discussion of figures of merit is only of educational interest, as only a limited number of instrument variations are commercially available due to the more economical mass production of a few instrument designs. In essence, technicians purchase what is available and the price of the instrument is correlated to its dispersion and resolution. As a note, there are manufacturers that will custom make a monochromator with specific blaze density and blaze angle, but these are relatively costly.

3.3.4 Refraction

Refraction goes "hand in hand" with dispersion. Dispersion of radiation in prisms is the result of refraction. The amount of refraction is described by Snell's law

$$\frac{\sin\theta_2}{\sin\theta_1} = \frac{\eta_1}{\eta_2} = \frac{\text{velocity 2}}{\text{velocity 1}} \tag{3.27}$$

where θ is the angle of refraction relative to the normal plane, η is the refractive index, and subscripts 1 and 2 are the two media. The effect of refraction on light moving through two materials is illustrated in Figure 3.11.

As mentioned in the previous sections (specifically Section 3.3.4), refraction is the basis of the separation (dispersion) of light in prisms, while diffraction is responsible for dispersion in grating monochromators.

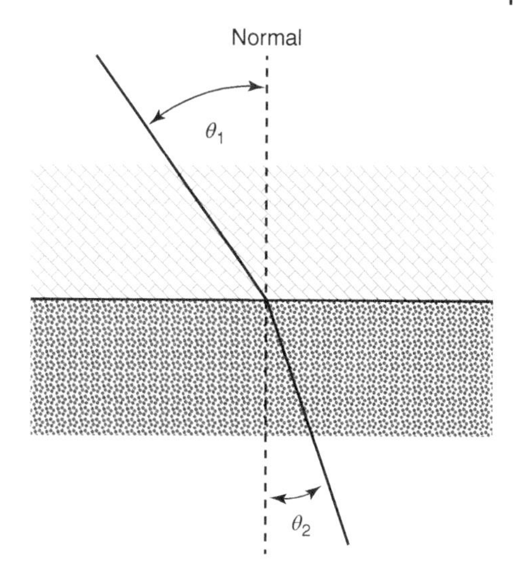

Figure 3.11 Illustration of the refraction of light passing through two media, one is less optically dense (the green-shaded region) and one more optically dense (the red-shaded region) (Source: Dunnivant and Ginsbach (Authors)).

3.4 Detectors in UV–visible Spectrometry

UV–visible measurements require that the energy contained in a photon be converted into a measurable electrical signal. Early detectors relied on a more solid-state version of the photoelectric effect that is best illustrated by a phototube (PT), one of the first detectors to convert radiant energy to electrical energy.

3.4.1 Phototubes

Figure 3.12 shows a diagram of a basic PT. A PT consists of an evacuated glass or quartz chamber containing an anode and a cathode. Cathode surfaces are composed of materials that readily give up electrons; Group I metals such as Cs work well for this purpose. A relatively large potential is placed across the anode and cathode, usually 90 V, and the gap is referred to as a dynode. Electrons contained in the cathode are released as photons with sufficient energy strike the surface. This causes electrons to move through the low-pressure gap to the anode, which produces a current. For a PT, a single photon causes only a single electron to be measured. For emission spectroscopy, the magnitude of current produced by the cascade of electrons in the detector is directly proportional to the concentration of analyte in the sample.

3.4.2 Photomultiplier Tubes

Photomultiplier tubes (PMTs) are an extension of the PT where numerous dynodes are aligned in a circular or linear manner. Here most of the electrodes act as both an anode and a cathode with each dynode (electrode pair) having a potential difference of +90 V; thus, the potential increases by 90 V as an electron goes from one electrode to the next (refer to Video S.3.1). When nine dynodes are used, a common feature in PMTs, the net result is a yield of 10^6–10^7 electrons from a single emitted photon. This causes considerable amplification of a weak signal compared to a photon tube that does not amplify the signal. PMTs were once excellent detectors for UV–vis, FAAS, and FAES measurements due to the low intensity of radiation in these systems and have dominated

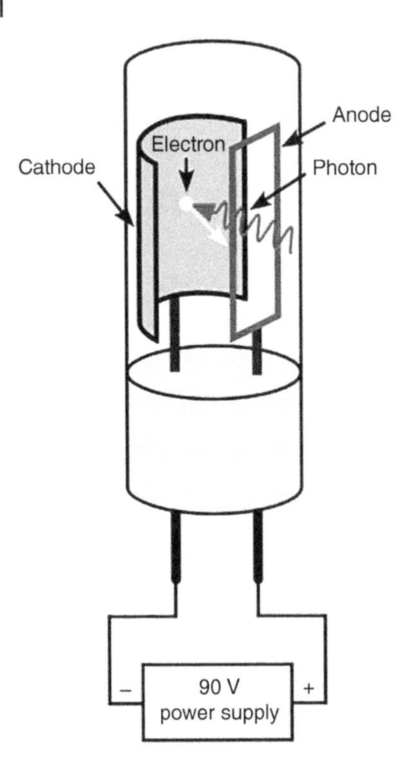

Figure 3.12 Diagram of a phototube (Source: Dunnivant and Ginsbach (Authors)).

these instruments for decades. But today they have been largely replaced by charge transfer devices (CTDs).

In Video S.3.1, only discrete packets photons are shown entering the detector from a chopped signal; this is only for illustration purposes since the stream of photons and the generated electrons are continuous producing an analog signal. Photons enter through a slit, strike the first cathode that is positively charged and a small number of electrons are produced. These electrons jump the gap to the next electrode where each arriving electron produces even more electrons due to the increased potential placed on the surface. This process is repeated over and over until a current is read at the anode (the final electrode).

Electron multipliers can also be of a continuous design as shown in Video S.3.2 for an organic cation entering an EM from a mass spectrometer.

3.4.3 p-n Junctions

In the mid-1990s, a new style of photon detector (and electron collector) was developed and implemented into instruments because of advancements in computer chip technology that decreased cost of producing these chips. Before charge-coupled devices (CCDs) can be understood and appreciated, the historical development of photosensitive computer chips must be presented.

The development of advanced detectors starts with a silicon diode device. A diode is an electronic device composed of two electrodes that can either act as a resistor or a conductor depending on the potential placed on the device. Such a device is shown in Figure 3.13a.

A silicon diode consists of two pieces of silicon, one positively doped (p-doped) and one negatively doped (n-doped), with the functional area (the depletion region) being at the interface between the two media (a p–n junction, the subject of this chapter's case study). The positively doped end was

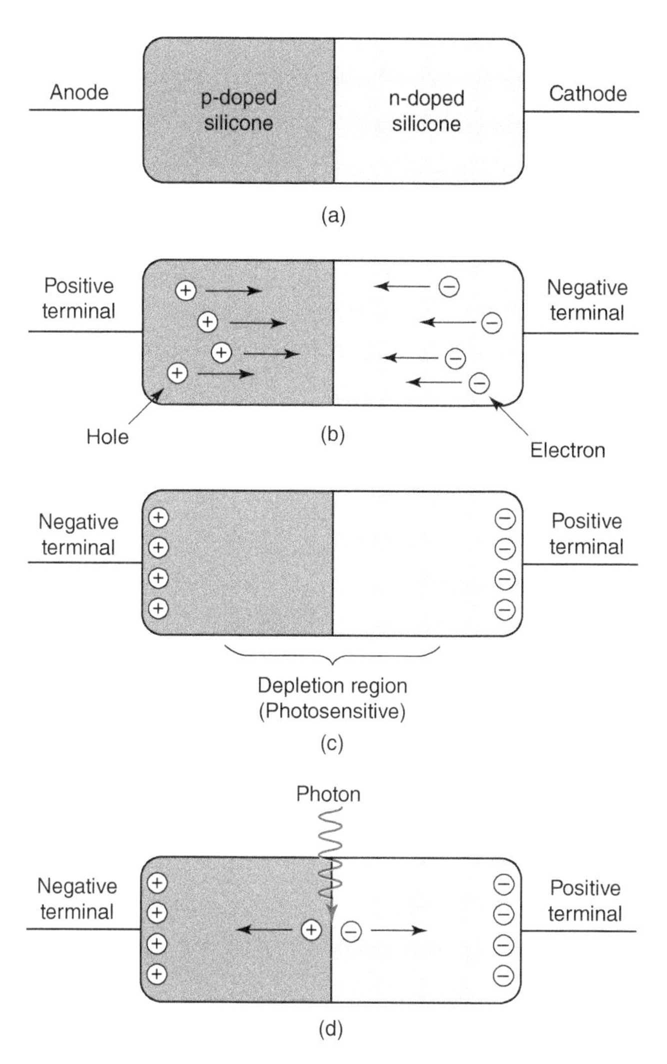

Figure 3.13 A basic silicon diode: (a) with no applied voltage, (b) with a forward (positive) bias placed on the electrodes, (c) with a reverse (negative) bias placed on the electrodes, and (d) the current generation resulting after a photon strikes the depletion region (Source: Dunnivant and Ginsbach (Authors)).

created by doping (replacing) an occasional Si atom with a Group III element, such as aluminum. The net result is referred to as a positively charged carrier that can move through the chip, or an electron hole since one bond is missing between Al and Si. In contrast, the negatively doped end of a diode is a similar piece of silicon where occasional Si atoms have been replaced with a Group V element, such as phosphorus. The net result is the creation of a negatively charged carrier since this material contains extra electrons.

Combining the p-doped and n-doped silica into one device (called a pn junction) creates a diode that can be operated in two modes; forward bias and reverse bias. In the forward bias mode, a positive potential is placed on the p-doped semiconductor and a negative potential is placed on the n-doped semiconductor (refer to Figure 3.13b). This results in the holes and electrons moving toward the center of the device where they can combine and reduce the resistance in the diode;

here the device acts as a conductor. The forward bias mode is of no interest in detectors utilized for spectrometry.

Figure 3.13c shows the silicon diode when a reverse bias is placed on the device. Here, the charges on the electrodes are reversed and the holes in the p-doped semiconductor are drawn toward the negative terminal, while the electrons in the n-doped side are drawn toward the positive terminal. This results in the creation of a depletion area at the interface of the two semiconductors, where there are few holes or electrons in the matrix. As a result of the voltage barrier, high resistance is created which allows minimal electric current to cross the p–n junction; in this mode, the device acts as a resistor. In this mode of operation, when a photon of sufficient energy (in the UV, visible, and near IR regions; photons with wavelengths from 190 to 1100 nm) strikes the depletion region, holes and electrons are created that then migrate to their respective terminals generating a current (Figure 3.13d). The amount of current generated is directly proportional to the number of reactive photons striking the depletion region. Thus, these devices can be used in the detection and quantification of radiant intensity. Silicon diodes are more sensitive than the older vacuum-style PTs but their performance is still poor when compared to PMTs described earlier (due to the lack of amplification). The advantage of a silicon diode is the large numbers that can be placed in a linear arrangement referred to as a linear diode array (LDA).

3.4.4 Linear Diode Arrays

An LDA allows the simultaneous detection of a wide number of wavelengths. When an LDA is utilized, UV or visible light is first separated by a monochromator and the separated wavelengths are aligned onto the surface of an LDA with a collimating mirror or lens. Figure 3.14 illustrates a section of a linear arrangement that can be up to 400 individual diodes. Each wavelength of interest is positioned on one of these many p–n junctions allowing the intensity of each wavelength to be measured simultaneously. The entire UV and visible spectrum can be measured with considerable wavelength resolution with one LDA.

LDAs offer the advantage of measuring all wavelengths in a sample at the same time, which is especially important in emission spectrometry. This type of detector is commonly used today in

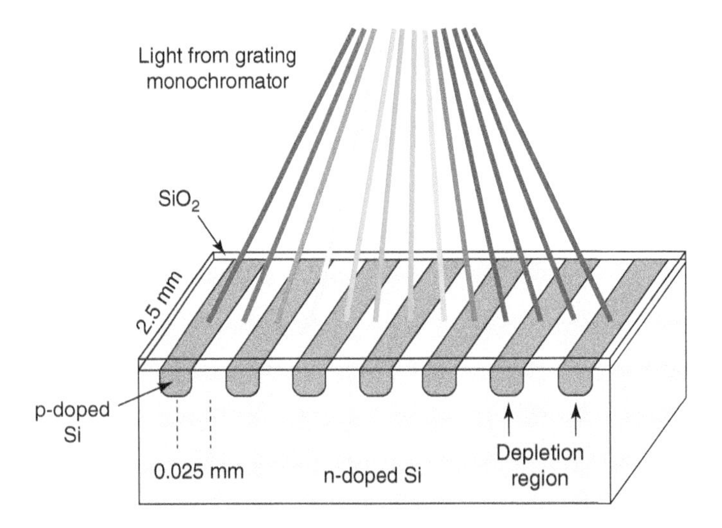

Figure 3.14 A linear diode array. Each p–n junction, the interface between shaded regions, is a diode (Source: Dunnivant and Ginsbach (Authors)).

instruments where detection limits are not a problem since LDAs do not amplify the original signal. Common uses include UV–visible detectors in inexpensive spectrometers and in many HPLC systems. The cost of production has greatly decreased in recent years and the cost of an LDA is now cheaper than a PMT. However, PMTs have higher sensitivity, a larger dynamic range, and lower signal to noise (S/N) ratios. In selecting between an LDA and a PMT detector, one must weigh the benefit of simultaneous wavelength detection versus detection limit. LDAs are most commonly used in absorption UV–vis spectrometry, but are not currently offered in common, more inexpensive FAAS or FAES systems. Diodes and LDAs are mentioned here to aid in the understanding of charge transfer devices, which are commonly used in UV–vis spectrometers today.

3.4.5 Charge Transfer Devices

There are presently two types of CTDs: charge injection devices (CIDs) and CCDs. Both of these overcome the disadvantages of LDAs (low dynamic range, poor sensitivity, and low S/Ns) by amplification (through timed storage or accumulation) of the original signal. As a result, they can match the performance of PMTs and actually exceed them by *simultaneously detecting multiple wavelengths* including separating different orders of diffraction. As a result of these advantages and due to the recent reductions in production costs and spatial constraints, CTDs are rapidly replacing PMTs.

CTDs are one- or two-dimensional arrays of detectors that can offer thousands of individual detector pixels. Both p-doped and n-doped semiconductors are used and individual pixels for each are shown in Figure 3.15.

CIDs use n-doped silicon (refer to Figure 3.15a). When a photon of sufficient energy strikes a bond in the silicon, a hole and an electron are created. The electron migrates to the positively charged substrate and is removed from the system. The hole migrates to a "potential well" that is created under the most negatively charged electrode. The system, as shown in Figure 3.13, acts as a capacitor because of the −5 and −10 V applied potential. It is important to note that each "well" can accumulate between 10^5 and 10^6 charges per electron; this closely matches the performance (amplification) of PMTs discussed earlier. Note that the silicon and electrode are separated by a nonconductive silicon dioxide layer. After a sufficient amount of time (when a large number of holes have been created and collected), the applied electrical potential on the −5 V electrode is removed and a voltmeter (potentiometer) is used to measure the capacitor potential under the −10 V electrode. The capacitor potential is directly related to the number of holes collected. If sufficient counts are measured, the holes are cleared from the CID and a new measurement is started. The advantage of CIDs is that counting can be continued by re-establishing the −5 V connection. Thus, amplification of the signal, or more specifically accumulation of holes, can be completed in two ways: (1) by initially measuring a significant amount of time and (2) by measuring and deciding to continue the collection process. This process is nondestructive; thus, the accumulated signal can be read repeatedly until a desired magnitude is obtained. When placed in a two-dimensional array, they can contain tens of thousands of pixels and each pixel can be positioned to collect data for a specific wavelength from the monochromator. Due to high thermal noise, CIDs are usually cooled to liquid nitrogen temperatures and therefore find little use in analytical AAS and HPLC instruments but are used in high-end astronomy telescopes.

CCDs, invented in 1969 at Bell Labs, are designed in an opposite manner as compared to CIDs. In CCDs, p-doped silicon is used and the electrodes atop the semiconductive material are positively charged. An individual pixel for a CCD is shown in Figure 3.9b. Here, photons strike the doped silicon, holes are neutralized and removed by the metal substrate below the pixel and electrons

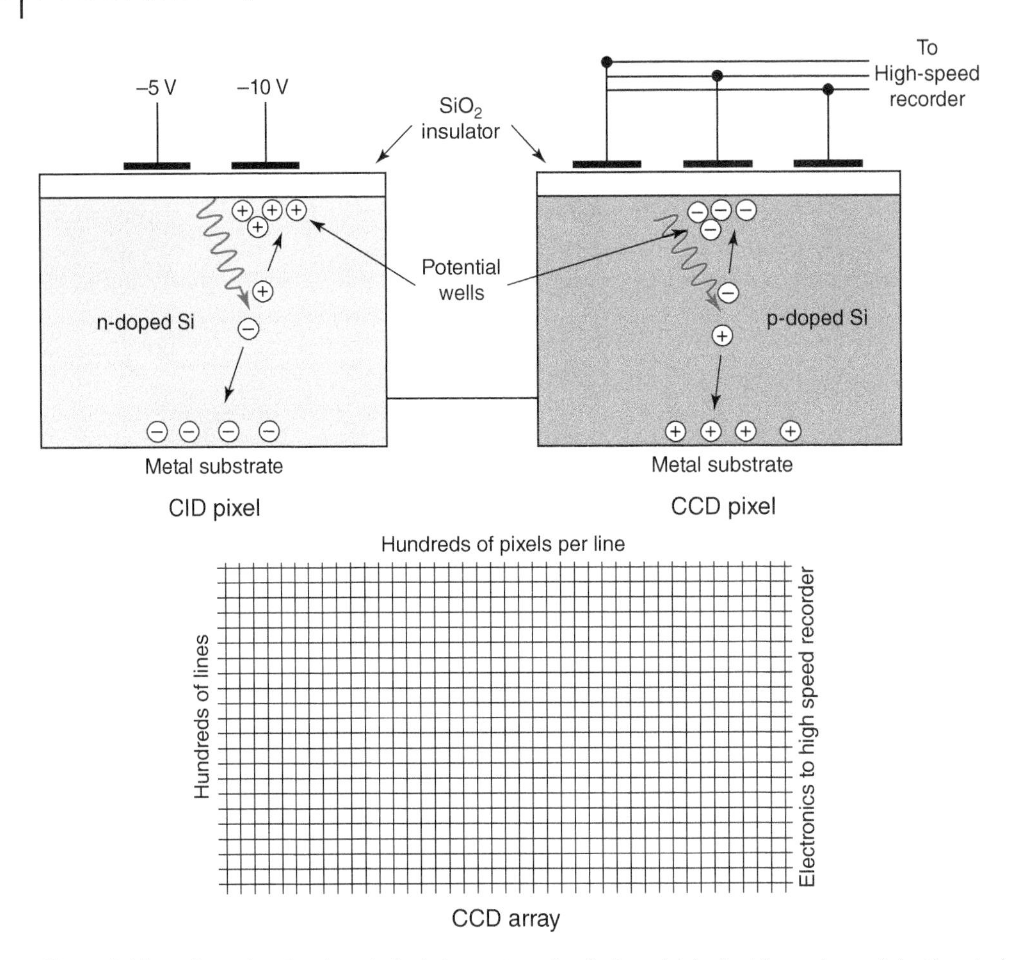

Figure 3.15 p-Doped and n-doped pixel charge transfer devices. (a) A pixel for a charge injection device (CID), (b) a pixel for a charge coupled device (CCD), and (c) an array of CCDs (each pixel is a detector) (Source: Dunnivant and Ginsbach (Authors)).

migrate to the potential well where they are stored. Arrays are read from left to right by shifting the potential on the electrons and measuring the potential charges transferred between the electrodes. Again the system (each pixel) acts as a capacitor where its potential can be measured by removing the applied potential and measuring the capacitor potential with a voltmeter. An individual pixel consists of three electrodes (a three-phase clock circuit) that are used to sweep the charges to the right where they are moved to the next pixel and onto a high-speed shift register and then onto a preamplifier and readout. This allows for a pixel-by-pixel and row-by-row two-dimensional detector surface to be created. However, the measurement is destructive, unlike in CID detectors, so only one measurement is possible. This is not a disadvantage in FAAS or FAES measurements since a manual measurement only takes seconds to obtain, and doubling or tripling the reading time is relatively insignificant. For automated sample introduction, the increased data collection time is trivial. For most purposes, typical accumulations of the signal by CCDs match the amplification performance of PMTs and CCDs are now much more inexpensive to produce.

For purposes of reference in spectrometry instruments, a CCD array (illustrated in Figure 3.15c) is about 1 cm × 1 cm in size and can contain tens of thousands of pixels, or greater. Contrast this to the surface area needed to obtain the same resolution; one would have to position 10 000 PMTs at the proper angle with respect to the monochromator with each PMT being about 1.5 cm in diameter and 8 cm in height. The small size of the CCD array allows it to be placed inside the monochromator after a collimating mirror or lens and an exit slit is not required. Instruments with CCDs are significantly smaller in size; thus, less bench space is needed in the laboratory.

The two-dimensional nature of CCDs will be of more importance in the next chapter where ICP–AES is discussed. In ICP–AES, two-dimensional reading devices are necessary to reduce interferences by allowing second and third diffraction orders that overlap in one dimension to be separated vertically onto the two-dimensional array of a CCD.

As an interesting side note, CCDs are the imaging detectors in older digital cameras, fax machines, advanced telescopes, and many other imaging instruments. Here, like in spectrometry, an analog (continuous) signal is segmented into a digital or pixilated image. Arrays used in high-end (expensive) cameras can offer the same resolution as extremely low ASA-rated emulsion camera films. The array can contain millions of pixels and a color grid on the array allows for color photography. As soon as an analytical instrument is placed into production, development of the next model is underway. For example, CCD, a remarkable detector in advanced FAES instruments, will soon be replaced by complementary metal oxide semiconductors (CMOS) detectors. The limitations of CCDs are the time requirements for amplification and the relatively slow and spatial-dependent readout of the signal. CMOS detectors have readout circuitry built into each pixel and each pixel can be randomly addressed and read. This enables relatively fast readout speeds. Lower-end CMOS detectors are already in modern cell phone cameras. More advanced CMOS detectors will have low noise, high-speed data collection and readout, and a broader dynamic range.

3.4.6 Noise in Detectors

Chemists have little difficulty when they are measuring high concentrations of an analyte, but this situation is relatively rare in environmental and geochemical applications. Most analysts attempt to determine whether an analyte is present or not and, in doing so, must measure the smallest amount of an analyte in a sample near the detection limit of the instrument. In many areas of science, such as environmental, forensics, and biological applications, the analyst is always struggling to lower the detection limit. This is referred to as “chasing the detection limit” since one of the goals of instrumental manufacture is to improve instruments to measure smaller and smaller quantities of analytes. The detection limit is basically the minimum concentration that one can measure above the “noise” in an instrument. To fully understand this concept, different types of noise must be discussed.

Environmental noise consists of factors in the immediate lab environment that will affect an instrument or sample. Obviously, if an instrument is sensitive to vibrations, such as an NMR, one would not place this instrument in an area of high vibration, such as next to an elevator. Similarly, an analyst would not want to make delicate ppm-, ppb-, and ppt-level concentration measurements in an environment where the analyte of interest is present in high concentrations (such as in the air near a metal smelter). Most environmental noise with respect to metal contamination can be avoided by locating the sample preparation area away from the instruments or locating each in a clean room or high-efficiency particulate air (HEPA) hood.

Chemical noise tends to be unavoidable and specific to each analyte. One of the simplest types of chemical noise, that is completely unavoidable, is the electronic and vibrational fluctuations present in an atom or compound due to the Maxwell distribution of energies. These result in broad

UV–visible absorption and emission spectra in aqueous samples but are of little consequence in gas phase transitions experienced in AA and ICP. Other forms of chemical noise include slight temperature and pressure fluctuations that affect measurements.

Instrumental noise is common and in many cases can be avoidable or managed. Instrumental noise is separated into three categories: thermal, shot, and flicker noise. Thermal noise (also referred to as Johnson noise) results from the thermal agitation of electrons in resistors, capacitors, and detectors. Most of the thermal noise can be overcome by cooling specific components of the instrument such as is done in advanced detectors (i.e. CIDs). Shot noise results from a current being generated by the premature movement of electrons across a junction. The best example is an electron being emitted from a photoemissive material, such as in photocells or electron multipliers that are used to measure the intensity of visible and UV light. There are a few electronic ways of minimizing shot noise, but these are beyond the scope of our discussion here. Flicker noise results from random fluctuations in current and is inversely related to frequency. Flicker noise is overcome by electronically modulating the detector output signal to a higher frequency where less noise is present (i.e. from 10^2 to 10^4 Hertz).

In practice, the analyst is concerned with distinguishing between a real signal and instrumental noise, quantified as the S/N ratio. S/N is mathematically defined as

$$\frac{S}{N} = \frac{\text{mean of signal}}{\text{s.d. of signal}} = \frac{\bar{x}}{\text{s.d.}} \tag{3.28}$$

where S is the mean of approximately 20 blank measurements and N is the standard deviation of these measurements. From a statistical standpoint, it should be noted that S/N is equal to the reciprocal of the relative standard deviation (RSD).

Two approaches are used to minimize noise: hardware and software. Common hardware approaches that are used to decrease noise are (1) grounding and shielding components and detectors, (2) using separate amplifiers for different signals, (3) placing frequency or wavelength filters "up line" from the detector, (4) modulating the instrument signal to a "clean" frequency, and (5) chopping the signal to obtain a reference reading of the background that can be subtracted from the sample signal. For spectroscopy, one of the most common software noise reduction techniques is to take as many readings as reasonably possible. The observed signal to noise ratio (S/N) is a function of the number of readings (n) taken as shown by

$$\frac{S}{N} = \sqrt{n}\frac{S_x}{N_x} \tag{3.29}$$

where S_x and N_x are the signal and noise readings for a specific setting. Note that as one takes more readings the S/N decreases by the square root of the number of measurements. By taking two measurements, one can increase the S/N by a factor of 1.4; or by taking four measurements the analyst can half the noise. For the topics covered in this book, molecular spectrometry, FAAS, ICP–AES, and ICP–MS instruments generally allow for multiple measurements over 5–10 seconds to be taken and averaged. And, of course, the analyst can always analyze a sample multiple times given the common presence of automatic samplers in the modern laboratory.

3.5 Summary

This chapter focused on the physics of light and how these principles can be used to separate and detect components of UV and visible light into specific wavelengths. While this has mostly been a review of college physics, it was necessary to review these concepts in regard to components of analytical instruments. In the following chapters, the optical physics and components presented in this chapter will be combined to construct instruments capable of measuring the absorbance or emission of gaseous atomic and molecular species with visible and UV radiation. These concepts will also be further extended to explain how a variety of additional instrumental components work. Chapter 4 will focus on molecular spectrometry, Chapter 5 will focus on flame atomic absorption and emission techniques, and Chapter 6 will focus on ICP combined with AES. Finally, Chapter 7 will extend atomic spectrometry to include mass spectrometry, the current gold standard in the analysis of metals in atmospheric, aqueous, soil, and biological tissues.

One of the most important technological discoveries in the past century was the p–n junction, the basis of every electronic component today. For example, your cell phone contains over five billion micro- (actually nano-) transistors. Research and read about this discovery at: https://en.wikipedia.org/wiki/P%E2%80%93n_junction

and https://en.wikipedia.org/wiki/Walter_Houser_Brattain.

Reference

Keliher, Peter N; Wohlers, Charles C. Echelle grating spectrometers in analytical spectrometry, *Analytical Chemistry (Washington)*, 1976, Vol. 48 (3), p. 333A–340A

3.6 Questions

1. What is the difference between spectrometry and spectroscopy?

2. What does each of the acronyms stand for?
   AAS
   ICP
   AES

OES
FAAS
FAES
MS

3 Draw a Jablonski diagram showing electronic energy transfer levels.

4 Name all types of electromagnetic radiation starting with the longest wavelength to the shortest wavelength. Correlate the frequency and energy levels with wavelength.

5 Name the color of the visible wavelengths.

6 What are the general wavelengths of each type of electromagnetic radiation?

7 What types of atomic and molecular energy transitions are associated with each type of electromagnetic radiation (Table 3.2)?

8 Show the mathematical relationships between the wavelength of an absorbed or emitted photon and the energy transitions (n values) of a valence electron.

9 How is absorbance related to transmission?

10 How is absorbance related to analyte concentration (Beer's law)?

11 Explain how fluorescence occurs with a Jablonski diagram and in words.

12 Explain how phosphorescence occurs with a Jablonski diagram and in words.

13 How do fluorescence and phosphorescence differ in transition time scales?

14 How does the quantification of fluorescence and phosphorescence differ from basic absorbance measurements?

15 Name and explain all the possible interactions between electromagnetic radiation and material surfaces, including reflection, diffraction (both types), dispersion, and refraction.

16 Dispersion can occur by diffraction and refraction. Explain how each occurs.

17 What is the refractive index of a substance?

18 What is the major difference between Echellette and Echelle gratings?

19 How are linear dispersion, reciprocal linear dispersion, and resolution used to judge the quality of a monochromator?

20 Draw a basic monochromator and explain how it works. What is its fundamental function?

21 What two components can result in dispersion in the monochromator?

22 Draw and explain how a phototube works.

23 Draw and explain how a photomultiplier tube works.

24 Contrast a phototube and a photomultiplier tube.

25 What is a dynode?

26 What is the difference between a discrete and a continuous PMT?

27 What is a charge-coupled device?

28 What is the difference between a forward and reverse bias in a CCD?

29 What is the elemental difference between the p-doped and n-doped areas of a diode?

30 What is a diode?

31 What is a linear diode array, how does it work, and what are its advantages?

32 Contrast a PMT and an LDA.

33 What are charge transfer devices (CTDs) and why they are replacing PMTs?

34 Contrast a CTD and a PMT.

35 Why are charge injection devices (CIDs) not common in FAAS and FAES?

36 How are complementary metal oxide semiconductor (CMOS) detectors superior to CCDs?

37 What are the sources of noise in detectors?

38 Explain how the signal to noise ratio is used and why it is important.

39 How have computers changed FAAS and FAES instruments?

Supporting Information

Additional supporting information may be found online in the Supporting Information section in the HTML rendition of this article.

4

Analytical Molecular Spectrometry

4.1 Introduction

Molecular spectrometry/spectroscopy covers a wide range of spectroscopic techniques to identify compounds and spectrometric techniques to quantify low concentrations of analytes. In Chapter 3, we discussed the components of typical UV–vis spectrometers and in this chapter, we will put these components together to build basic and high-end spectrometers. Molecular spectrometer covers a wide range of the electromagnetic spectrum (EMS) shown in Figure 3.2 and reproduced here in Figure 4.1.

Typical categories of the EMS of use in chemistry include gamma radioactivity detection, X-ray diffraction (XRD) for powder and crystal structure determination, UV for vacuum UV wavelength detection and promoting reactions, UV–vis for spectroscopic detection and quantification of trace concentrations of analytes, near-IR for Raman spectroscopy (identification) of organic molecules, IR for the detection of organic functional groups and compound identification, microwaves for promoting reactions, microwave region for electron paramagnetic resonance spectroscopy, and nuclear magnetic resonance (NMR) for organic compound identification. The most important with respect to analytical chemistry (trace compound concentration determination) discussed in this book are gamma rays (the subject of Chapter 15) and UV–vis that is used for a variety of organic analytes containing a chromophore (a UV–vis absorption active group). This chapter will cover a variety of spectrometers used in the detection of low-concentration analytes (parts per million and lower) undergoing absorption of UV–vis photons or emission of these photons via fluorescence or phosphorescence.

4.2 Basic UV–Vis Spectrometer

Figure 4.2 illustrates the basic components of UV–vis, fluorescence, and phosphorescence spectrometers.

The standard UV light source is a deuterium lamp for UV wavelengths and a tungsten filament lamp for visible wavelengths. The intensity of the wavelengths is regulated by an adjustable physical slit opening. The wavelength selector, depending on the cost and quality of the instrument, can be a prism for inexpensive instruments, or Echelle or Echelette monochromators for more high-end instruments. The sample, usually aqueous in analytical chemistry, is contained in a 1-cm cuvette made of plastic or standard glass for visible wavelengths and quartz for UV wavelengths. The detectors for absorbance (subject to Beer's law discussed in Section 3.2.1), fluorescence, and

Essential Methods of Instrumental Analysis, First Edition. Frank M. Dunnivant and Jake W. Ginsbach.

Companion Website: www.wiley.com/go/essentialmethodsofinstrumentalanalysis1e

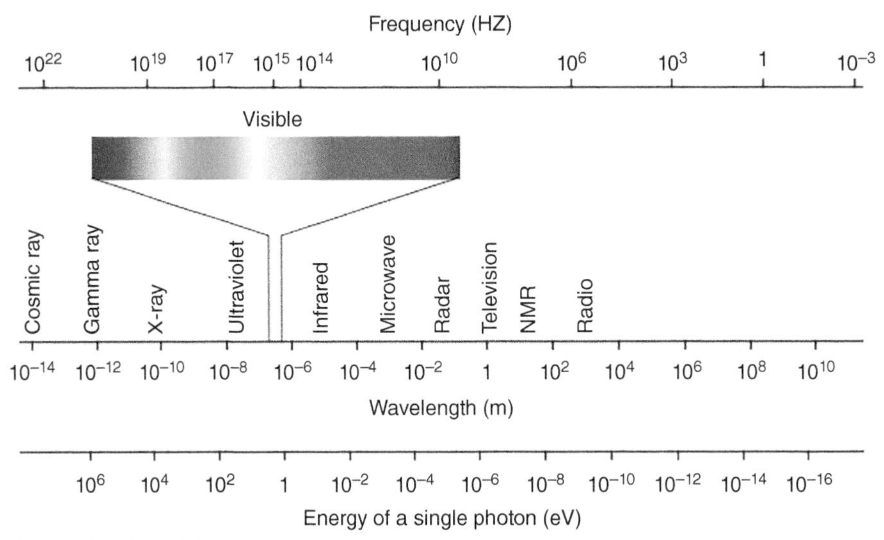

Figure 4.1 The electromagnetic spectrum (Source: Authors Dunnivant and Ginsbach)

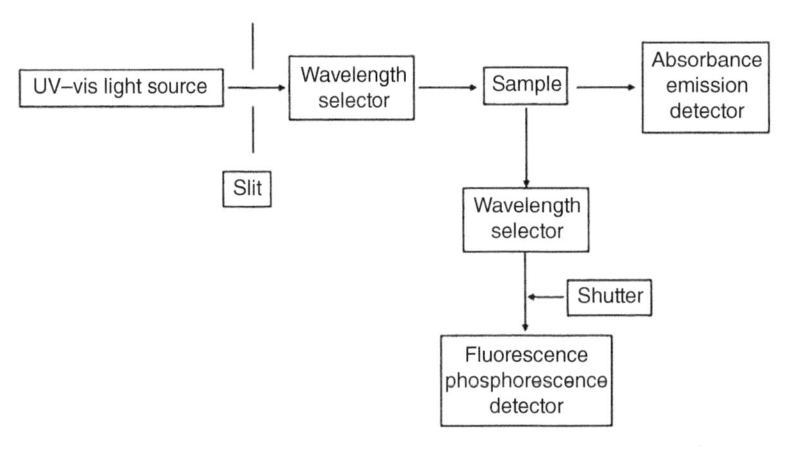

Figure 4.2 A basic single-beam UV–vis spectrometer (Source: Authors Dunnivant and Ginsbach)

phosphorescence wavelengths can be a phototube (PT), linear diode array (LDA), photomultiplier tube (PMT), or charged coupled device (CCD) depending on the age, cost, and quality of the instrument. The time-sensitive shutter is used to distinguish between fluorescence and phosphorescence emissions.

4.3 From Simple to Complex UV–Visible Spectrometers

There are a variety of commercially available spectrometers for measuring inorganic and organic analytes, including field and stationary laboratory instruments. One of the most commonly used single-beam, field instruments is the Hach® unit. Figure 4.3 shows a Hach DR300 colorimeter used in field applications including the analysis of aluminum, ammonia nitrogen, bromine, chlorine dioxide, chlorine pH, iron, manganese, molybdenum, monochloramine and free ammonium,

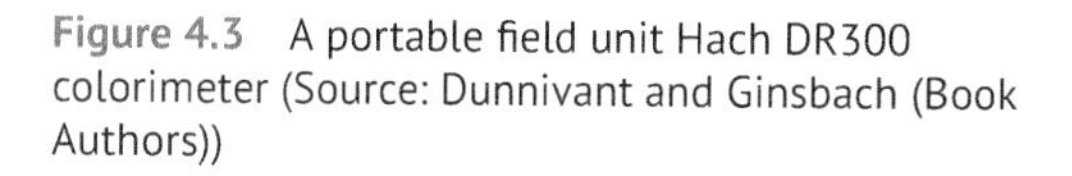

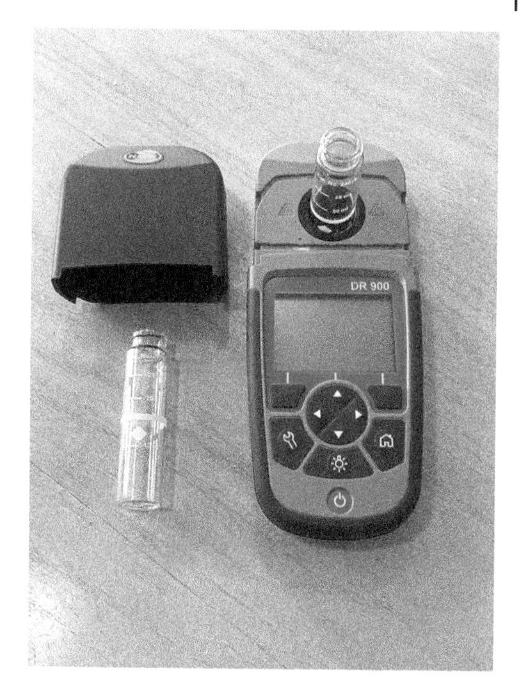

Figure 4.3 A portable field unit Hach DR300 colorimeter (Source: Dunnivant and Ginsbach (Book Authors))

nitrate, dissolved oxygen, ozone, phosphate, and zinc. Most of these analytes have a precalibrated standard curve (line) programmed into the instrument for relatively rapid analysis using powdered reagent packets.

For decades, the workhorse of UV–visible instruments was the spectrophotometer-20 shown in Figure 4.4. This single-beam instrument was originally manufactured with analogue (needle) readout but was later upgraded with a digital readout which significantly improved the accuracy and recorded significant figures of the absorbance readings. Aqueous samples are taken, chromophores are present or developed with reagents, the solution is placed in a cuvette, and an absorbance reading is taken and compared to a standard curve (line) from external standards that are treated the same as the samples.

Today most teaching laboratories use more simple and compact systems such as the StellarNet® and Vernier® instruments. These are also single-beam instruments, but in addition, these units allow for wavelength scanning to determine a complete UV spectrum in order to determine the maximum absorbance wavelength of a sample and allow for measuring the absorbance as a function of time to study reaction kinetics.

More advanced research-grade instruments use a double-beam technology shown in Figure 4.5 where the UV–vis wavelengths are split between a reference cell and the sample cell. This design produces more stable (less noise) and reproducible absorbance reading from the detector.

4.4 Fluorescence and Phosphorescence Instruments

Fluorescence and phosphorescence instruments are slightly more complicated than a standard UV–vis spectrometer. The main difference is the radiation source and the emitted wavelengths must be refined with individual monochromators. A basic diagram is shown in Figure 4.6.

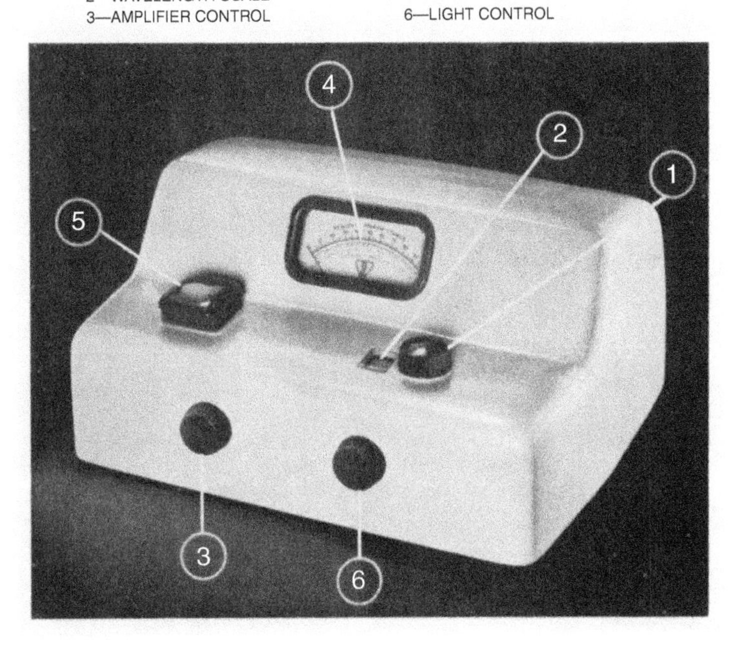

Figure 4.4 An original Spect-20 spectrophotometer (Source: With permission of Science History Institute)

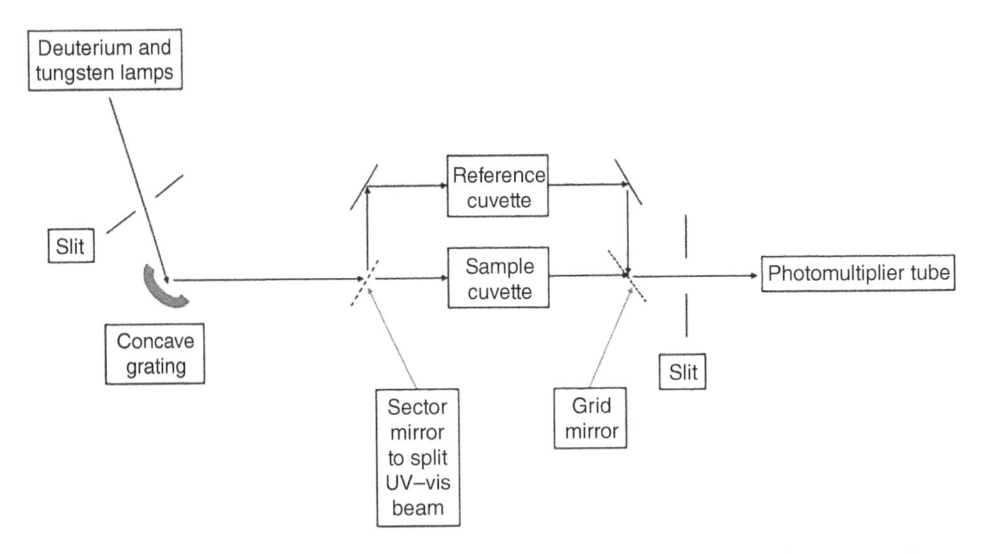

Figure 4.5 A double-beam UV–vis spectrometer design used in research-grade instruments (Source: Authors Dunnivant and Ginsbach)

Standard deuterium (for UV wavelengths) and tungsten (for visible wavelengths) lamps are used for source radiation. The first monochromator selects the excitation wavelength that is directed to the sample. Fluorescence and phosphorescence are emitted in all directions, but the detector emission monochromator is aligned at 90° to the sample in order to remove radiation (and noise) from the source lamp. Recall from Chapter 3 that in fluorescence, an excited molecule then undergoes a vibrational relaxation (again, in polyatomic compounds) to a lower vibrational state within the same electronic state. Fluorescence emission occurs when the electron falls from one excited electronic state back to the ground electronic state (a singlet-to-singlet transition). This type of transition occurs since most vibrational transitions occur faster (10^{-15} to 10^{-12} seconds) than electronic transitions (around 10^{-8} seconds). Since the vibrational relaxation emits energy in the form of heat, the energy of the fluorescence emission is lower (and consequentially the wavelength is longer) than the absorbed energy.

Again from Chapter 3, phosphorescence occurs when a radiation-less vibrational transition occurs between two different spin states (for example, from a singlet (S_1) to a triplet (T_1) excited state). After this spin-forbidden transition, another spin-forbidden transition from the excited state to the singlet state must occur (from a triplet to a singlet state in Figure 3.5). Phosphorescence transitions are easy to distinguish from fluorescence since they occur much more slowly; fluorescence occurs over time scales of 10^{-5} seconds while phosphorescence occurs over time scales of minutes to hours since spin-forbidden transitions have a low probability of occurring (a low transition rate).

A very useful technique is possible with phosphorescence since emission can still be detected after the source lamp is turned off or blocked from the detector. It is possible to place a shutter between the excited sample and the detector. The shutter is opened after the source lamp is turned off and phosphorescence is occurring. This allows the detector to measure only phosphorescence. Phosphorescence has no use in atomic spectrometry but can be used for the analysis of metals in polyatomic states, such as those occurring for uranium oxide species. Phosphorescence has multiple uses in organic chemistry. In general, fluorescence and phosphorescence have much better detection limits, typically three orders of magnitude better.

4.5 Instrument Maintenance

UV–vis, fluorescence, and phosphorescence require little maintenance other than keeping a clean and dust-free workplace. Most optical components are fixed and monochromators are calibrated when the instrument is turned on. Source lamps typically last for at least a decade. The most common problem is keeping the sample cell holder clean from solution spillage.

4.6 Summary

Molecular spectrometer is an important tool in the detection and study of inorganic and organic chemicals but as discussed in Chapter 1, all instruments require calibration with a set of external standards. The dynamic range of absorption spectrometry is limited by Beer's law while fluorescence and phosphorescence have several orders of magnitude of linearity.

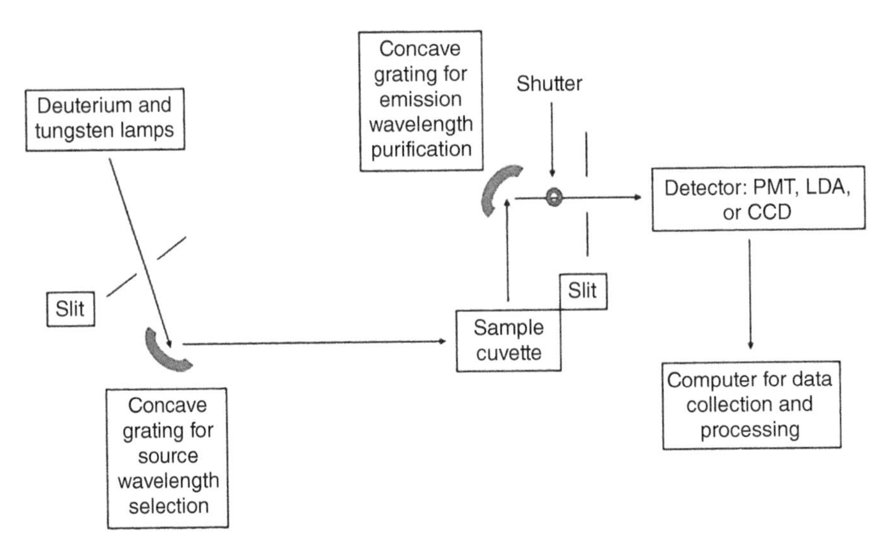

Figure 4.6 A combined fluorescence and phosphorescence instrument (Source: Authors Dunnivant and Ginsbach)

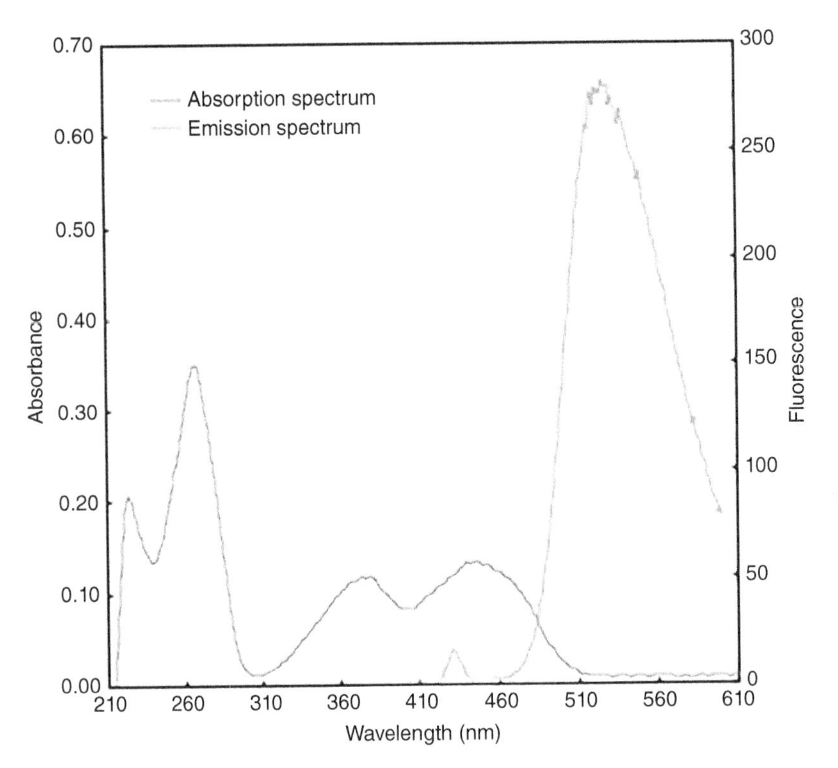

Figure 4.7 A UV–visible wavelength scan to determine the maximum for Riboflavin absorbance and fluorescence (Source: Used with permission from Nate Boland)

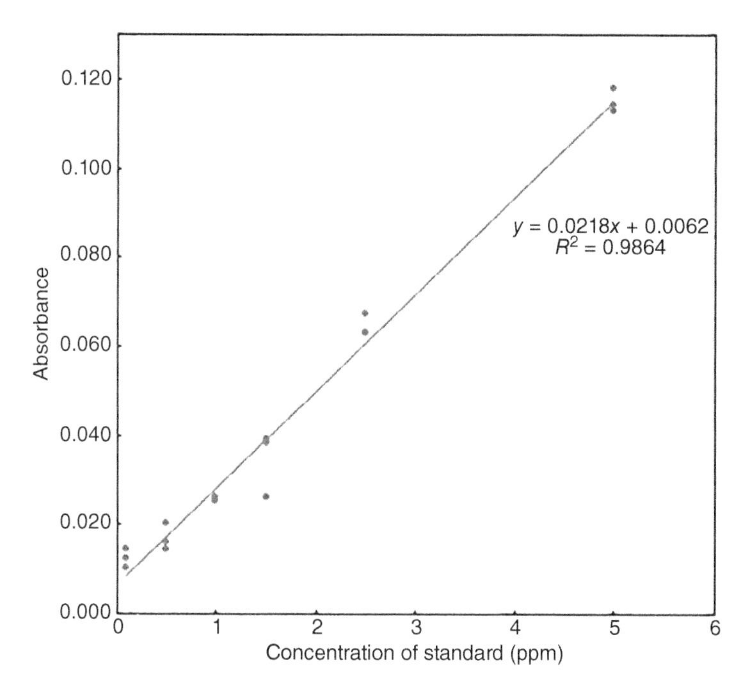

Figure 4.8 An absorbance calibration curve (line) for Riboflavin at 375.28 nm (Source: Used with permission from Nate Boland)

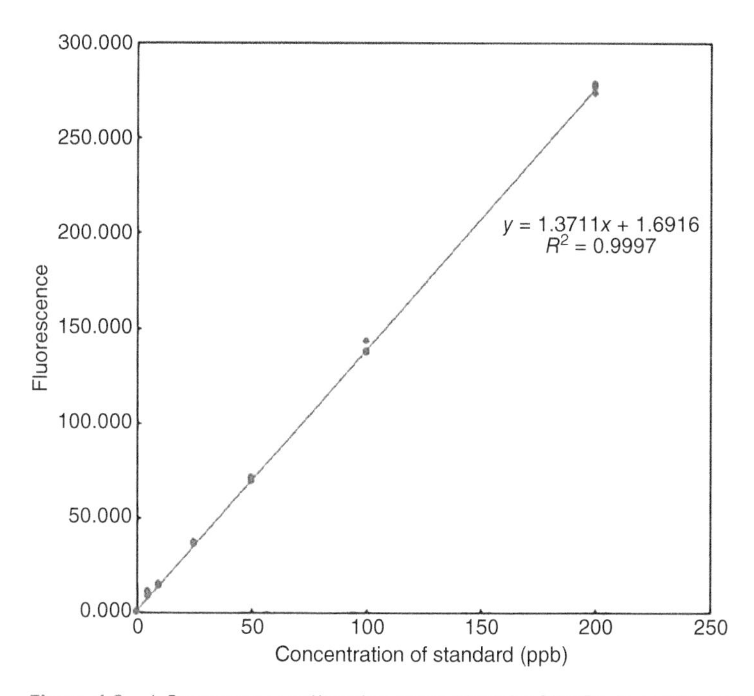

Figure 4.9 A fluorescence calibration curve (line) for Riboflavin at 526 nm (Source: Used with permission from Nate Boland)

4.7 Case Study: Quantitation of Riboflavin by UV–Vis and Fluorescence Spectrometry by Nate Boland

Riboflavin (vitamin B2) is bright yellow and fluorescent green, which provides two means for quantitative detection: both absorption and fluorescence spectroscopy. In a quantitative methods laboratory, Professor Nate Boland of Whitman College had his students compare the means of detecting and quantifying riboflavin by both these methods and use each method to quantify the amount of riboflavin in over-the-counter dietary supplements.

Figure 4.7 shows a UV–vis wavelength scan to determine the λ-max for Riboflavin absorbance and fluorescence. Figure 4.8 shows the absorbance calibration curve (line) while Figure 4.9 shows the fluorescence calibration curve (line) created using external standards.

The two methods produced slightly different concentrations of Riboflavin in vitamin supplement tablets. Twenty tablets were analyzed at once and the results are given for the average concentration in one tablet. The absorbance method resulted in a tablet concentration of 176 mg $\pm$ 5% while the fluorescence method resulted in 140 mg $\pm$ 1.6%. The manufacturer's label quoted a concentration of 100 mg. Note: dietary supplements are not regulated by the Food and Drug Administration and routinely vary in concentration and dose.

5

Flame Atomic Absorption and Flame Emission Spectrometry

5.1 Introduction and History of Atomic Absorption Spectrometry (AAS)

The first observation of atomic emission dates back to at least the first campfire where hominoids/humans observed a yellow color in the flame. This color was caused by the relaxation of the 3p electron to a 3s orbital in sodium (refer to the energy level diagram in Figure 3.3 given earlier), and in part by carbene ions. Slightly more advanced but still unexplained observations were responsible for the first development of colorful fireworks in China over 2000 years ago. A few of the more relevant discoveries for atomic spectroscopy were the first observations by Newton of the separation of white light into different colors by a prism in 1740, the development of the first spectroscope (a device for studying small concentrations of elements) in 1859 by Kirchhoff and Bunsen, and the first quantitative analysis (of sodium) by flame emission by Champion, Pellet, and Grenier in 1873. The birth of atomic spectrometry began with the first patent of atomic absorption spectrometry by Walsh in 1955. In the same year, flames were employed to atomize and excite atoms of several elements. The first atomic absorption instrument was made commercially available in 1962. Since then, there have been a series of rapid developments that are ongoing in atomic and emission spectrometry including a variety of fuels and oxidants that can be used for the flame, the replacement of prisms with grating monochromators, a variety of novel sample introduction techniques (hydride, graphite furnace, cold vapor, and glow discharge), advances in electronics (especially microprocessors to control the instrument and for the collection and processing of data), and the development of atomic fluorescence spectrometry. Surprisingly, detection limits for the basic instruments used in flame atomic absorption and emission spectrometry have improved little since the 1960s, but specialty sample introduction techniques such as hydride generation and graphite furnaces have greatly improved detection limits for a few elements.

5.2 Components of a Flame Atomic Absorption/Emission Spectrometer System

5.2.1 Overview

The general layout of optical components for a flame atomic absorption and emission spectrophotometer is shown in Figure 5.1. In FAAS, a source of pure light (wavelength) is needed to excite the analytes without causing excessive instrumental noise. Most instruments today use a hollow cathode lamp (HCL) that is specific to each element being analyzed to emit a very narrow bandwidth of

Essential Methods of Instrumental Analysis, First Edition. Frank M. Dunnivant and Jake W. Ginsbach.

Companion Website: www.wiley.com/go/essentialmethodsofinstrumentalanalysis1e

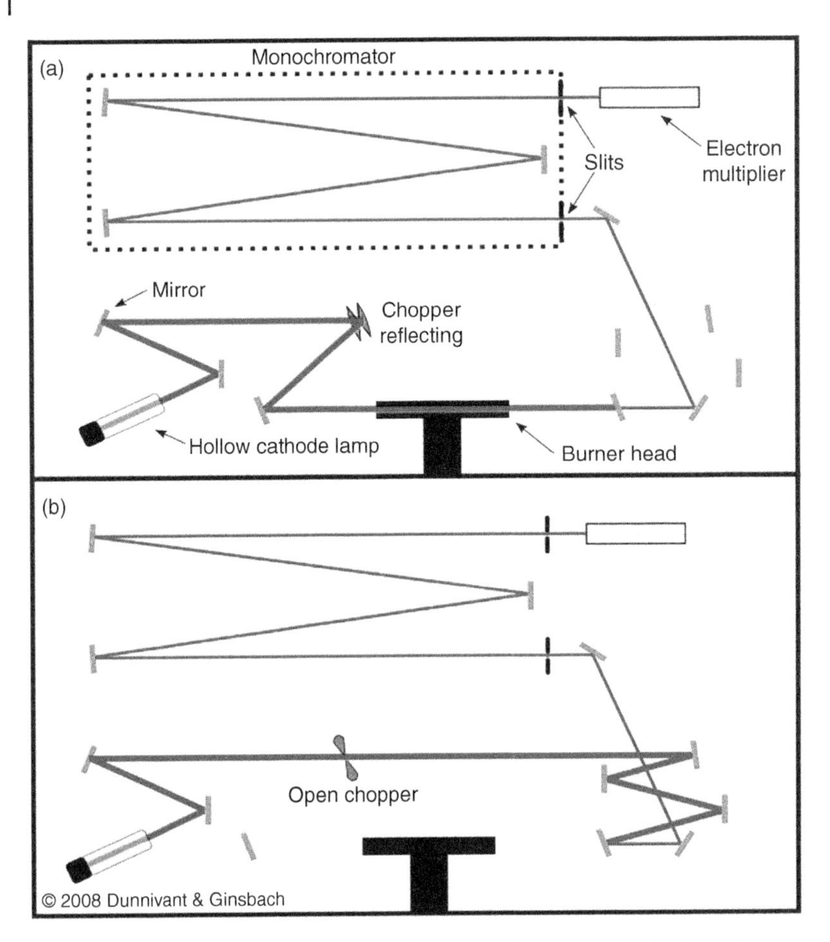

Figure 5.1 An overview of a flame atomic spectrophotometer. (a) In detection mode where the source beam goes through the flame sample cell and (b) in reference mode where the source beam bypasses the sample cell (Source: Dunnivant and Ginsbach)

ultraviolet (UV) or visible radiation into the instrument for detection. All modern and some older atomic absorption systems use double-beam technology where the instrument splits the beam of source light, with respect to time, into two paths. One of these beams does not pass through the sample and is used to measure radiation intensity and fluctuations in the source lamp, while the other beam is used to measure the radiation that interacts with the analyte. The splitting of the source beam is accomplished with a chopper, as illustrated in Figure 5.2, where the chopper is in the reflection position in the center of the figure. The other half of the time, radiation is passed through the sample cell: in this case, the flame that contains the atomized gaseous metal analytes. A portion of the metal atoms absorbs a specific wavelength of radiation (matching the wavelength to four significant figures emitted by an HCL) that results in a quantitative reduction in the intensity of radiation leaving the sample cell. After interaction with the sample, or in the case of the reference beam bypassing the sample in the burner head, the beam of light is reflected by mirrors into the monochromator. This reflected light that contains various wavelengths past through a small slit that is size (width) adjustable. Then it reflects off a focusing mirror to travel to the dispersing device (today a grating monochromator is used). Finally, the separated wavelengths of light are focused toward the exit slit with another focusing mirror. By changing the angle of the monochromator,

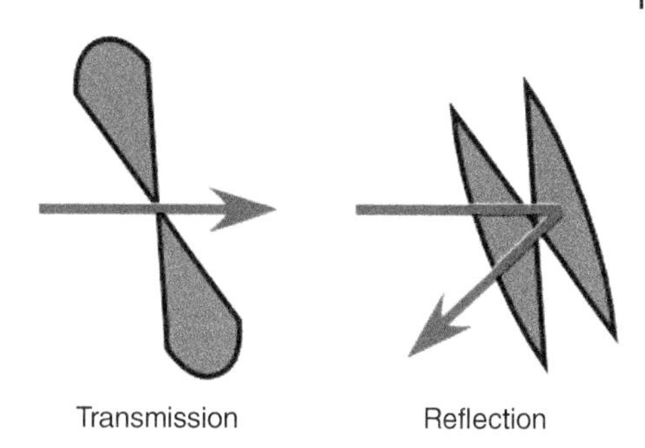

Figure 5.2 A chopper, a device used to split source light with respect to time (Source: Dunnivant and Ginsbach)

different wavelengths of light enter the detector through the exit slit of the monochromator. The detector in older AAS units was a photoelectron multiplier tube (PMT) that amplifies and converts the signal of photons to electrons that are measured as an electrical current. Today, various forms of advanced photon-multiplying devices, described later, are used.

An emission spectrometer can be identical to an absorption system except that no external radiant light source is used to excite the atoms. In flame emission spectroscopy, the electrons in the analyte atoms are excited by the thermal energy in the flame. Thus, the sample is the source of photon emissions through relaxation via resonance fluorescence (Section 3.3). Note that this results in emission systems that are only single beam in design.

One modern and very important component not shown in Figure 5.1 is an automatic sampler. An automatic sampler is a device that is used to analyze numerous samples without the constant attention of an analyst. Modern automatic samplers can hold over a hundred samples. They can be used for most of the modes of FAAS and FAES. Instrument setup for running a set of samples with an automatic sampler is similar to the normal instrument setup: the correct lamp must be installed and aligned and the flame must be lit and optimized with respect to burner height. Software and instrument settings may also need to be adjusted to ensure a smooth run of samples. Of course, automatic samplers require the use of a computer to both run the automatic sampler and collect the relatively large amount of data produced from various samples. Automatic samplers greatly reduce the cost of analysis when large numbers of samples need to be analyzed for the same element. Usually, an analyst will spend a normal day shift processing (digesting and diluting) samples. An instrument equipped with an automatic sampler can be set up at the end of the day and allowed to run all night without paying an analyst to manually run the samples or "babysit" the instrument. In the morning, the analyst arrives with a collection of data to process.

5.2.2 Optical Radiation Sources

For FAAS and FAES, the wavelengths of interest are in the UV and visible range. There are three basic types of radiation sources that are utilized in these instruments: continuous sources, line sources, and laser sources. A continuous source, also referred to as a broadband source, emits radiation containing a broad range of wavelengths. A plot of intensity on the *y*-axis and wavelength on the *x*-axis is shaped like a broad Gaussian distribution with a few small peaks and shallow valleys. The emission wavelengths of a continuous source can range over hundreds of nanometers. Examples of lamps considered to be continuous sources are deuterium, mercury, xenon, and tungsten lamps. These various lamps are used as background correction lamps (signal to noise (S/N) correction devices) in AAS and AES instruments and not as source lamps for analyte detection.

Line sources are lamps that emit very narrow bands of radiation, but this source of radiation is not as pure as radiation from a laser. The most common line source radiation generator used in AAS is the HCL. These lamps are encased in a cylinder made out of glass walls and a quartz end cap. Glass end caps can be used for visible wavelength emitting materials, while quartz must be used for UV-emitting HCLs. These cylinders are filled with a noble gas (Ne or Ar) to sub-atmospheric pressures of 1–5 Torr. HCLs also contain a tungsten anode, a cathode composed of the metal of interest, and various insulators (usually made out of mica). Lamps containing more than one element in their cathode are also available, but most FAAS and FAES instruments can only measure one element at a time.

An HCL works by placing a 300 V potential across the electrodes that generates a current of 5–15 mA. As the electrons migrate toward the cathode, they collide with the noble gas atoms and ionize them. These charged noble gas atoms collide with the cathode. If the kinetic energy of the noble gas cations is high enough, some of the metal atoms on the cathode will be dislodged, producing an atomic cloud of metal in the gaseous phase. This process is referred to as sputtering. A portion of the collision energy will cause the gaseous phase metal atoms to enter into an excited electronic state. These excited gaseous-phase atoms relax from the excited state to the ground state through resonance fluorescence that emits a specific wavelength of UV or visible radiation (Section 3.3). This emission usually corresponds to only one or a few energy gaps that results in the generation of a relatively pure source of light. Higher currents can be used to generate more photons, but too much current results in self-absorption of the photons and Doppler broadening of the bandwidth (a degradation of the spectral purity of the signal). A proper balance of current and generated radiation intensity allows for maximum intensity of the lamp and maximum analyte sensitivity for absorption spectrometry.

While the use of an HCL is most common in AAS systems, other line source generators are available including lasers and electrode-less discharge lamps (EDLs) such as microwave EDLs and radiofrequency EDLs. Lasers produce the most pure form of radiation but are of little to no use in AAS or AES, and thus, will not be discussed here because they are not tunable to the range of wavelengths needed in AAS. Lasers are utilized for sample introduction for inductively coupled plasma systems; this laser ablation technique is discussed in Chapter 6. EDLs are also relatively rare in AAS instruments and are only used for a few selected elements that are too volatile or unstable at pressures and amperages used in HCLs. In these lamps, the metal atoms are excited using microwave or radiofrequency generators. EDLs produce higher intensity radiation than an HCL but are generally not as reliable or commercially available for all elements of interests. As a result, most analytical technicians use HCLs in their FAAS systems. Examples of common elements used in EDLs are Hg, As, and Se.

5.2.3 Mirrors

Mirrors are important components of all spectrophotometers. Mirrors are used to direct radiation by reflecting it in a specific direction. Most atomic absorption units employ mirrors that reflect only UV and visible wavelength radiation. Mirrors are usually made of flat- or concave-shaped glass coated with a metal surface, usually aluminum. In some cases, a plastic support can be used as the coating substrate. Metals are plated on the surface by a technique known as thin-layer vacuum coating. A variety of mirrors exist, including plane, concave, and collimating (a form of focusing).

5.2.4 Choppers

A chopper is a device that splits a single beam of radiant light into two directions. A chopper is a circular-shaped device split into quarters (refer to the figure below). Opposing quarters of the chopper are open (contain no material as in the left-hand side of Figure 5.2) while the other quarters of the chopper contain a mirrored surface (as in the right-hand side of Figure 5.2) in order to direct the light to the sample cell (as shown in Figure 5.1). The chopper rapidly spins in the beam of light, directing the light by reflection as the mirror interacts with the light or by allowing the radiation to pass as the empty portion of the chopper moves past the beam of light. Thus, choppers split the beam of light with respect to time (as opposed to space where two adjacent mirrors would direct the light in two different directions). If the chopper spins faster than the fluctuation of noise in the source signal, an accurate measurement of the background noise can be obtained and corrected for in the sample readings.

5.2.5 Burner Head

The burner head in FAAS and FAES systems is where all of the chemical reactions take place. The burner head, as shown in Video S.5.1, consists of an inlet tube, fuel and air inlets, a nebulizer, mixing cell, and the flame (the reaction and sample cell). Aqueous samples move through the inlet tube into the nebulizer which atomizes the liquid into small droplets using a Teflon or glass impact bead (not shown in this figure) placed at the entrance to the nebulizer to help break up and aspirate the inlet fluid. The mixture of sample droplets created by the impact bead, oxidant, and fuel is homogenized by the mixing fins in the mixing cell before this mixture is atomized in the flame. The sample liquid is drawn into the nebulizer by a phenomenon known as the Bernoulli effect where a compressible fluid (the fuel and oxidant gases) is passed through a constriction in a pipe.

The Bernoulli principle is the pressure differential created when gases flow through a constriction. The gaseous flow, where the velocity is below the speed of sound, creates streamlines along the path of flow (represented by the horizontal lines in Figure 5.3). A streamline is an imaginary line that describes the path of a gaseous molecule through a system operating under laminar flow (a flow system with little random motion or mixing). The Bernoulli principle states that the sum of the mechanical energies along a streamline is the same at all points on that streamline. This requires that the sum of kinetic energy and potential energy remain constant along the streamline. If the gas is flowing out of one reservoir (the reservoir with a larger radius) into a constricted reservoir,

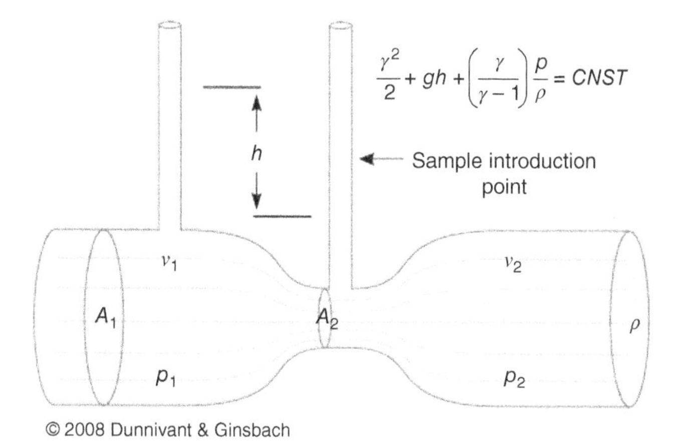

Figure 5.3 Diagram explaining the Bernoulli principle (A is cross-sectional area, v is fluid velocity, p is pressure, ρ is fluid density, and h is head pressure (or difference in pressure) (Source: Dunnivant and Ginsbach)

the sum of all forms of energy along the way is the same on all streamlines. The total energy at any point can be described by the following equation:

$$\text{Total Energy} = \frac{v^2}{2} + gh + \left(\frac{\gamma}{\gamma - 1}\right)\frac{p}{\rho} \tag{5.1}$$

where γ is the ratio of the specific heats of the fluid, p is the pressure at a point, ρ is the density at the point, v is the speed of the fluid at the point, g is the acceleration due to gravity, and h is the height of the point above a reference plane. Since the total energy of the system must be conserved, the total energy must equal a constant for the system.

Consider a gas molecule moving from left to right along one of the horizontal streamlines in Figure 5.3. While all of the variables in the total energy equation change, it is necessary to only focus on the pressure and the velocity at a particular point. This is valid because the pressure and the velocity at a single point are the dominant contributors to the overall total energy. When this molecule moves from left to right it encounters less pressure and subsequently h becomes smaller. In order for energy to be conserved, the velocity (v) must become larger. The converse is true when the particle moves farther right into an area of higher pressure. In this instance, the velocity must become smaller so the total energy of the molecule never changes over the entire system. As a result of Bernoulli's equation, the highest speed occurs at the lowest pressure, and the lowest speed occurs at the highest pressure.

The Bernoulli discussion (and Figure 5.3) is only illustrative of the general concept of pressure and flow balances. The design of the inlet chamber in AAS units is slightly different, but the principles of the Bernoulli equation cause the sample to enter into the mixing chamber in FAAS and FAES units. For the situations occurring in FAAS and FAES, if a fluid reservoir (the aqueous sample inlet tube) is connected to the low-pressure region of the fuel and oxidant gas inlet constriction, the lower pressure present in the constriction will draw fluid into the system (nebulizer chamber). A compression valve located on the sample inlet pipe is used to regulate the flow. As a result, the Bernoulli effect causes the liquid to move freely into the flame systems without the use of a pump.

Next, the sample enters into the nebulizer, a mixing chamber where the sample is broken into an aerosol mist by the impact bead. The droplet size of this aerosol formed in the nebulizer is of importance since this directly affects how much analyte reaches the flame. Droplets with diameters greater than 20 μm are trapped in the spray chamber by attaching to surfaces and flow to the waste container. Only about 10% of the water that enters the nebulizer reaches the flame. The empirically determined governing equation for the determination of droplet size is

$$d_o = \frac{585}{v}\sqrt{\frac{\gamma}{\rho}} + 597\left[\frac{\eta}{\sqrt{\gamma\rho}}\right]^{0.45} 1000\left(\frac{Q_{\text{liquid}}}{Q_{\text{gas}}}\right)^{1.5} \tag{5.2}$$

where d_o is the droplet size which is a function of viscosity (η), density (ρ), and surface tension (γ) of the sample solution, the flow rate of the nebulizer gas (Q_{gas}) and the aspirated solution (Q_{liq}), and the velocity of the nebulizing gas (v).

After the sample is pulled into the nebulizer and turned into an aerosol mist, it is mixed with the fuel and oxidant by two mixing fins. Common fuels used in FAAS and FAES units are acetylene (for hotter flames) and hydrogen (for cooler flames). Oxygen in the compressed air or nitrous oxide is used as the oxidant to regulate the temperature of the flame. Different elements require different flame conditions, including the choice of fuel and oxidant and the ratio of the fuel to oxidant mixtures. Hydrogen-air flames produce temperatures of about 2000 °C, while acetylene-air flames yield temperatures of approximately 2300 °C and acetylene-nitrous oxide yields temperatures of 2900 °C. Within these fuel types, fuel-rich mixtures yield cooler flames and oxidant-rich mixtures yield hotter temperatures. Temperatures are optimized for a particular analyte since different metal elements are excited or atomized under different conditions. In addition, some metals readily form oxides in an oxygen-rich atmosphere, a reducing (fuel-rich) environment is necessary to produce atomic instead of molecular species (such as oxides). Other elements are stable in the atomic state under any fuel/oxidant mixture. After this specific mixture of fuel and oxidant are mixed together with the sample, they exit the burner head and pass into the flame.

Several processes and reactions occur rapidly when the sample molecules enter into the flame. First, the water is evaporated and removed from the metal complex. Next, the heat of the flame degrades organic and dehydrated inorganic complexes into gaseous atomic states (ground electronic states) that are then excited by the thermal energy in the flame. In the lower portion of the flame, absorption of photons occurs by the electronic ground state gaseous atoms. As the analytes rise into cooler regions of the flame, the excited atoms relax and emit a photon for emission spectrometry. Finally, the fumes and metals from the flame are removed from the laboratory by a fume hood exhaust system.

5.2.6 Instrumental Noise in the Source Lamp and Flame

Now that the source lamps and flames have been introduced, it is time to discuss sources of noise in AAS measurements; some of these also apply to AES measurements. FAAS and FAES instruments usually start with a pure source of light, and it is desirable to end with that same wavelength in as pure of a form as possible. Noise results when this process breaks down and the intensity of the wavelength of interest decreases or impure radiation reaches the detector. Decreases in the quality of light occur as the radiation passes through air and at the interfaces of surfaces. In addition, there are three common causes of line broadening: natural, Doppler, and pressure broadening.

5.2.6.1 Natural Broadening and the Uncertainty Effect

Natural broadening of pure spectral lines occurs due to the finite amount of time an atom spends in its excited electronic state. As the absolute time of the two states (ground and excited) approaches infinity, the width of the line resulting from a transition approaches zero; this is a direct result of the Heisenberg uncertainty principle for time and energy.

$$\Delta E \Delta t \geq \frac{\hbar}{2} \tag{5.3}$$

For example, the time required for absorption of a photon by an atomic species is approximately 10^{-15} seconds, while the lifetime of the excited state is about 10^{-9} seconds. This excited state transition is sufficiently short enough that the uncertainty in energy is greater than 10^{-25} J. For UV and visible wavelengths in the system discussed here, the line broadening from this uncertainty in energy affects the wavelength by 10^{-5} to 10^{-6} nm, and is considered negligible compared to other forms of line broadening.

5.2.6.2 Doppler Broadening

The Doppler shift of a wavelength is an important observation in physics. This broadening is caused when an object is moving with respect to a detector while it simultaneously is emitting a wave such as a photon or sound. The observed wavelength will be slightly different when the emitter is moving toward or away from the detector. Everyone who has listened to a train whistle at a railroad crossing has observed this principle; as the train approaches a stationary observer, the sound frequencies are compressed and a slightly higher frequency is heard. In contrast, as the train passes the observer, the frequency is broadened and a lower frequency pitch results. For the instruments discussed in this text, the Doppler effect is observed only in an HCL. If an excited atom is moving toward the sample cell and detector, a slightly shorter wavelength will be observed, while an atom moving away from the detector will emit a longer wavelength. This is also referred to as thermal motion. Even though atomic speeds are significantly less than the speed of light (1000 m/s), this effect can result in spectral broadening since the wavelength of interest may now overlay with another wavelength present in the sample or flame. The net result is an increase in noise and possibly an overlap with another absorbing or emitting atomic species. For the conditions in common FAAS flames, the width of a spectral line is about two orders of magnitude greater than the breadth present in the naturally occurring line due to natural broadening. This is calculated for individual wavelengths by

$$\Delta \nu_D = \frac{\upsilon}{c}\left(\frac{2RT}{M}\right)^{0.5} \tag{5.4}$$

where ν is the frequency of interest, R is the ideal gas law constant, T is temperature, and M is the atomic mass of the element. For typical operating conditions, the broadening caused by the Doppler effect is about 10^{-4} nm. This effect accounts for most of the line width broadening in flame-based instruments.

5.2.6.3 Pressure Broadening

Pressure broadening, also known as collisional or Lorentzian broadening, results from collisions between the gaseous atom of interest and any other atom. Collisions result in radiation-less relaxations by distributing electronic energy into vibrational and rotational energy that lengthens the wavelength of the line as compared to its central frequency (unaffected frequency or wavelength in a vacuum). Pressure broadening can occur in the lamp and the flame in an AAS instrument. In a source lamp, such as an HCL, most collisions are between gaseous sputtered metal atoms and the noble gas. The pressure of Ar or Ne in the source lamp is very low to decrease the frequency of these collisions; most of these collisions result from other gaseous sputtered metal atoms present in the source lamp. The net result is a line broadening of approximately 10^{-6} nm and is much less significant than the Doppler Effect (approximately 10^{-4} nm). The observed two-order-of-magnitude difference illustrates the lack of importance of pressure broadening in HCLs. However, in high-pressure background source lamps, such as deuterium, Hg, and Xe lamps, collisions are more common and the resulting broadening is capitalized upon to emit a broad range of wavelengths in the UV and visible regions. In the flame, the reaction cell used in most AAS units, collisions occur between the analyte of interest, fuel and oxidant molecules, and other ions.

5.2.7 Slits

After the radiant light has passed through the sample, it is directed by mirrors to the entrance slit of the monochromator. A slit is nothing more than a hole or slot in a black surface that allows a narrow beam of light to pass through it. The purpose of an entrance slit is to only allow a fine beam of light to enter the monochromator. After the monochromator separates the white light into its components, a narrow band of wavelengths is directed through the exit slit. This allows only a narrow band of wavelengths to exit the system and reach the detector.

5.2.8 Monochromators

A variety of monochromators, a device that is used to separate wavelengths of light through dispersion, were discussed in Sections 3.3.3 and 5.2.8. There are two types of monochromators: prisms and grating systems. Despite achieving the same goals, as noted in Section 3.3.3, prisms and grating systems separate various wavelengths of light in different fashions. Prisms refract light at the interface of two surfaces with differing refractive indexes creating angular dispersion. Historically, prisms were the first monochromators to be developed, but they have limitations. Their resolution is significantly lower than a grating system, and their separation technique is nonlinear (with respect to distance along the exit slit) which creates mechanical problems with focusing a specific wavelength on the exit slit. The one advantage that prisms possess over grating systems is their low manufacturing cost.

Diffraction gratings are materials with a large number of parallel and closely spaced slits or ridges. Diffraction causes constructive interference at unique points for each wavelength. In Section 3.3.3, the theory behind the two governing equations for diffraction yielding constructive interference was described by

$$n\lambda = d(\sin i \pm \sin r) \tag{5.5}$$

with each of the variables shown again in Figure 5.4. Despite their higher cost grating monochromators are used in all modern medium- to high-end spectrometry systems.

This equation was simplified for an Echelle-style grating to

$$n\lambda = 2d \sin \beta \tag{5.6}$$

The first grating monochromators used were of the Czerney–Turner style illustrated in Figure 5.5. This common form of monochromator was used for decades when PMTs (Section 3.4.2) were the detector of choice. UV and visible wavelengths enter the monochromator through an entrance slit where they are reflected onto the grating device where spectral separation occurs. The separated wavelengths were collimated (focused by wavelength) with a concave mirror toward the exit slit. The tilt angle of the grating device determined the band of wavelengths exiting the monochromator and reaching the detector: usually a PMT.

Today, with the replacement of PMTs in higher-end instruments by more modern microelectronic circuitry (charge transfer and injection devices described in Section 3.4.3), only an entrance

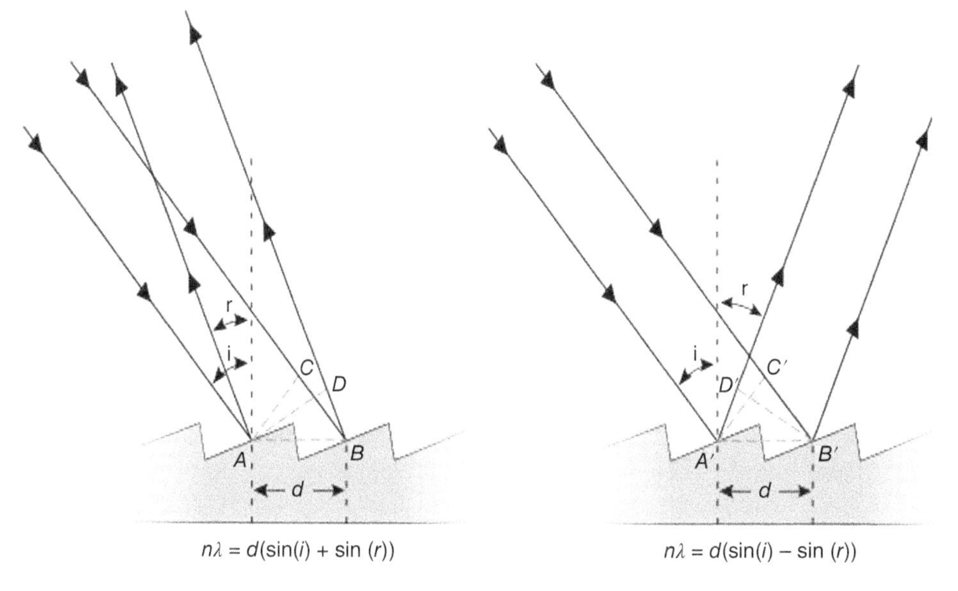

Figure 5.4 Diffraction resulting from a typical Echellette-type grating (Source: Dunnivant and Ginsbach)

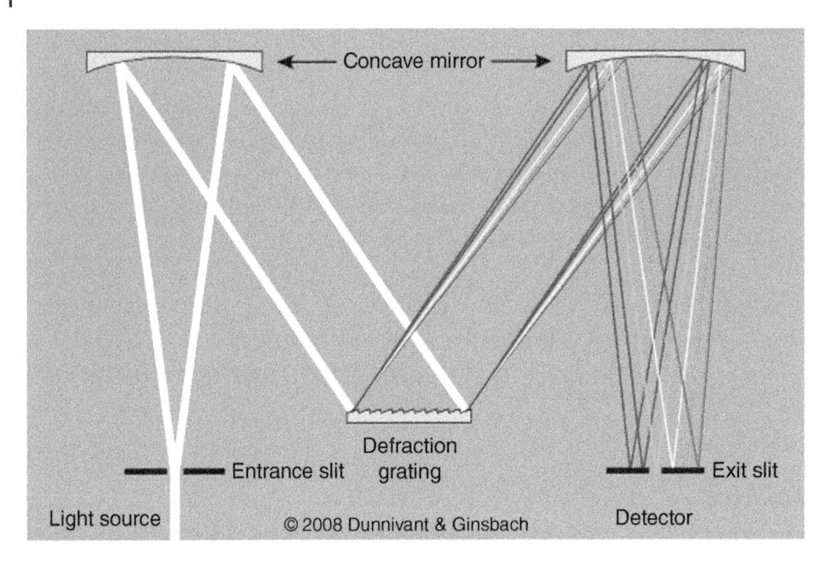

Figure 5.5 A Czerney–Turner-style grating monochromator. Note that only two diffracted beams of light are shown leaving the diffraction grating but identical beams leave each blazed surface and are collimated by a concave mirror onto the exit slit as one rainbow of wavelengths (Source: Dunnivant and Ginsbach)

slit is necessary. In some higher-end FAAS and FAES units, newer monochromator/detector systems have an Echelle grating monochromator and charge transfer device placed together where all wavelengths are measured simultaneously with a charge transfer device without the need for an exit slit (illustrated in the next section). This can also be used in absorption spectrometry where individual source lamps are required (only one element is still detected at one time) but is mostly of importance in emission spectrometry where all elements present in the sample cell are undergoing wavelength-specific relaxations.

5.2.9 Detectors

FAAS and FAES measurements require that the energy contained in a photon be converted into a measurable electrical signal. Early detectors relied on a more solid-state version of the photoelectric effect that is best illustrated by a phototube, one of the first detectors to convert radiant energy to electrical energy.

A variety of UV–visible detectors used in optical instruments were discussed in Chapter 3, including Phototubes (Section 3.4.1), PhotoMultiplier Tubes (Section 3.4.2), Charge Transfer Devices (Sections 3.4.3 and 3.4.5), and Linear Diode Arrays (Section 3.4.4).

5.2.10 A Review of Instrumental Noise in Detectors from Chapter 3

Chemists have little difficulty when they are measuring high concentrations of an analyte, but this situation is relatively rare. Most analysts attempt to determine if an analyte is present or not, and in doing so, must measure the smallest amount of an analyte in a sample near the detection limit of the instrument. In many areas of science, such as environmental, forensics, and biological applications, the analyst is always struggling to lower the detection limit. This is referred to as "chasing the detection limit" since one of the goals of instrumental manufacture is to improve instruments to measure smaller and smaller quantities of analytes. The detection limit is basically the minimum concentration that one can measure above the "noise" in an instrument. To fully understand this concept, different types of noise must be discussed.

Environmental noise consists of factors in the immediate lab environment that will affect an instrument or sample. Obviously, if an instrument is sensitive to vibrations, such as an NMR, one would not place this instrument in an area of high vibration, such as next to an elevator. Similarly, an analyst would not want to make delicate ppm-, ppb-, and ppt-level concentration measurements in an environment where the analyte of interest is present in high concentrations (such as in the air near a metal smelter). Most environmental noise with respect to metal contamination can be avoided by locating the sample preparation area away from the instruments or locating each in a clean room or HEPA hood.

Chemical noise tends to be unavoidable and specific to each analyte. One of the simplest types of chemical noise, that is completely unavoidable, is the electronic and vibrational fluctuations present in an atom or compound due to the Maxwell distribution of energies. These result in broad UV–visible absorption and emission spectra in aqueous samples but are of little consequence in gas phase transitions experienced in AA and ICP. Other forms of chemical noise include slight temperature and pressure fluctuations that affect measurements.

Instrumental noise is common and in many cases, can be avoidable or managed. Instrumental noise is separated into three categories: thermal, shot, and flicker noise. Thermal noise (also referred to as Johnson noise) results from the thermal agitation of electrons in resistors, capacitors, and detectors. Most of the thermal noise can be overcome by cooling specific components of the instrument such as is done in advanced detectors (i.e. charge injection devices). Shot noise results from a current being generated by the premature movement of electrons across a junction. The best example is an electron being emitted from a photoemissive material, such as in photocells or electron multipliers that are used to measure the intensity of visible and UV light. There are a few electronic ways of minimizing shot noise, but these are beyond the scope of our discussion here. Flicker noise results from random fluctuations in current and is inversely related to frequency. Flicker noise is overcome by electronically modulating the detector output signal to a higher frequency where less noise is present (i.e. from 10^2 to 10^4 Hertz).

In practice, the analyst is concerned with distinguishing between a real signal and instrumental noise, quantified as the S/N ratio. S/N ratio is mathematically defined as

$$\frac{S}{N} = \frac{\text{mean of signal}}{\text{s.d. of signal}} = \frac{x}{\text{s.d.}} \tag{5.7}$$

where S is the mean of approximately 20 blank measurements and N is the standard deviation of these measurements. From a statistical standpoint, it should be noted that S/N is equal to the reciprocal of the relative standard deviation (RSD).

Two approaches are used to minimize noise: hardware and software. Common hardware approaches that are used to decrease noise are (1) grounding and shielding components and detectors, (2) using separate amplifiers for different signals, (3) placing frequency or wavelength filters "up line" from the detector, (4) modulating the instrument signal to a "clean" frequency, and (5) chopping the signal to obtain a reference reading of the background that can be subtracted from the sample signal. For spectroscopy, one of the most common software noise reduction techniques is to take as many readings as reasonably possible. The observed S/N ratio is a function of the number of readings (n) taken as shown by

$$\frac{S}{N} = \sqrt{n}\frac{S_x}{N_x} \tag{5.8}$$

where S_x and N_x are the signal and noise readings for a specific setting. Note that as one takes more readings, the S/N decreases by the square root of the number of measurements. By taking two measurements, one can increase the S/N by a factor of 1.4; or by taking four measurements, the analyst can half the noise. For the topics covered in this Etextbook, FAAS, ICP–AES, and ICP–MS instruments generally allow for multiple measurements over 5–10 seconds to be taken and averaged. And, of course, the analyst can always analyze a sample multiple times, given the common presence of automatic samplers in the modern laboratory.

5.2.11 Data Processing

Data collection has greatly advanced with the aid of computer technology that has replaced the strip charts of the decades before. But with this change, other adjustments have been necessary. For example, almost all instrumentation signals are analog in nature (a continuous stream of data), while computers require data to be in a digital form (segmented). One of the best analogies used to explain the difference between analog and digital comes from the music recording industry. Vinyl records from the pre-1985 era were analog recordings, the music (groves in the vinyl) was a continuous source of sound just as a guitar string continuously vibrates. Digital recording takes the analog input and breaks it into very short bits (segments), so short that most human ears cannot detect the individual segments (others argue, such as Neal Young, that compact discs (CDs) are the worst thing to happen to the recording industry due to a decrease in sound quality and vinyl records are making a comeback). This same analog to digital (A–D) conversion must take place before a computer can record and process data from an analytical instrument.

Computers have unquestionably allowed more control and extended capabilities of analytical instrumentation. Examples include the automated nature of the instrument, operation of automatic samplers, automatic collection and processing of data including automatically generating calibration lines, and reporting concentrations of analytes in samples as opposed to detector response. However, in some ways, computers have decreased operator knowledge of the instrument; two examples are the lack of knowledge concerning the "nulling or zeroing" of an instrument and "balancing" electronics of a system to optimum values. Prior to the addition of computers to instrumentation, these functions were manually adjusted, while today they are part of the automatic setup.

As an example, the operation of a typical FAAS blank and sample measurement will be used. Recall that in this system, a source lamp is split into two beams, one reference and one passing through the sample. Both beams go through the monochromator and to the detection, where the reference and sample signal is processed separately. Finally, an absorbance reading is displayed and recorded for each blank and sample. But notice, blank and reference readings allow all of the source radiation to pass through the instrument unhindered (defined as 100% transmission), thus generating a maximum response from the PMT detector, yet the reading displayed on the instrument panel is 0.000 absorbance units. Hence, there are unnoticed calculations and corrections going on "behind the scenes" in the instrument. What actually happens is the instrument sets the reference or blank sample to read a maximum signal (maximum number of electrons being generated by the PMT). Recall that the signal is rapidly chopped into reference and sample readings. Next, the instrument measures the signal for the sample, subtracts the reading of the blank from the sample, and the difference is the absorbed signal. Thus, readings go from low to high absorbance values.

It is also important to note that for systems such as FAAS units, electrical components that interact with each other must be "balanced" where signals, voltages, and currents must be optimized and matched between components. For example, when an HCL is turned on, the current placed on the electrodes is set to give out the maximum radiant output that the other instrument components can take but not damage the lamp by overheating or removing too much of the cathode material. The intensity of this radiant signal is adjusted further by increasing or decreasing the size of the entrance and exit slit on the monochromator. Finally, the "gain," the electrical potential applied to the PMT dynodes, is adjusted to provide maximum amplification but not overload the PMT. Today, this is all adjusted automatically, but the process should still be understood.

Most computer-controlled systems operate the instrument using two computer programs: a method and a sequence. The method program controls physical conditions of the instrument such as lamp current, fuel and oxidant flows, and electronic gain placed on the PMT. Once the instrument is operating at its maximum performance, a sequence is started. A sequence is a program that tells the instrument what samples to run, where they are located in the automatic sampler, and what order to run the samples. The sequence also collects and stores the data for each sample in a separate file. Finally, the method is used again to calibrate the instrument using the data from the reference standards and to calculate the concentration of analytes in the samples (convert detector response to concentration).

Overall, computers are a significant asset to analytical instrumentation. They have increased the capabilities of instruments and significantly decreased the cost of analysis through the operation of automatic samplers and advanced data processing. An additional aspect of data processing is the elimination of the need to retype data into a spreadsheet or report form, which reduces typographical errors and other mistakes.

5.3 Specialized Sample Introduction Techniques and Analysis

5.3.1 Aqueous Sample Introduction

Most of the FAAS and FAES systems use aqueous sample introduction through the nebulizer and into the burner head. This works well for samples that are already in the aqueous phase or can be digested in acid such as soil, atmospheric particles, and tissue samples. All aqueous samples for FAAS and FAES contain some amount of strong acid, usually in the 1–5%. This acid acts to keep the metal analytes in the dissolved phase and to avoid adsorption of metal ions to sample containers and instrument surfaces. FAAS and FAES operated under these conditions suffer from relatively poor detection limits that in some cases, can be improved upon with specialized sample introduction techniques. These techniques, described below, are physical attachments to the basic AAS unit that may or may not replace the burner head and in some cases, allow solid samples and high particulate-containing samples to be analyzed. A few of the techniques allow the analysis of elements not commonly measured by FAAS or FAES.

5.3.2 Mercury Cold Vapor

Inorganic and organic forms of mercury are ubiquitous in the environment, including water and food, and come primarily from the burning of coal. As a result, it is necessary to detect the concentration of mercury to assess the danger caused by this toxin. Mercury is a neurotoxin and extremely small concentrations (ppb or ppt) and can have detrimental effects due to bioaccumulation in the food chain (increases in concentration as one goes from one tropic level to the next). Several fish species, located in streams downwind from coal-burning regions, contain significant, and in some cases, dangerous concentrations of Hg. Flame AAS techniques only yield detection limits of approximately one part per million (ppm), which is inadequate for environmental and food monitoring.

The cold vapor technique described below yields detection limits in the parts-per-trillion (ppt) range. Equal or even lower detection limits can be obtained by ICP–MS (the subject of Chapter 7).

An overview of the cold vapor system is shown in Video S.5.2 where an external glass vessel is used to generate elemental (and volatile) mercury that is passed through a Pyrex/quartz cell placed on the standard burner head. No flame is needed to atomize the mercury, hence the name "cold" vapor technique.

A sample containing digested water, sediment, or tissue that contains cationic mercury (Hg^{2+}) is added to the external glass vessel, which is then closed and purged with argon to remove any oxygen. Next, $SnCl_2$ is added via a syringe to reduce Hg^{2+} to elemental Hg. The elemental Hg is stripped from the water solution and passes as a pulse of vapor through the sample cell. The instrument is operated in the absorption mode; a hollow cathode Hg source lamp provides a specific wavelength to be absorbed by $Hg_{(g)}$ in the cell. After detection, the mercury vapor passes through a potassium permanganate solution to convert the mercury vapor back to ionic mercury so that no mercury is released into the laboratory environment or into the natural atmosphere. All blanks, external standards, and samples must be analyzed in the same manner. A drawback to this method is that samples must be processed individually, without automation. For each new sample, the argon stream must be interrupted to allow addition of a new sample to the glass container, which must then be purged with argon.

The advantage of the technique is a three order of magnitude improvement in the detection limit. Disadvantages are labor costs associated with digestion and manual instrumental analysis. When numerous samples are routinely analyzed this technique has been replaced with ICP–AES (with a detection limit of 1 part per billion) and ICP–MS (with a detection limit of less than 10 ppt). Cold vapor mercury analysis is still commonly used when mercury is the only metal of interest and economics does not support the purchase and maintenance of an ICP–MS system.

5.3.3 Hydride Generation

Another external attachment to a FAAS instrument is the hydride generation system that is used to analyze for arsenic (As), bismuth (Bi), mercury (Hg), antimony (Sb), selenium (Se), and tellurium (Te); some of these are notable toxins. This technique works in a similar manner to the cold vapor technique but sodium borohydride is used as the reducing agent to generate a volatile metal hydride complex. In addition, a flame is used to decompose the metal hydride. Again water, soil, and tissue samples must be digested to free the metal from organic and inorganic complexes and place it in its cationic state. The generated metal hydride is passed through the flame as a pulse input where it is degraded by heat to its gaseous elemental state. In this state, the metal will absorb the source radiation (again from an HCL) and the absorbance reading is directly proportional to the concentration of metal in the sample. Instrumental calibration and data processing is identical to the cold vapor mercury technique.

An example reaction that occurs in the reaction vessel for As is shown below

$$3BH_4^-{}_{(aq)} + 3H^+{}_{(aq)} + 4H_3AsO_{3(aq)} \rightarrow 3H_3BO_{3(aq)} + 4AsH_{3(g)} + 3H_2O_{(l)}$$

A video illustration of this technique can be viewed in Video S.5.3. As with the cold vapor technique, the advantage is an improved detection limit, with the disadvantage of high labor costs associated with digestion and individual and manual instrumental analysis.

5.3.4 Electrothermal Vaporization (Graphite Furnace Atomic Absorption, GFAA)

The graphite furnace, formally known as an electrothermal vaporization unit, uses a typical FAAS unit but replaces the burner head with a furnace system. No flame is used in the operation of this system; instead, the metal in the sample is atomized by heating the cell with electrical resistance to temperatures not obtainable in flame systems. The heart of the system is illustrated in Video S.5.4. All systems use an automatic sampler to ensure reproducible results between replicate analyses. In GFAA, samples are usually digested (in acid) to assure homogeneity of the injected solution. A sample is placed into the graphite furnace cell through a small hole in the side. Cells range in size depending on the brand of the instrument but are usually about the diameter of a standard writing pencil (~0.5 cm) and 2–3 cm in length. Argon gas is passed through the cell to pass vapor and analytes into the radiant beam and the furnace and sample are then cycled through a three-step heating process. First, the water is driven off by resistance heating at 107 °C (which partially blocks the beam of the HLC source but this is not recorded). Next, the sample is "ashed" at 480 °C to degrade any organic material in the sample (again this absorbance signal reduction is not recorded). Finally, the cell is rapidly heated to 2000 °C where the analytes are placed in their volatile elemental state where they absorb radiant light from the HCL; this signal is recorded. The system is then prepared for another run by heating the cell to 2500 °C to remove excess material before cooling the chamber with tap water back to room temperature. Then another standard or sample can be added and the process is repeated.

One obvious advantage of this system over the cold vapor and hydride technique is automation, which reduces the cost of analysis. Another advantage of the GFAA technique over FAAS or FAES is the improvement in detection limits, typically in the low parts-per-billion (ppb) or high ppt ranges. Most of the elements that are analyzed by FAAS can be analyzed by GFAA but high background concentrations of a few rare earth and alkaline earth elements in the graphite tubes limit their detection limits.

An illustration of sample introduction and the heating steps is given in Video S.5.4.

The Zeeman Effect (Correction): In the graphite furnace samples containing high amounts of solids and organic matter can be introduced which leads to high background levels and spectral interferences when these compounds are heated and degraded during the atomization step. These problems can be overcome by using a Zeeman correction calculation. The Zeeman Effect capitalizes on the observation that the absorption profile of an element is split into several polarized components in the presence of a strong magnetic field. When absorption measurements using radiation from a deuterium or mercury vapor lamp are taken with and without a magnetic field present in the reaction cell, the presence of spectral interferences can be corrected for (subtracted out) and this difference in signals is referred to as the "Zeeman background corrected" atomic absorption signal. The physics behind this process is beyond the scope of this text and it is only important to note that the Zeeman correction is available and used in most graphite furnace systems.

5.3.5 Glow-Discharge Atomization for Solid Samples

The glow-discharge device is a highly specialized sample introduction system since it is mostly used for electrically conductive samples, such as a piece of metal or semiconductors. Like the GFAA unit, it is an attachment that replaces the burner head. A sample is placed in a low-pressure argon chamber where a potential is placed between the container and the sample. Excited argon atoms sputter atoms from the surface of the sample, similar to the operation of the HCLs. Gaseous phase metal atoms rise into the path of the source radiation and absorb their characteristic wavelength. The largest advantage of the glow discharge technique is the direct analysis of solid samples.

5.3.5.1 Fluorescence

The above discussions have mostly focused on absorption and emission processes and instruments. Recent advances in atomic fluorescence spectrometry make this technique possible for a few elements (Hg, As, Se, Te, Sb, and Bi). Again, fluorescence occurs when an electron is excited to a higher electronic state and decays by resonance, direct line, or stepwise fluorescence. Instrument components are similar to those discussed above and in the next chapter, but the key difference is that the source lamp is located at a 90° angle with respect to the detector in order to prevent the source radiation from being measured by the detector. Lamps used to excite electrons include HCLs, EDLs, lasers, plasmas, and xenon arc lamps. Atomizers include flames, plasmas, electrothermal atomizers, and glow discharge chambers; thus, samples can be introduced as cold vapors, liquids, hydrides, and solids. Sub-ppb detection limits are obtainable from these instruments.

5.4 General Operation of FAAS and FAES Instruments

For most FAAS and FAES measurements, aqueous samples are required. Once samples are prepared, which can take hours, the instrument is prepared for operation. Instrument setup usually requires:

- checking the fuel and oxidant gas pressures and flow rates,
- turning on the instrument and lamp (and aligning it, and allowing the electronics to warm up and stabilize for at least 10 minutes),
- checking the drainage system to ensure waste flow,
- lighting the flame and allow it to thermally stabilize for 5–10 minutes,
- adjusting ("balancing") the electronics (this is automatically performed on most modern instruments),
- testing the stability of the signal,
- loading the method and creating a sequence for the samples, and
- starting the sequence.

5.5 Maintenance

FAAS and FAES instruments are fairly low maintenance. HCLs have a limited life depending on the manufacturer and will need to be updated. Optics are fixed and the monochromator will be calibrated with respect to wavelength when the instrument is turned on. Samples containing particles must be filtered through a 0.45 mm glass fiber filter prior to analysis to avoid clogging the inlet tube, entrance to the spray chamber, and slit on the burner head. If high salt or organic matter samples are analyzed, the burner head will have to be cleaned after each experiment.

5.6 Summary

FAAS and FAES dominated the market for decades but are slowly being replaced by ICP systems. Today, FAAS and FAES systems are limited to situations where only a few elements are analyzed occasionally. "Mass production" analyses in FAAS–FAES are most economical with automatic samplers but even then these techniques are being replaced by ICP–AES and ICP–MS because of superior detection limits and analysis of multiple elements at one time. These advanced techniques are the subject of the next two chapters.

5.7 Questions

1. Give the complete name of the following acronyms:
   AES
   AAS
   FAAS
   FAES
   HCL
   PT
   PMT
   CCD
   LDA
   CTD
   GFAA

2. Draw and label double-beam flame atomic spectrometer.

3. What types of electromagnetic wavelengths do FAAS and FAES instruments use?

4. What is the difference between line and continuous spectrum sources?

5. Draw and explain how an HCL works.

6. What is the purpose of the chopper in an AAS or AES instrument?

7. How can an instrument perform both AAS and AES?

8. Explain the Bernoulli principle with respect to the FAAS burner head.

9. What is the optimum droplet size that makes it into and through the burner head and what factors influence this size?

10. What types of fuels and oxidants are used in FAAS and FAES?

11. Name and discuss all of the common sources of noise in the lamp and flame.

12. What is the purpose of a slit in the various locations in an instrument?

13. What two components can result in dispersion in the monochromator?

14. What are the sources of noise in detectors?

15 Explain how the *S*/*N* ratio is used and why it is important.

16 How have computers changed FAAS and FAES instruments?

17 Draw and explain how the mercury cold vapor technique works.

18 Draw and explain how the hydride generation unit works in FAAS. What elements is it used to analyze?

19 Explain how electrothermal vaporization (graphite furnace atomic absorption, GFAA) works.

20 How is the GFAA unit superior to the FAAS or FAES instruments?

21 How does the glow-discharge unit work? In what industry is it mostly used?

Supporting Information

Additional supporting information may be found online in the Supporting Information section in the HTML rendition of this article.

6

Inductively Coupled Plasma

Inductively Coupled Plasma: Atomic Emission Spectrometry

6.1 Introduction and History

Greenfield et al. developed plasma-based instruments in the mid-1960s about the same time flame-based instruments such as FAAS and FAES (Chapter 5) became prominent (1964). These first plasma-based instruments used direct current (DC) and microwave-induced (MI) systems to generate the plasma. Interference effects and plasma instability limited the utility of plasma instruments during analysis; consequently, flame-based spectrometry instruments (such as FAAS) dominated the analytical market for metals analysis and remain effective today.

The limitations of the first plasma instruments were overcome by utilizing an inductively coupled plasma (ICP) instead of DC or MI-generated plasma. ICP optical systems became popular in the 1980s due to their decreased cost, lower time investment during analysis, and labor-saving advantages. FAAS/FAES instruments require a unique radiation source (lamp) for the approximately 35 elements they can measure. Because the lamp must be changed between each element of interest, FAAS/FAES techniques analyze a single element at a time and are unable to easily analyze metalloids. ICP optical systems, by contrast, can analyze about 60 different elements at the same time with a single source (the plasma). The most common instruments today are inductively coupled plasma–atomic emission spectrometers (ICP–AES of OES; optical emission spectrometry) and inductively coupled plasma–mass spectrometers (ICP–MS). ICP–AES/OES will be discussed in this chapter, while ICP–MS will be the subject of Chapter 7.

6.2 Atomic Emission Spectrometry Theory

The operation of an ICP–AES system relies upon the same interaction of molecules with electromagnetic radiation that was presented in Chapter 2. The two emission systems, FAES and ICP–AES, differ in the way atomic species are created and excited. Because of the relatively low temperatures (~2000 to 2500 °C) in a flame-based system, not all of the atoms or elements present in the sample are excited, particularly if they exist in a polyatomic compound. Some elements readily form nonemitting and refractory oxides that result in an underestimation of their concentration. In plasma-based systems, the temperature is considerably hotter (~6000 to 10 000 K), which results in more effective excitation of atoms (generally greater than 90%) of approximately 60 elements including some metalloids and nonmetals. This intense heat prevents polyatomic species from forming, thus increasing the detection limits for many elements. Atoms are excited, and in many cases ionized, by the intense heat of the plasma, and the emission of a photon

Essential Methods of Instrumental Analysis, First Edition. Frank M. Dunnivant and Jake W. Ginsbach.

Companion Website: www.wiley.com/go/essentialmethodsofinstrumentalanalysis1e

occurs via resonance fluorescence (normal valance electron relaxation by photon emission). While plasma-based systems eliminate many problems, they are not free of interferences due to the excitation and subsequent emission of spectral lines for every element in the sample as well as the Ar added to facilitate plasma generation. The spectral overlay that results from these possible emissions is overcome in modern instruments with specialized sequential monochromators (Section 3.3.2). ICP–AES, compared to FAAS/FAES, offers high selectivity between elements, high sensitivity, and a large dynamic range, especially as compared to FAAS, which is limited by Beer's law, lower detection limits, multielement detection, and fewer matrix interferences.

6.3 Components of an Inductively Coupled Plasma: Atomic Emission Spectrometry System (ICP–AES)

6.3.1 Overview

An ICP–AES system can be divided into two basic parts: the ICP source and the atomic emission spectrometry detector. Figure 6.1 shows the common components of an ICP–AES system from the late 1980s to the 1990s. The ICP source has mostly been unchanged since its invention with the exception of innovations in monochromator type, which enables greater suppression of interference phenomena. Modifications of this common system will be explained in the following sections.

Sample solutions include digested soil or other solid material or natural water. Typically, the sample solution is acidified up to 2–3% in HNO_3 to prevent adsorption of metals onto polypropylene sample bottles or onto instrument tubing or glassware prior to introduction into the plasma. In Figure 6.2, the sample is introduced to the nebulizer chamber via a peristaltic pump and Tygon® tubing attached to an automatic sampler. A peristaltic pump operates by sequentially compressing flexible tubing with evenly spaced and rotating rollers that pull/push the liquid through the system. The rate of sample introduction into the plasma changes as the rotation rate of the peristaltic rollers increases or decreases. The flow of sample and Ar gas through the small aperture of the nebulizer creates very small droplets that form a mist of μm-sized particles in the nebulizer chamber. Larger sample droplets collect on the chamber walls and are removed through a drain, while smaller particles travel with the Ar flow and enter the torch. Evaporation, atomization, and excitations/ionizations occur in the plasma at temperatures reaching 10 000 K. Ar not related to the sample is also excited and ionized because this gas both carries the sample aerosol and confines the location of the plasma to prevent damage to the rest of the instrument. As the excited/ionized atoms leave the hot portion of the plasma, excited valence electrons relax and emit a photon characteristic of the electron transition. This photon is specific to the element but does not yield any information about the isotopic state of the element, unlike in mass spectrometry (Chapter 7). Visible and ultraviolet (UV) radiation emitted from the sample constituents enters the monochromator through a small slit where the wavelengths are separated by grating(s) and/or prism(s) before being captured and measured by a wide variety of detectors.

Because spectral interferences may still occur, the choice and configuration of the monochromators in the instrument are important and have been the target of innovation. In Figure 6.1, the most common form of a monochromator in the 1980s and 1990s (a Rowland circle) and detector (photomultiplier; PMT) is shown: The Rowland system utilizes a concave Echellette-style grating monochromator to separate the various emission lines and simultaneously focus individual wavelengths on to a series of slits, with each slit aligned to allow a specific wavelength of radiation to pass to a detector. The standard detector, a photomultiplier tube (PMT), was discussed in Section 3.4.2.

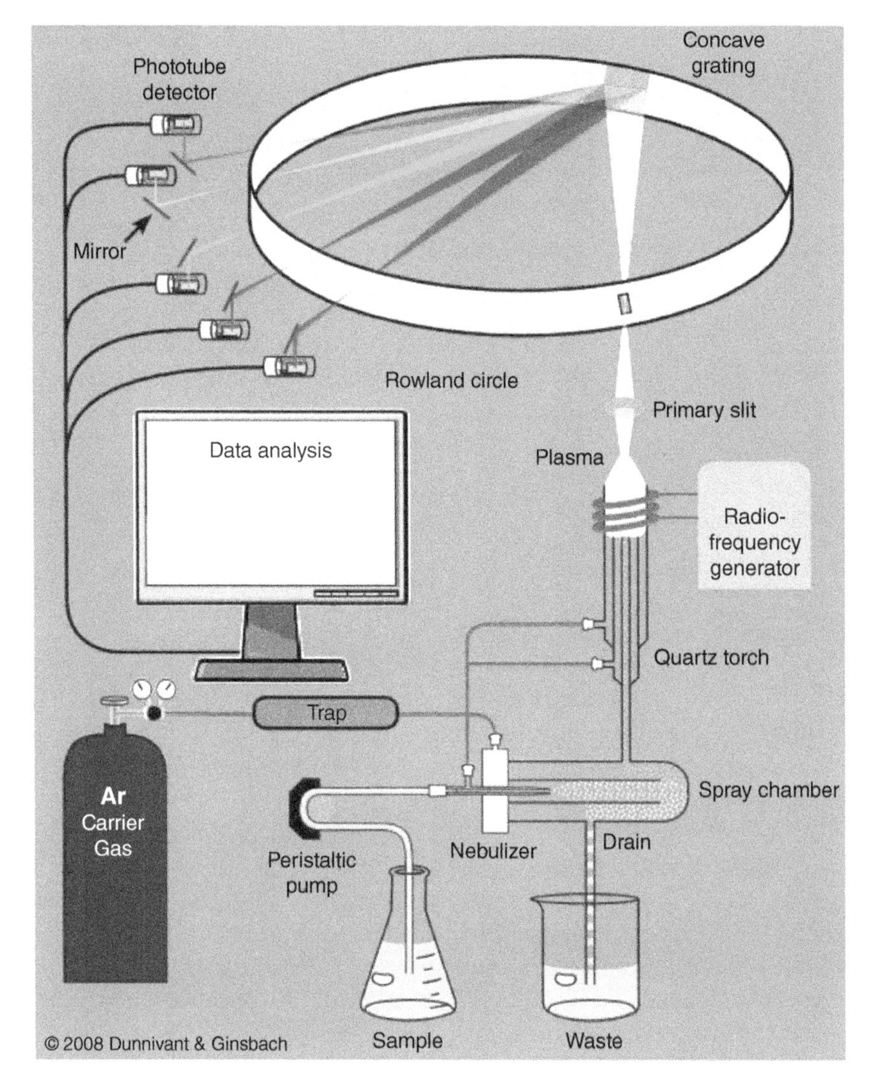

Figure 6.1 Overview of a basic inductively coupled plasma–atomic emission spectrometry (ICP–AES) from the 1990s (Source: Dunnivant and Ginsbach).

Some systems use multiple PMTs at fixed locations to monitor each wavelength simultaneously (Figure 6.1) whereas other systems use a single PMT and move it to different locations to detect each wavelength. Data from these detectors are processed by a computer because multiple wavelengths are measured in an ICP–AES system at the same time.

6.3.2 Sample Introduction and Optimization

The predominant form of sample matrix in ICP–AES today is a liquid sample: acidified water or solids digested into aqueous forms. Given the automated nature of the ICP analysis, all modern systems are purchased with automatic samplers where a computer-controlled robotic sampling arm takes liquids from each sample via a peristaltic pump from plastic tubes located in specific locations in a sampling tray. Liquid samples are pumped into the nebulizer and sample chamber

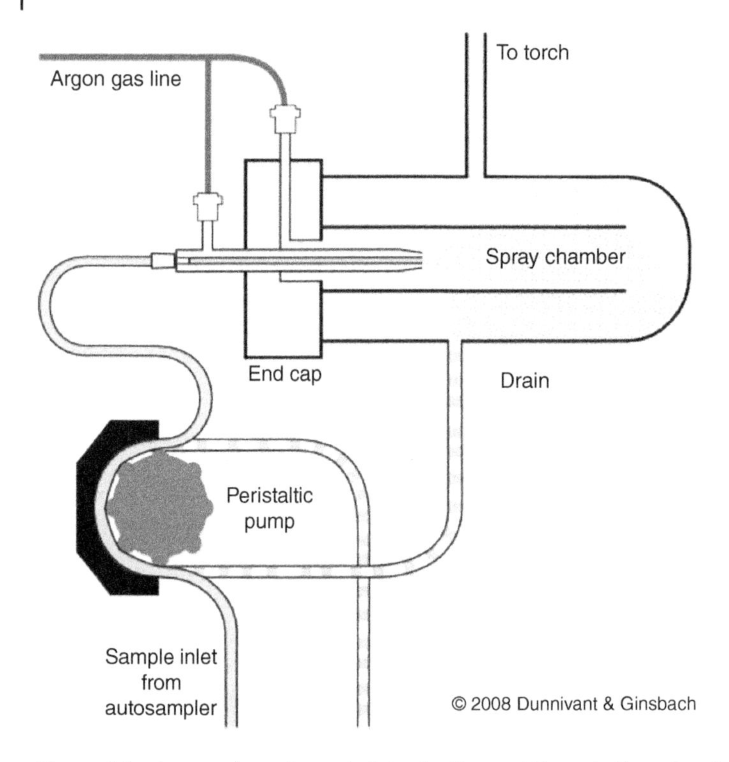

Figure 6.2 An overview of sample introduction and the nebulizer chamber (The nebulizer shown here is a pneumatic style, described below) (Source: Dunnivant and Ginsbach).

via a peristaltic pump as shown below. Then the samples pass through a nebulizer that creates a fine mist of liquid particles. Larger water droplets condense on the sides of the spray chamber and are removed via the drain (pumped out of the chamber also by the same peristaltic pump), while finer water droplets move with the argon flow and enter the plasma. Nebulizers help ensure that the sample enters into the plasma at a uniform flow rate and specific droplet size. Droplets that are greater than 5 μm in diameter are likely to interfere with plasma stability.

While there are numerous types of nebulizers for a variety of specific applications, the three most common types are (1) pneumatic, (2) ultrasonic, and (3) grid. Because argon is used in generating the plasma (discussed below), it is most often used as the gas in these various nebulizers, but other gases can be used. The most common pneumatic nebulizer for samples containing low concentrations of total dissolved solids (TDS) is the concentric nebulizer shown in Figure 6.3, but higher suspended solid and dissolved solid samples are commonly introduced to the plasma via the Babington nebulizer (Figure 6.4)

Ultrasonic nebulizers are used to provide more sample delivery to the plasma and thus improve detection limits. An ultrasonic generator surface, usually a piezoelectric crystal that rapidly vibrates

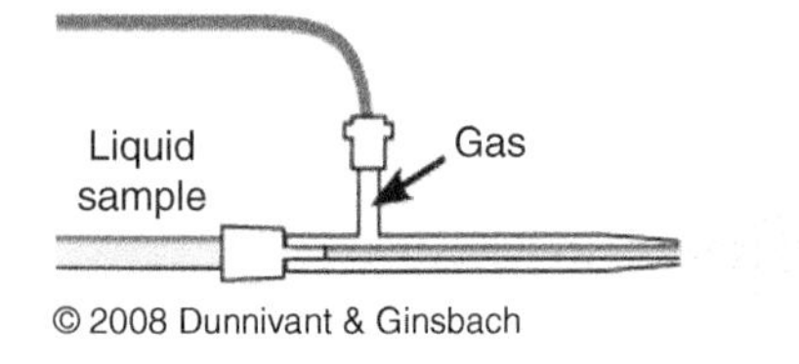

Figure 6.3 Diagram of a pneumatic concentric nebulizer (Source: Dunnivant and Ginsbach).

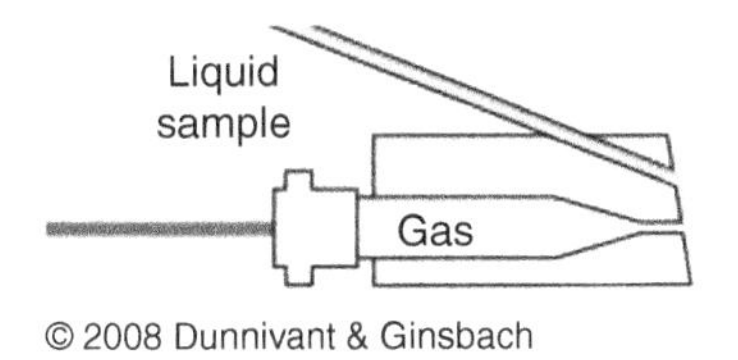

Figure 6.4 Diagram of a pneumatic Babington nebulizer (Source: Dunnivant and Ginsbach).

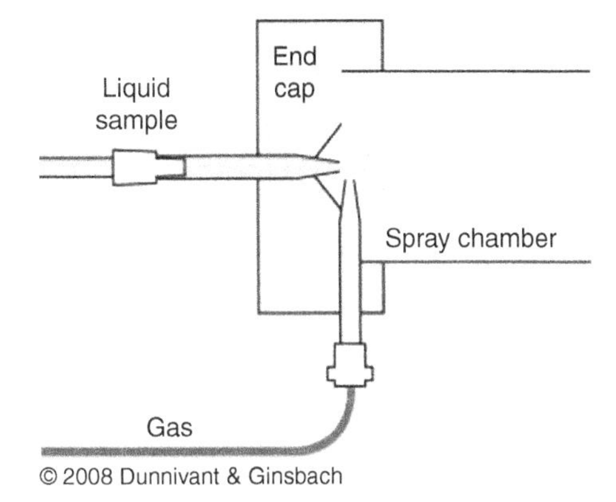

Figure 6.5 Diagram of a pneumatic cross-flow nebulizer (Source: Dunnivant and Ginsbach).

to generate sonic energy, is used to create extremely fine droplets that, at low flow rates, are completely transferred to the plasma (unlike considerably lower efficiencies from pneumatic nebulizers). Grid nebulizers (Figure 6.5) create a fine mist by placing a grid in front of the argon flow. The liquid sample is allowed to flow down the grid, and as argon passes through the grid, it creates fine droplets. Again, the most common commercially available nebulizers use the pneumatic design.

ICP systems have a few sample requirements or limitations. Micrometer-sized particles must be removed by filtration or centrifuge or dissolved prior to introduction to a sample vial because the particles can easily clog the tip of the nebulizer. A clogged tip will not be automatically detected by the instrument, but will be detected by a trained operator in the absence of any sample, internal standard, or standard analyte concentrations in the spectra.

Another limitation of the nebulizer is the presence of high TDS that will eventually cause the accumulation of solids in the tip of the nebulizer, as well as in the sample chamber and on the sampler cones located just after the plasma (in ICP–MS instruments only). Most ICP instruments limit the sample TDS level to approximately 0.1–0.2% salts by weight. Higher salt contents can enhance atomization and ionization of some elements and suppress or interfere with others. Likewise, the acid concentration of liquids entering the plasma should be limited to approximately 2–3% for nitric and hydrochloric acid. Rarely, hydrofluoric acid may be used if a particularly refractory element is of interest, but then the acid concentrations are less than 0.5% because it will degrade the quartz torch with time. Some samples, such as food-based beverages or plant extracts, can contain high concentrations of organic matter. It is best to oxidize this organic matter in an acid or peroxide digestion prior to analysis since carbon will rapidly accumulate on the components of the instrument if high organic matter-containing samples are directly injected into the plasma. This carbon can reduce sample flow and capture elements and ions as the carbon builds up, specifically at the cone apertures that facilitate the transition from ambient pressure in the plasma region to the high vacuum detector region (in ICP–MS systems). When a portion of a carbon plaque drops back into the analytical stream, an anomalously high, and false, measurement can occur.

Other forms of sample introduction are used but are not as common as liquid injection. These include (1) direct insertion of 10–30 mg of sample on a graphite probe into the plasma via the normal nebulizer entry port, (2) the injection of the effluent from chromatographic separation systems, especially from high-performance liquid chromatography (HPLC), supercritical chromatography, and ion chromatography (IC) units where compounds containing elemental analytes of interest are first separated by chromatography and then introduced into the plasma for elemental analysis, (3) where the cold vapor techniques described in Section 5.3.2 for mercury analysis is connected to the plasma inlet via the nebulizer port, and (4) where the hydride generation system from Section 5.3.3 for selected metals can also be connected to the nebulizer port.

6.3.3 Inductively Coupled Plasma Torch

The torch unit of an ICP is used to create and sustain a plasma. A plasma is an electrically conducting gaseous mixture containing enough cations and electrons (though the plasma has a neutral charge overall) to maintain the conductance. One common example of a simple plasma is a regular flame which will conduct an electrical current across it; cations and electrons are created upon ignition of the fuel and travel upward in the flame until they are cooled above the flame. Another common plasma is used in scanning electron microscopy (SEM) where a sample is coated with graphite or metal in a vacuum chamber in order to make the surface conductive (a standard requirement for obtaining high-quality images). DC arc and microwave plasmas can also be used to generate plasmas, but for purposes of metal analysis, the ICP system described below is most important. The purpose of the torch is to (1) evaporate the solvent (usually water) from the analyte salts, (2) atomize the atoms in the salt (break the ionic bonds and form gaseous state atoms), and (3) excite or ionize the atoms. In the case of ICP–AES, only excitation is needed as in FAES, but given the extreme temperatures (up to 10 000 K) of the argon plasma used in modern ICP systems, the excitation of atoms is virtually complete for most elements. (Chapter 7 describes how the ICP–MS capitalizes on the efficient ionization of atoms in the plasma.) A modern torch system is illustrated later in Videos S.6.1 and S.6.2. The torch described here is composed of three concentric quartz tubes. The samples and the argon gas used to aspirate it pass through the center tube. The plasma-generating gas (argon) passes through the middle tube, and the argon passing through the outer tube is used to cool the quartz torch.

The plasma is sustained by a radiofrequency (RF) generator that creates an oscillating magnetic field around the torch that results in ohmic (inductive) heating of the charged gases at the end of the torch. Ohmic heating occurs when an electrical current is passed through a conductor, which in turn creates more heat. Three types of RF generators have been used. Older types are based on piezoelectric crystal oscillators and "free-running" generators in which the oscillation is set according to the combinations of the components of the circuit; examples include the Armstrong, Hartley, Colpitts, and tuned-anode-tuned gate oscillator electronic circuits. Most modern analytical ICP systems use solid-state semiconductor generators where the circuit consists of (1) a capacitor used to store a high electrical charge (thus requiring the 220–240 V electrical power requirements) and (2) an inductor coil to deliver the oscillating current to the torch and generate the magnetic field around the torch. The capacitor is hidden from view in ICP systems and buried in the electronics of the instrument. The inductor coil is visible and is the approximately 3-mm hollow copper coil wrapped three times around the end of the torch. The capacitor responsible for generating the plasma from argon gas oscillates an electrical field at a rate between approximately 27 and 41 MHz (a frequency regulated by the Federation Communications Commission [FCC]) and through induction creates a magnetic field in the plasma. The intensity of the frequency, measured in Watts, is sufficient to promote the valance electrons in some of the Ar atoms but not ionize them sufficiently to initiate or sustain a plasma. The creation of plasma occurs when a spark (from a Tesla coil; basically an automatic gas grill lighter) introduces free electrons at the end of the torch when the electrical field is being oscillated at a specific frequency by a RF generator. The seed electrons from the Tesla coil oscillate in an angular path and periodically collide with argon gas atoms and ionize them, releasing more electrons. Due to their kinetic energy and collisions with other atoms, a large amount of heat is generated, enough to generate and sustain a plasma at temperatures up to 10 000 K. In terms of an electrical analogy, the term "inductively coupled" in ICP is a result of the coupling of the induction coil and the electrons. The copper induction coil serves as the "primary winding" of the RF transformer and the "secondary winding" is the oscillating electrons and cations in the plasma; the two "windings" are thus coupled together because the second winding depends on the presence of the first.

The RF generator also contains a feedback loop where the interaction of the electronic/magnetic field in the capacitor/induction coil is monitored. If the flow rate of argon gas is too low or if small amounts of O_2 or high amounts of water vapor are present in the torch during the initiation of the plasma, the RF generator will sense their presence by a change in the feedback between the oscillation and the RF generator. When this happens, the RF generator shuts down to protect the electronics in the system from overheating. Thus, atmospheric O_2 cannot be present in the Ar and the flow rate of nebulizer sample (water) must also be controlled to successfully light the torch. The process of lighting the plasma is visualized in Video S.6.1. In this animation, the argon is turned on and the pressures are allowed to equilibrate without the introduction of any sample or blank solutions to the nebulizer. The RF generator then ramps the wattage through a series of cycles. A typical ramp cycle begins by turning the RF generator to 200 W. Subsequently, the electronics go through a preliminary check that adjusts the electronics and minimizes the resistance between the RF generator and the induction coil. Next, the RF wattage is ramped to 900 W where excitation of valence electrons in some Ar atoms occurs, and the Tesla coil initiates the excitation and ionization of Ar atoms and the plasma is lit. The circulation of electrons and Ar cations in close association results in ohmic heating that generates more electrons, collisions between Ar atoms, and an increase in temperature. The wattage is then increased to 1200 and the resistance is minimized again via electronic adjustments. Finally, the wattage is adjusted to the suggested energy level for the elements of interest; this is usually between 1250 and 1550 W.

If the region of ohmic heating is not controlled, the plasma will continue to heat until it melts the quartz torch (and the other components of the ICP spectrometer). The torch is cooled in two ways: the first is by the tangential introduction of relatively large volumes of argon gas through the outer tube of the torch. This argon flow spirals around the middle tube resulting in uniform cooling. Furthermore, cooled water flows through the copper induction coil of the RF generator that is wrapped around the end of the torch. These combined cooling systems promote an equilibrium maximum temperature of approximately 10 000 K in the hottest portion of the plasma. The portion of the plasma that ICP–AES measurements are concerned with is about 5000–6000 K and is located in a cone-shaped region outside the quartz torch.

After successful ignition of the torch, samples are introduced into the system through the nebulizer (Video S.6.2). Upon entry into the plasma, the solvent evaporates and salts form. Then, these compounds decompose as they move farther into the hotter portions of the plasma. Next, the valence electrons on the analyte atoms are excited (for ICP–AES measurements at lower temperatures) or completely removed (ionized for ICP–MS measurements at higher temperatures). The intense heat that sample molecules encounter in the plasma is sufficient to decompose most refractory compounds. Thus, only atoms or atomic ions are present in the plasma. This heat also causes complete excitation of atoms and leads to higher detection limits, compared to FAES where the extent of excitation is element- and flame-temperature-dependent. In ICP–AES, as the atoms exit, the plasma and cool to approximately 6000 K, they relax and emit a characteristic photon that enters one of the monochromator/detector systems described in the next section. In ICP–MS (Chapter 7), the plasma enters a vacuum chamber where the cations are separated by a mass filter (mass analyzer) and detected by a specialized PMT. An illustration of the drying, atomization, excitation, and ionization is shown in Video S.6.2.

The consumption rate of Ar gas is an important issue and accounts for most of the operational costs of an ICP. Older systems can use up to 20 L/minute, while modern systems have reduced the flow to below 10 L/minute. Systems are being designed that are portable and use far smaller flow rates of Ar gas. Many laboratories use cryogenic sources of Ar to reduce operational costs. Current prices for *liquid* Ar (equivalent to approximately 5000 cubic feet at STP) are approximately $325, whereas the cost of three T-size cylinders of *gaseous* Ar (equivalent to approximately 334 cubic feet at STP) is approximately $420. Thus, use of a cryogenic tank is far more economical than individual gas cylinders. Gases generated from a cryogenic tank are often more pure than those obtained from a standard T-sized gas cylinder.

6.3.4 Separation and Detection

After the atoms are excited or ionized in the plasma, they exit the high-temperature region and electronically relax. This results in the emission of at least one element-specific photon available for detection. The first choice to be made by the analyst is the position of the monochromator entrance relative to the plasma source. Two choices are available, axial and radial alignment. A radial design allows photons to move from the side of the plasma (at the end of the torch) to the entrance slit, while an axial design gives higher intensities (and better detection limits) since the photons come from the center and end of the plasma. Thus, depending on the concentration of analytes in a given sample, the analyst may choose to monitor the emissions from the radial angle for higher concentrations; this avoids burning out the detector, which is costly to replace. Or the analyst may choose to use the axial monochromator alignment for lower concentration samples. Changing the angle of observation can potentially avoid additional sample preparation such as dilutions. Many brands of instruments allow a selection between radial and axial monitoring.

Three broad categories of detection are available for analyzing the emitted photons: sequential, simultaneous multichannel, and Fourier transform systems. In the first two, all wavelengths enter a monochromator where they are dispersed by prisms and/or grating monochromators and are then transmitted to the detector (most commonly the PMT presented in Video S.3.1, or a CCD presented in Figure 3.15). Fourier transform systems produce spectral separations by capitalizing on constructive interference techniques as first developed for Michelson's interferometer in 1881. In Fourier transform systems, no slits or monochromators are required, and this creates better detector limits because more intense radiation reaches the detector. Fourier transform systems also have higher spectral resolution (and thus have fewer spectral interferences) and can simultaneously monitor all wavelengths for longer times. This text will only cover sequential and simultaneous multichannel systems that use monochromators because Fourier transform systems extend past the basic ICP instrumentation goal of this textbook and are considered "higher-end" systems, and are rarely seen at an undergraduate level of instruction.

The first ICP–AES systems used a sequential monochromator to separate the analyte-emitted photons by wavelength before detecting them with a single detector. This approach is analogous to the FAES system where only one element could be analyzed at a time. In a sequential system, if a sample needed to be analyzed for multiple elements, the grating system was rotated to direct the appropriate wavelength to the exit slit of the monochromator and into a PMT (described in Section 3.4.2). While ICP–AES provided superior detection limits compared to FAES instruments, it was just as slow and labor-intensive as FAES.

Simultaneous multichannel systems rapidly became economical as component production costs decreased and labor costs increased with time. In these systems, multiple elements can be detected at the same time using one of two designs. One way is to use a standard grating monochromator that separates the photons based on wavelength and directs the photons of interest to specific exit slits. Then, a single detector was rapidly moved from one slit to the other to analyze photons of various wavelengths. This type of system is shown in Figure 6.6. Another way to accomplish multiple-element detection is by placing an individual detector, again usually a PMT, at each exit slit. Such a system, the Rowland circle, was shown in Figure 6.1 and is shown again here in Figure 6.7. The Rowland design increases the upfront cost of an instrument and the running cost as PMT are replaced but this design significantly decreases the analysis time.

More recent advances use a combination of two monochromators (one prism and one Echelle grating, or one conventional grating monochromator and one Echelle grating), remove the exit slit, and place a multiple element detection system, such as a charge-coupled device (CCD described in Section 3.4.3) or charge injection device (CID described in Section 2.4.3) in the path of the separated wavelengths (refer to Figure 6.8).

Some instruments use one detector for UV wavelengths and another for visible wavelengths. There are a variety of designs for these latter systems illustrated below. These more advanced instruments have an increased cost but are more economical recently due to decrease in the production costs of CCDs and CIDs versus PMTs. They also significantly reduce the lab bench space requirements. Instruments such as the one illustrated in Figure 6.9 (1) offer superior resolution compared to a Rowland circle due to the advanced monochromator systems, (2) offer the immediate option to monitor secondary or greater diffraction that can be used in high-concentration samples, (3) reduce

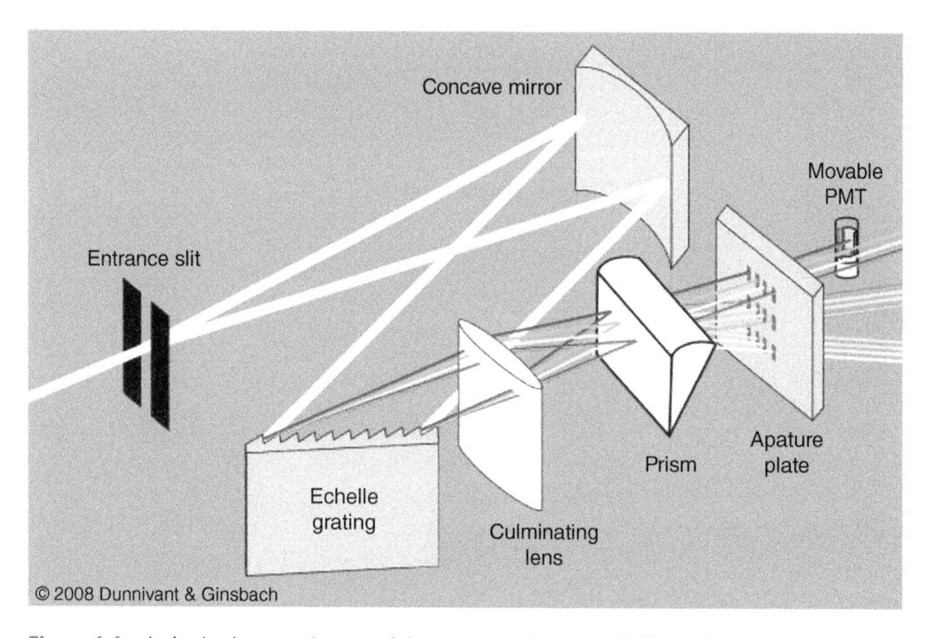

Figure 6.6 A single detector (sequential type monochromator) ICP–AES system (Source: Dunnivant and Ginsbach).

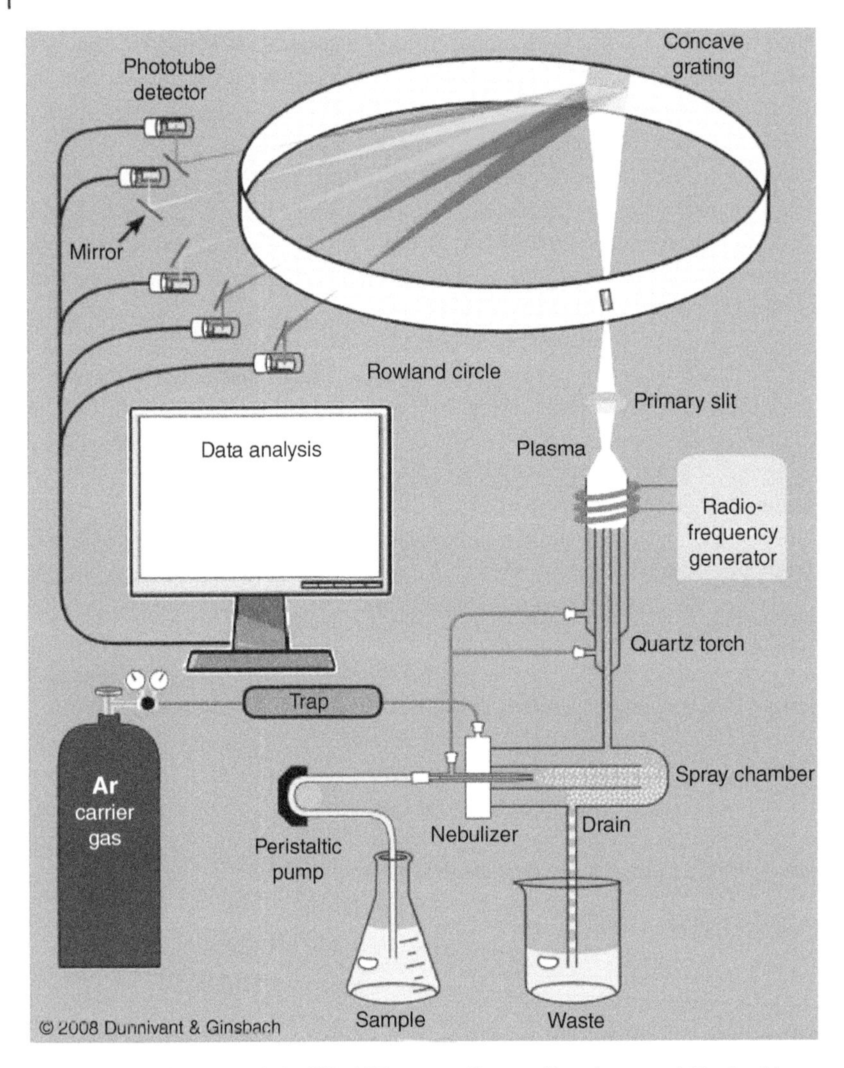

Figure 6.7 A Rowland circle ICP–AES system (Source: Dunnivant and Ginsbach).

the size of the instrument due to the reduced size of the detector(s), and (4) offer more inexpensive instruments as the cost of CCDs and CIDs become more economical.

6.3.5 Data Collection

When single-element detection was the goal of instrumentation, a simple chart recorder or digital display was sufficient for data collection. However, with the advances in ICP–AES technology today, computers are a necessity, not only to control the automatic sampler and the instrument control but also for data collection. Data collection systems are divided into a "method file" to run the instrument and "sequence file" to tell the instrument where a sample is in the sample tray, when to run it, and where to store the collected data file.

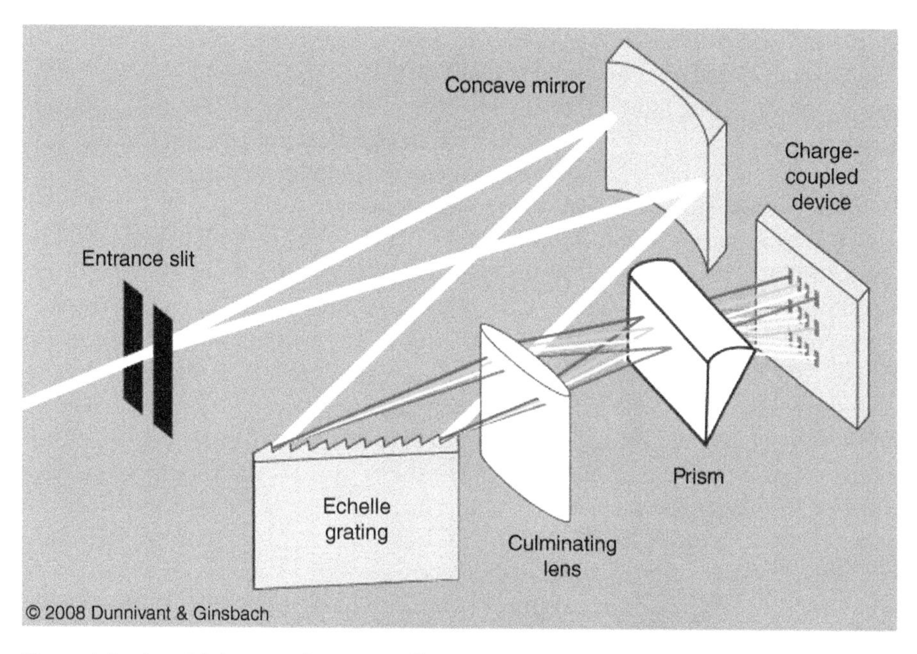

Figure 6.8 A multiple monochromator ICP–AES instrument (Source: Dunnivant and Ginsbach).

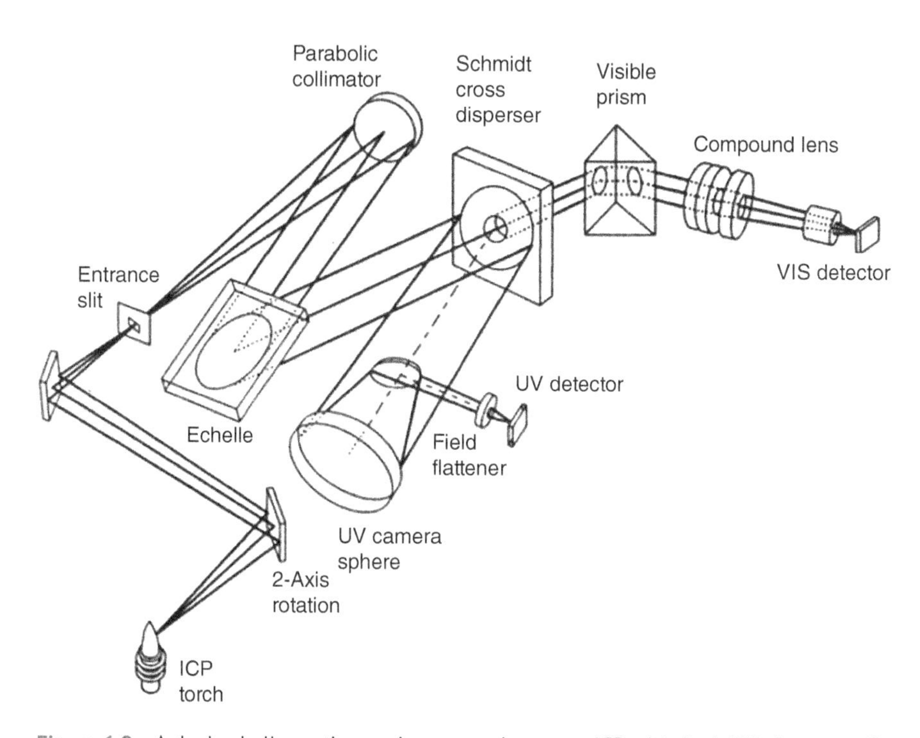

Figure 6.9 A dual echelle grating–prism monochromator ICP with dual CCD detectors (Source: Barnard et al. (1993) / with permission of American Chemical Society).

6.4 Interferences

ICP systems greatly reduce the number of interferences over those created in flame-based systems. Nebulizer, chemical, ionization, and spectral interferences are all present in ICP systems, but spectral interferences are most prominent. Nebulizer interferences (also known as matrix effects) can arise from physical and chemical differences between reference standards and samples, or between samples, such as the inconsistent presence of matrix salts and organic compounds or different viscosities and surface tension of the liquid. Each of these can be overcome by the use of standard addition calibration techniques discussed in Chapter 1, Section 1.1.2.2, but at a significant increase in the cost of analysis (primarily due to labor costs). For low ionic strength samples, nebulizer interferences are less prominent. Chemical interferences are common in FAAS and FAES but are less common or practically nonexistent in ICP–AES due to the relatively high temperature of the plasma, long residence time in the plasma, and inert atmosphere of the Ar plasma. Ionization interferences, in direct arc- and microwave-produced plasmas, usually only occur for easily ionized elements such as alkali and alkaline earth elements. The net result of ionization interferences is an increase or decrease in the intensity of emission lines for these elements. Few ionization interferences occur for these elements in ICP–AES.

Spectral interferences can be common in high-temperature plasmas as opposed to flame-based systems given the complete excitation and subsequent emission of all compounds in the sample (including the argon). Spectral interferences can be divided into three different classes. The first type is spectral line coincidence when resolution of the monochromator of the system is too poor to separate the analyte line from a matrix line. The use of Echelle monochromators with higher resolution eliminates spectral interferences. The second cause of interferences occurs when a wavelength of interest overlaps completely with a nearby "broadened line wing." This can be solved by monitoring a secondary emission line for that particular element. Most instruments use a background correction technique to overcome this type of interference. The final type of spectra interference referred to as spectral continuum occurs when stray light results from the recombination of electrons with Ar ions in the plasma that emits multiple intense lines. This can be avoided in some instruments by adjusting the temperature of the plasma or monitoring alternate lines. Stray light from matrix emissions can be avoided by use of high-end optical components such as "solar-blind PMTs."

6.5 Maintenance

ICP systems require considerable maintenance that includes monitoring and changing the Ar gas source given the high use rates, checking to ensure that the nebulizer is not clogged, and checking torch stability.

6.6 Case Study: Quantitation of Heavy Metals in Consumer Products by Dan Burgard

Polyvinyl chloride (PVC) packaging is a ubiquitous polymer used in building materials, medical devices, electronics, automotive parts, footwear, and packaging. PVC packaging can contain zinc, lead, and cadmium for use as UV light and heat stabilization. Several states have restrictions

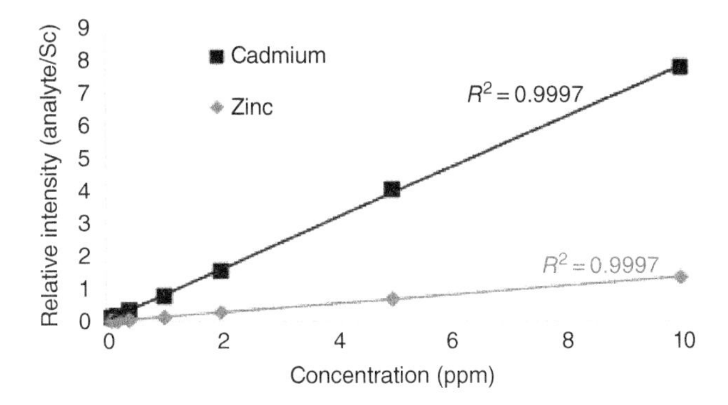

Figure 6.10 Emission calibration curves for zinc (206.200 nm) and cadmium (214.440) (Source: Dan Burgard).

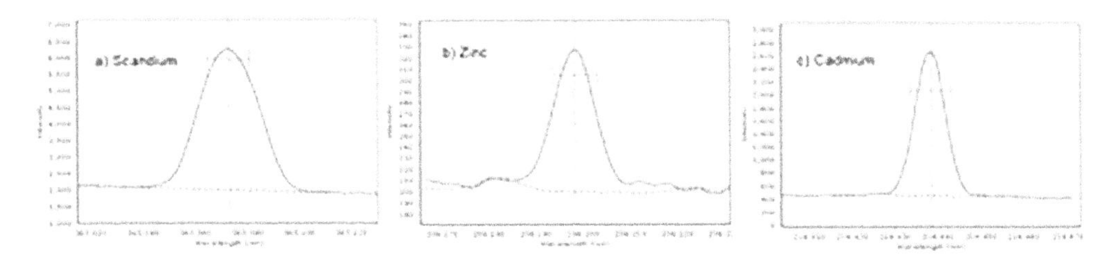

Figure 6.11 The emission spectra from an extracted consumer plastic packaging sample (Source: Dan Burgard).

against these elements in consumer packaging, yet their use remains. In an Instrumental Analysis laboratory, Professor Dan Burgard of University of Puget Sound had his students extract metals from samples of PVC-based consumer packaging and quantify the levels with Inductively Coupled Plasma–Atomic Emission Spectroscopy (ICP–AES). An internal standard of scandium was used in addition to a surrogate standard of yttrium.

Figure 6.10 shows an external calibration curve for cadmium and zinc solutions ranging from 100 to 10 ppm. The plotted intensity values are relative to the internal standard Sc. Figure 6.11 shows the intensity versus wavelength emission spectra from an extracted sample.

The plastic packaging sample was found to contain 8.42 ppm cadmium and 1.38 ppm zinc by weight. The instrumental deviation from triplicate measurements of the same sample varied by 0.7% and 3%, respectively.

6.7 Summary

Optical spectrometry has made considerable advances since the first flame-based instruments were introduced in the 1960s. Recent advances in plasma technology, monochromator layout, and computer-based detector systems such as charge transfer devices have given a new meaning to "state-of-the-art" technology. Today, low-end FAAS instruments are only used in situations where only one or two elements are analyzed infrequently. High sample throughput situations, such as those found in industry and environmental monitoring, require automated systems that can perform multiple analyses in a minimum amount of time. But these demands are paid for with instruments of significantly higher cost. While ICP–AES detection limits are significantly better than flame-based techniques, mass spectrometry yields even better detection limits and can distinguish between different isotopes. This is the subject of Chapter 7.

6.8 Questions

1 Approximately how many elements can FAAS be used to analyze?

2 Approximately how many elements can ICP–AES or ICP–MS be used to analyze?

3 Why can ICP be used to analyze more elements than FAES?

4 Why is acid added to all samples during sampling or for sample digestion in FAAS, FAES, and ICP analysis?

5 Draw and explain how a Rowland circle of PMTs is used in ICP–AES.

6 Explain the purpose and operation of the nebulizer in the ICP.

7 What is the most common type of nebulizer in ICP?

8 List all of the types of sample introduction to plasma.

9 What is a plasma?

10 What is ohmic heating?

11 Write a detailed discussion explaining how an ICP plasma is "light" and maintained.

12 Explain the name "inductively coupled plasma."

13 Why does the plasma not melt the instrument? How is it cooled?

14 Explain the three zones present in the plasma.

15 Contrast the separation of wavelengths by the two different systems in Figures 3.7 and 3.8.

16 List and explain the types of interferences in ICP–AES.

Supporting Information

Additional supporting information may be found online in the Supporting Information section in the HTML rendition of this article.

References

Barnard, T.W., Crockett, M.I., Ivaldi, J.C., et al. (1993). Solid-state detector for ICP-OES. *Analytical Chemistry* 65 (9): 1231–1239.

Greenfield, S., Jones, I., and Berry, C.T. (1964). High-pressure plasma as spectroscopic emission sources. *Analyst (London)* 89 (1064): 713.

7

Inductively Coupled Plasma–Mass Spectrometry

7.1 Introduction and History

The earliest forms of mass spectrometry (MS) go back to the observation of canal rays by Goldstein in 1886 and again by Wien in 1899. Thompson's later discovery of the electron also used one of the simplest mass spectrometers to bend the path of the cathode rays (electrons) and determine their charge to mass ratio. Later, in 1928, the first isotopic measurements were made by Aston. These basic experiments and instruments were presented to most readers in first-year general chemistry. More modern aspects of MS are attributed to Arthur Jeffrey Dempster and F.W. Aston in 1918 and 1919. Since this time, there has been a flurry of activity (not only concerning minor advances in components of mass spectrometers such as different types of instrument interfaces [direct injection, gas chromatography (GC) and high-performance liquid chromatography HPLC]) to different ionization sources (electron and chemical ionization) but also new types of ion separators. For example, double-focusing magnetic sector mass filters were developed by Mattauch and Herzog in 1934 (and recently revised into a new type of mass filter), time-of-flight MS (TOFMS) by Stephens in 1946, ion cyclotron resonance MS by Hipple and Thomas in 1949, quadrupole MS by Steinwedel in 1953, and ion trap MS by Paul and Dehmelt in the 1960s. MS was coupled with ICP as a means of sample introduction in 1980. Although not specific to ICP, even as recently as 1985, Hillenkamp and Michael Karas developed the matrix-assisted laser desorption/ionization (MALDI) technique (a laser-based sample introduction device) that radically advanced the analysis of protein structures and more types of mass analyzers will certainly be developed. Ion mobility spectrometer capabilities have recently advanced and are the basis of luggage scanning at airports. This chapter will deal only with basic mass spectrometer instruments that can be used in the analysis of atomic cations.

7.2 Components of a Mass Spectrometer

7.2.1 Overview

The sample introduction systems (automatic sampler to torch) are almost identical on optical and inductively coupled plasma mass spectrometry (ICP–MS) units (Section 6.3). While the inductively coupled plasma atomic emission spectroscopy (ICP–AES) is interfaced with an optical grating system, the plasma in an ICP–MS instrument must enter into a vacuum so that atomic cations can be separated by a mass filter. The common components of a modern ICP–MS are shown in Figure 7.1 (the sampling interface is not shown). The torch and the plasma were discussed in Section 6.3 (Videos S.6.1 and S.6.2). For MS systems, the detector is axially aligned with the plasma to follow

Essential Methods of Instrumental Analysis, First Edition. Frank M. Dunnivant and Jake W. Ginsbach.

Companion Website: www.wiley.com/go/essentialmethodsofinstrumentalanalysis1e

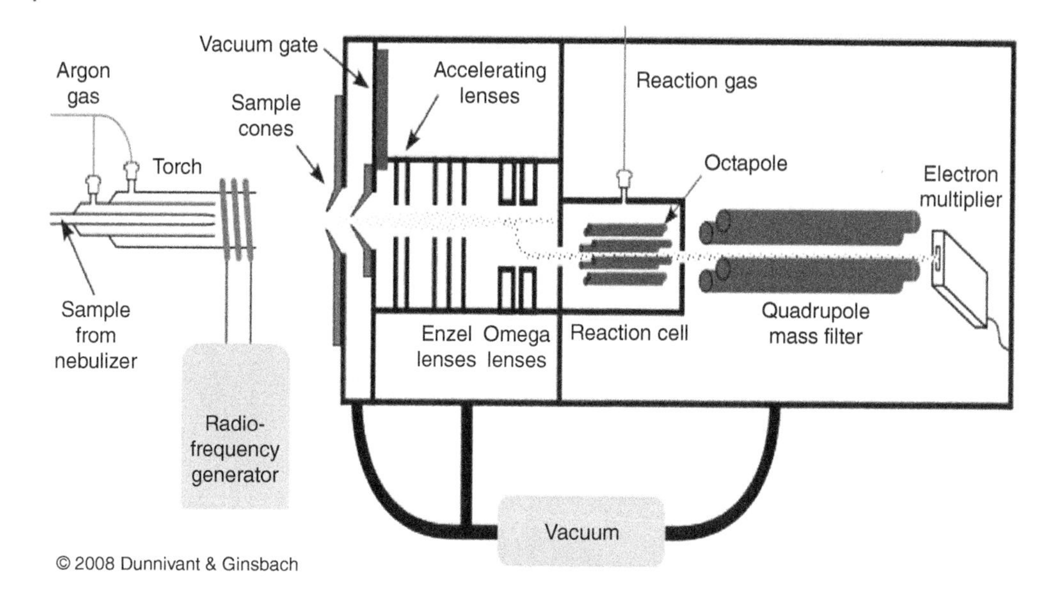

Figure 7.1 Illustration of the gas flow, RF wattage adjustment, and lighting of the plasma torch commonly used in ICP systems (Source: Dunnivant and Ginsbach).

the flow trajectory of the argon. After the analytes are ionized in the plasma at atmospheric pressures, they must enter into a low-pressure system before they can be accelerated and separated by mass to charge (m/z) ratios. This pressure difference is accomplished with a series of cones (small holes) between the plasma and the mass analyzer. The first cone, the sample cone, is a protruding cone, usually made of Ni, that has a small hole (1.0 mm in diameter) at its tip to allow the cations and Ar to pass. The next chamber interface contains another cone (the skimmer cone) with an even smaller diameter hole (equal to or less than 0.1 mm in diameter) that allows less sample to enter into the low vacuum chamber ($\sim 10^{-5}$ Torr which is about 10^{-8} atm). The smaller hole in the skimmer cone helps maintain a lower vacuum in the mass filter chamber. As the cations enter this second chamber, they are exposed to an accelerating lens (negatively charged plates) that places a fairly uniform amount of kinetic energy on the cations. Then the neutral particles and photons are filtered out by a second type of electronic lens (Einzel and Omega lens). The specific design of the lens varies among different manufacturers despite the fact that they accomplish the same goals. Most higher-end systems have a reaction cell placed just before the mass filter. This cell removes polyatomic interferences (that have the same mass as the analyte of interest) by gas-phase chemical reactions. Then, the cations enter into the mass filter that separates the different atoms with respect to their mass to charge ratio (m/z) before they eventually enter into a detector. Mass analyzers that have higher than unit amu resolution, such as a double-focusing mass filter, bypass the reaction cell since polyatomic interferences have different masses at three or four significant figures. Given the large amount of data and the extremely short scan times of the MS (100 ms), computer operation and computer-enhanced data collection are required. The most variation between various ICP–MS manufacturers is the presence or absence of a reaction cell and the type of mass filter.

7.2.2 Sample Introduction

The most common sample introduction system for an ICP–MS is made up of a nebulizer and spray chamber like an ICP–AES system (Section 6.3). While the majority of applications use this setup, there are some specialized applications that allow solid samples to be analyzed. Solid samples can be placed directly into the ICP with a graphite rod that contains a small quantity of sample. Other sample introduction procedures cause solid samples to sublimate before they enter the plasma. One common form of sample introduction not presented here is the glow discharge system (Section 5.3.5), which is used heavily by the semiconductor and metallurgy industries.

Another solid introduction technique is laser ablation which is becoming more common, especially for geological and materials science applications. In laser ablation, the automatic sampler and peristaltic pump for liquid samples are replaced with the working components of the laser ablation system. This consists of a small chamber to hold the solid sample on a movable stage, a laser to ablate (heat and vaporize the solid), a viewing window or charge-coupled device (CCD) camera to align the laser to a specific spot on the sample, and an argon gas stream to purge the ablation chamber and rapidly transport the vaporized sample to the inlet of the plasma. The laser is focused on a 10–25-μm section of the sample and a pulse of energy from the laser vaporizes the sample. The sample is transported to the plasma as a short pulse of vapor that is atomized and ionized in the plasma, and the generated cations are analyzed by the MS unit. Given the relatively small sampling area of the laser, numerous analyses can be conducted for a given sample and an average of analyte concentrations are determined. Common laser types include Nd-YAG, ruby, CO_2, and N_2. The only requirement of the laser is that it has sufficient power to ablate and vaporize refractory (high bond energy) sample matrices. Obviously, one of the quantitative limitations of the laser ablation technique is obtaining reference standards. While solid reference standards can be relatively easily made and obtained, it is almost impossible to match the matrix of all samples. Detection limits for this technique are in the range of 0.1–10 ppm which is much higher (poorer) than aqueous sample detection limits in an ICP–MS. As a result of the difficulties encountered with instrument calibration, qualitative analysis is commonly performed.

7.2.3 Mass Analyzer Interface

Sample cations enter into the mass filter from the plasma through a series of nickel or platinum cones that contain a small hole (from 1.0 mm to less than 0.1 mm) in the center. These cones are necessary to achieve the pressure drop that is required for the mass analyzer. The low pressure of the system, 10^{-5} Torr, minimizes the collision of the chemically reactive analyte cations with ambient gases, thus maintaining a confined beam of ions. A photograph of typical cones is shown in Figure 7.2. Cones are one of the most maintenance-intensive components of an ICP–MS since they require frequent cleaning, but the cleaning process is easy and relatively fast. After cleaning, cones must be conditioned prior to use by exposing the clean cones to a mid- to high-range reference standard via the plasma for 20–30 minutes. This conditioning process will avoid analyte loss on the cone and memory effects (the persistent presence of an analyte in a blank run typically occurring after a high concentration standard or sample has been analyzed).

The low pressure in the spectrometer chamber is maintained with two vacuum pumps. First, an external rotary vacuum pump is used to remove gas molecules down to a pressure of 10^{-1} to 10^{-4} Torr; a rotary vacuum pump is a positive-displacement pump that consists of vanes mounted to a spinning rotor. After a sufficient vacuum has been reached, a turbo molecular pump takes the vacuum down to 10^{-5} to 10^{-6} Torr. A molecular pump operates by using high-speed (50 000 rpms today but older pumps operated at 90 000 rpms) rotating blades to literally knock gaseous molecules out of the chamber. Low vacuum pressures are needed to minimize secondary collisions between analyte cations and ambient atmospheric molecules that would deflect the cation path away from the mass filter and detector and interfere with the desired trajectory in the mass filter.

Figure 7.2 Photograph of a Sampler (the larger one) and a Skimmer Cone for ICP-MS. Source: Courtesy of Agilent Technologies, Inc.

7.2.4 Lenses

After entering into the evacuated region, a number of electronic lenses are used to manipulate the path of the ions flowing from the plasma. First and foremost are the accelerator lenses. An accelerator lens consists of two to three plates with a relatively large hole in them (larger than the hole in the cones). Each plate has an increasingly negative charge placed across it that results in the attraction of the cations toward the plate increasing their kinetic energy. The hole in the center allows most of the cations to pass directly through the plate. The imposed kinetic energy is needed to pass the cations through the subsequent reaction cell, mass filter, and onto the detector with sufficient energy to dislodge electrons on the surface of the detector (an electron multiplier (EM) device).

The next type of lens used in the MS is a focusing lens that centers the cations into a small beam. This lens is used to focus ions into the center of the reaction cell (if present) and the mass filter. One such electrical lens is the Einzel lens that is analogous to a focusing lens in an optical spectrophotometer. An Einzel lens contains six parallel plates, three on each side of a rectangular box, that are exposed to various electric potentials (Figure 7.3). These potentials create an electrical field that bends the cations near the outside of the plates toward the focal point. The lens stretches the length of a given beam of ions since ions on the outside (near the plates) have to travel a longer distance to reach the focal point.

The final class of lenses removes neutral (elemental) atoms and photons that enter through the cones. Both photons and neutrals would be detected by the universal detector (an EM) and would give false signals and increase the instrumental noise if they are not filtered. Besides causing increased noise, neutrals passing through the mass filter can become adsorbed onto metal components that can interfere with their proper function. There are two major types of lenses that remove neutral particles and photons: a Bessel box and Omega lens. A Bessel box, also referred to

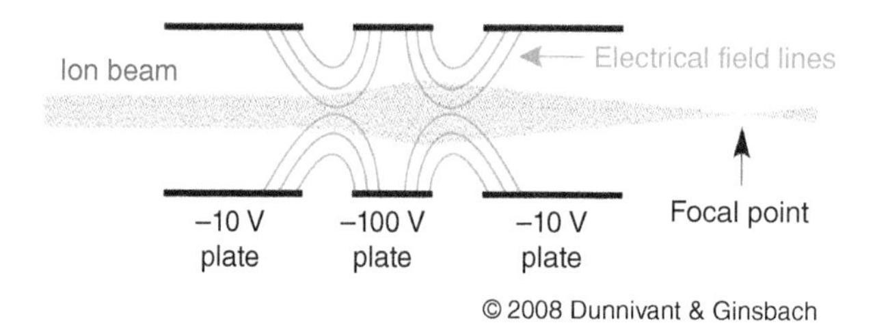

Figure 7.3 Diagram of an Einzel Lens used to focus a beam of ionic particles. Source: Dunnivant and Ginsbach.

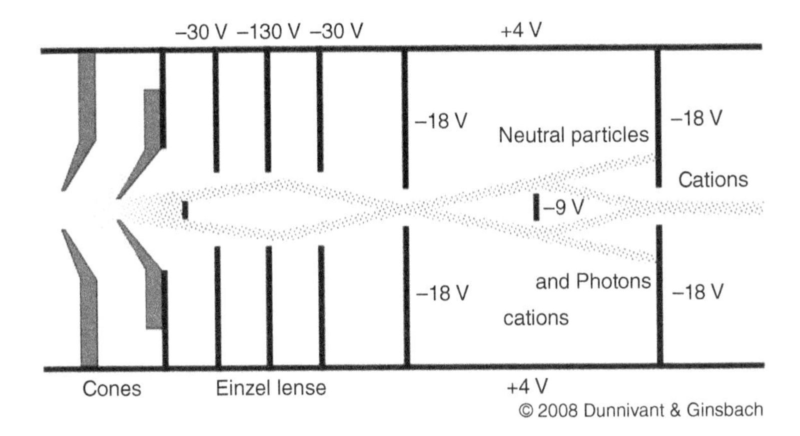

Figure 7.4 A Bessel Box photon stop. Source: Dunnivant and Ginsbach.

as a photon stop, comprises two photon stops, an Einzel lens, and a set of three lenses (Figure 7.4). The first photon stop (located before the Einzel lens) prevents particles from flowing directly down the evacuated chamber. The Einzel lens focuses the particles into the Bessel box and around the second photon stop. The positive voltage (+4 V) on the outside of the Bessel box and the negative voltage on the second photon stop (−9 V) direct the cations back to the exit slit. Neutral particles and photons are unaffected by the electrical field and are removed.

Another type of lens, an Omega lens, filters out the photons and neutral particles. A cross section of an Omega lens consists of four electrodes, two near the top and two near the bottom of the ion beam, and is presented in Figure 7.5. The lens works by carefully balancing the charges of the electrodes to deflect the beam of cations, but not the neutral species or the photons from the plasma. This deflection is accomplished by placing a positive charge on the first top electrode and a negative charge on the first bottom electrode that acts to deflect the beam of cations downward in the front of the lens (refer to Figure 7.5). Next, the beam needs to be stabilized with respect to the horizontal direction to guide the beam into the reaction cell or mass filter. So, an opposite set of electrodes is present, one with a negative charge on the top and one with a positive charge on the bottom. The net result is the deflection of the cations toward the mass analyzer in the absence of particles and photons that continue straight and collide with the end plate. Both the Einzel and Bessel systems are subject to contamination and need to be maintained (usually every six months or so depending on use).

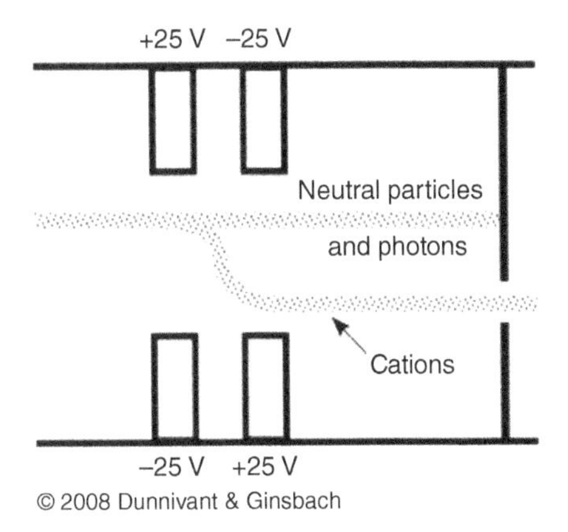

Figure 7.5 An Omega Lens. Source: Dunnivant and Ginsbach.

7.2.5 Mass Interferences and Reaction Cells

ICP–MS instruments separate and detect analytes based on the atom's mass to charge ratio. Since the plasma in an ICP system is adjusted to maximize singularly charged species, sample identity is directly related to atomic mass. While ICP–AES can be relatively free from spectral interferences (with monochromator systems that produce nm resolutions to three-decimal places), certain elements have problematic interferences in ICP–MS analysis due to the limited unit resolution (one amu) of some mass filters (especially the most common quadrupole mass filters). All ICP systems are subject to the nebulizer interferences given in the previous chapter. Spectral interferences are divided into three categories: isobaric, polyatomic, and doubly charged species. Isobaric interferences occur in mass analyzers that only have a unit resolution. For example, $^{40}Ar^+$ will interfere with $^{40}Ca^+$ and $^{114}Sn^+$ will interfere with $^{114}Cd^+$. High-resolution instruments will resolve more significant figures of the cation's mass and will easily distinguish between these elements. Polyatomic interferences result when molecular species form in the plasma that has the same mass as the analyte of interest. Their formation can be dependent on the presence of trace amounts of O_2 and N_2 in the Ar or sample, certain salts in the sample, and the energy of the plasma. For example, $^{40}Ca^{16}O^+$ can overlay with $^{56}Fe^+$, $^{40}Ar^{23}Na^+$ with $^{63}Cu^+$, and $^{80}Ar_2{}^+$ and $^{80}Ca_2{}^+$ with $^{80}Se^+$. The final type of interference occurs with doubly charged species. Since mass analyzers separate atoms based on their mass to charge ratio, $^{136}Ba^{2+}$ interferes with the quantification of $^{68}Zn^+$ since their mass to charge ratios are identical. The presence of any of these types of interferences will result in overestimation of the analyte concentration. Fortunately, there are several ways of overcoming these interferences.

The easiest, but most expensive, way to overcome all three spectral interferences is to use a high-resolution mass analyzer (discussed later), but, at a minimum, this can double to quadruple the cost of an instrument. Most inexpensive alternatives include the use of interference equations to estimate the concentration of the interfering element or polyatomic species, the use of a cool plasma technique to minimize the formation of polyatomic interferences, and the use of collision and/or reaction cells prior to the entry to the mass filter. These three techniques will be discussed in detail below.

7.2.5.1 Interference Equations

Most elements are present on the Earth in their known solar abundance (the isotopic composition of each element that was created during the formation of our solar system). Important exceptions are elements in the uranium and thorium decay series, most notably lead. For these elements, the isotopic ratios are dependent upon the source of the sample. For example, lead isotope ratios found in the environment can be attributed to at least three possible sources: geologic lead, leaded gasoline, and mined lead shot from bullets.

Interference equations are mathematical relationships based on the known abundances of each element that are used to calculate the total concentration of all of the isotopes of a particular ion. Isobaric correction is relatively easy when two or more isotopes of each element (the analyte and the interfering isotope) are present in the solar abundance. There are two ways to correct for this type of interference: (1) the analyte of interest can be monitored at a different mass unit (different isotope) or (2) the interfering element can be quantified as a different isotope (mass unit), and the result can be subtracted from the analyte concentration. Polyatomic interferences can be corrected for in the same manner but to a less effective degree. This type of correction is illustrated in the following example taken from the ICP–MS primer from Agilent Technologies Company, a manufacturer of ICP–MS systems.

Example 7.1 Arsenic is an important and common pollutant in groundwater and an industrial and agricultural pollutant. The analyte of interest is ^{75}As, but $^{40}Ar^{35}Cl$ has an identical mass on a low-resolution mass filter system, and since most water samples contain chloride, this interfering ion will be present in varying concentrations. These can be corrected for by doing the following instrumental and mathematical analysis. Note that all analyses suggested below require external standard calibration or for the instrument to be operated in semi-quantitative mode (a way of estimating analyte concentrations based on the calibration of a different element or isotope).

1) Acquire data at masses 75, 77, 82, and 83.
2) Assume the signal at mass 83 is from ^{83}Kr and use this to estimate the signal from ^{82}Kr (based on solar abundances).
3) Subtract the estimated contribution from ^{82}Kr from the signal at 82. The residual value should be the counts per second for ^{82}Se.
4) Use the estimated ^{82}Se data to predict the size of the signal from ^{77}Se on mass 77 (again, based on solar abundances).
5) Subtract the estimated ^{77}Se contribution from the counts per second signal at mass 77. The residual value should be from $^{40}Ar^{37}Cl$.
6) Use the calculated $^{40}Ar^{37}Cl$ data to estimate the contribution on mass 75 from $^{40}Ar^{35}Cl$.
7) Subtract the estimated contribution from $^{40}Ar^{35}Cl$ on mass 75. The residual is ^{75}As.

This process may seem complicated but is necessary to obtain accurate concentration data for As in the absence of a high-resolution mass filter. It should also be noted that this type of analysis has limitations. (1) If another interference appears at any of the alternative mass units used, the process will not work. (2) If the intensity of interference is large, then a large error in the analyte concentration will result. But as a side note, As only occurs as isotope 75. Superior detection limits can be obtained with a hydride generation FAAS mentioned in Section 5.3.3.

7.2.5.2 Cool Plasma Technique

The ionization of Ar-based polyatomic species in the normal "hot" plasma can be overcome by operating the radiofrequency at a lower wattage (from 600 to 900 W) and therefore lowering the temperature of the plasma. This technique, a function on all modern ICP–MS systems, allows for the removal of polyatomic interferences in the analysis of K, Ca, and Fe. One downside is the tendency to form more matrix-induced oxide cations.

7.2.5.3 Collision/Reactor Cells

The limitations of the two techniques described above, and the price of high-resolution MS, led to the development of collision and reaction cells in the late 1990s and early 21st century to remove these interferences. Numerous PhD dissertations, as well as research and development programs in industry, are active in this area and there are books specifically devoted to this topic. Two basic types of approaches have been used: (1) a collision cell that uses He to select for an optimum kinetic energy by slowing interfering ions relative to the analyte and only allowing the passage of the higher energy analyte and (2) reaction cells that promote reactions between a reagent gas and the interferences in order to remove them from detection.

The actual collision/reaction cell (CRC) is a quadra-, hexa-, or octa-pole that is considerably smaller than the subsequent quadrupole (mass filter) and is enclosed in a chamber that can contain higher pressures than the surrounding vacuum chamber. No mass separation occurs in the multipole since only a direct current (DC) is applied to the poles. Instead, the main purpose of the multipole is to keep the beam focused/contained to provide a space for the necessary collisions or chemical reactions to occur. While the number of poles in the reaction cell varies with different instruments, the larger number of poles allow for a more effective cell since the cross sectional area of the ion beam is larger for an octa-pole over a hexa- or quadrupole. The majority of CRCs can be operated in either mode by altering the gas utilized by the system. The price of the instruments increases slightly with the addition of these cells. However, removing interferences with a collision cell is still less expensive than the alternative: a high-resolution mass filter.

7.2.5.4 Collision Cells

In a collision cell, a nonreactive gas, usually He, is used to remove polyatomic ions that have the same mass to charge ratio as the analyte of interest. These multipole collision cells are relatively small as compared to the mass-filtering quadrupole and confine the ion beam from the plasma. Helium gas is added to the cell, while the analyte of interest (an atomic species) and the interferent (a polyatomic species) move through the chamber. Polyatomic species are larger than atomic species and therefore collide with the He gas more often. The net result of these collisions is a greater reduction in the kinetic energy (measured in eVs) of the polyatomic species in relationship to the atomic species. As the polyatomic and analyte ions exit the collision cell, they are screened by a discriminator voltage. A discriminator voltage is the counterpart to an accelerating lens and contains a slit with a positive voltage; this process is commonly referred to as kinetic energy discrimination. When a positive voltage is applied to this gate, only cations possessing sufficient kinetic energy will pass through the slit. Smaller cations retaining more of their energy, after being subjected to the collisions with He, will pass through the slit, while larger polyatomic cations that have been slowed by the He collisions, will be repelled by the voltage. The polyatomic species that do not pass into the mass analyzer collide with the walls of the chamber and are neutralized and removed by the vacuum system. Common interferences that are removed in this manner are sample matrix-based interferences such as $^{35}Cl^{16}O^+$ from interfering with $^{51}V^+$, $^{40}Ar^{12}C^+$ from interfering with $^{52}Cr^+$, $^{23}Na^{40}Ar^+$ from interfering with $^{63}Cu^+$, $^{40}Ar^{35}Cl^+$ from interfering with $^{75}As^+$, and plasma-based interferences such as $^{40}Ar^{16}O^+$ and $^{40}Ar^{38}Ar^+$. Interfering polyatomic species can be reduced down to ppt levels through kinetic energy discrimination. An animation for a typical collision cell is shown in Video S.7.1.

7.2.5.5 Reaction Cells

The physical structure and design of a collision cell, depending on the manufacturer, is similar or identical to that of a reaction cell. However, instead of utilizing an inert gas such as helium, more reactive gases are introduced into the cell. H_2 is the most common reactive gas but CH_4, O_2, and NH_3 are also used. Table 7.1 shows a variety of reaction gases and their intended use.

The purpose of the reactive gas is to break up or create chemical species, through a set of chemical reactions, and change their polyatomic masses to one that does not coincide with the mass of the analyte of interest. These cells have significantly extended the elemental range of ICP–MS to include some very important elements; the most important being $^{39}K^+$, $^{40}Ca^+$, and $^{56}Fe^+$ which had previously been difficult to measure due to the interferences of $^{38}Ar^1H^+$, $^{40}Ar^+$, and $^{40}Ar^{16}O^+$, respectively. The removal of interferences can be divided into three general categories: charge exchange, atom transfer, and adduct formation (i.e. condensation reactions).

A generic reaction for a charge transfer reaction would be

$$A^+ + B^+ + R \rightarrow A^+ + B + R^+ \tag{7.1}$$

Table 7.1 Reagent gases used in collision and reaction cell ICP–MS systems.

Collision Gas	Purpose
He, Ar, Ne, Xe	Used as a collision gas to decrease the kinetic energy of the polyatomic interference
H_2, NH_3, Xe, CH_4, N_2	Used in charge exchange reactions
O_2, N_2O, NO, CO_2	Used to oxidize the interference or analyte
H_2, CO	Used to reduce the interference
CH_4, C_2H_6, C_2H_4, CH_3F, SF_6, CH_3OH	Used in adduction reactions to remove interferences

Source: Data from Koppenaal et al. (2004)

where A^+ is the analyte, B^+ is the isobaric interferent, and R is the reagent gas. An example of a charge exchange reaction is removal of the cationic Ar dimer in the analysis of selenium.

$$^{80}Se^+ + {}^{80}Ar_2{}^+ + H_2 \rightarrow {}^{80}Se^+ + {}^{40}Ar^{40}Ar + H_2{}^{+-} \quad (7.2)$$

The neutral Ar dimer is now removed by the photon stop and vacuum and $^{80}Se^+$ is easily transported through the mass filter. Another specific case would be the interference of $^{40}Ar^+$ with the measurement of $^{40}Ca^+$. The reaction is

$$^{40}Ca^+ + {}^{40}Ar^+ + H_2 \rightarrow {}^{40}Ca^+ + {}^{40}Ar + H_2{}^+ \quad (7.3)$$

In this reaction, the interfering cationic species is neutralized and removed by the vacuum and does not enter the MS. It should be noted that in charge exchange reactions, the ionization potential of the reagent gas must lie between the ionization potentials of the interfering ion and the analytes in order to promote charge transfer from the interfering ion instead of the analyte. Such a requirement is not necessary for atom transfer and adduction formation/condensation reactions. Two such reactions follow.

Atom Addition Reaction

$$A^+ + B^+ + R \rightarrow AR^+ + B \quad (7.4)$$

$$Fe^+ + ArO^+ + N_2O \rightarrow FeO^+ + ArO^+ + N_2 \quad (7.5)$$

In this case, the interference of ArO^+ with the measurement of Fe is removed by oxidizing the Fe to its oxide that has a different mass from the argon oxide and quantifying Fe as FeO^+. Another example is given below for the removal of ^{90}Zr interference in the detection of ^{90}Sr.

$$^{90}Sr^+ + {}^{90}Zr + 1/2O_2 \rightarrow {}^{90}Sr^+ + {}^{90}ZrO^+ \quad (7.6)$$

These chemical reactions in the cell create cations that can potentially interfere with other analytes. Hence, it is not uncommon for these problematic analytes to be measured singularly (no multielemental analysis). As a result, the reaction cell mode is frequently utilized for singular applications or for argon interferences (e.g. $^{40}Ar^+$ and $^{40}Ar\,{}^{16}O^+$) with hydrogen since the products of the reaction do not interfere with other analytes of interest. If possible, operating the CRC in the collision mode is preferable since the interferences are removed from the system. After the spectral interferences have been removed by either process, the ion beam is separated by mass to charge ratio with the mass filter. An animation of a typical reaction occurring in a reaction cell with H_2 is shown in Video S.7.2.

7.2.6 Mass Filters (Mass Analyzers)

Mass analyzers separate the cations based on ion velocity, mass, or mass to charge ratio. A number of mass filters/analyzers are available for combination with ICP systems. These can be used individually or coupled in a series of mass analyzers to improve mass resolution and provide more conclusive analyte identification. This text will discuss the most commonly available mass filters, but not all are commercially available for ICP systems.

The measure of "power" of a mass analyzer is resolution, the ratio of the average mass (m) of the two adjacent ion peaks being separated to the mass difference (Δm) of the adjacent peaks, represented by

$$R_S = m/\Delta m \tag{7.1}$$

Resolution (R_S) is achieved when the midpoint between two adjacent peaks is within 10% of the baseline just before and after the peaks of interest (the valley between the two peaks). Resolution requirements can range from high-resolution instruments that may require discrimination of a few ten thousands (1/10 000) of a gram molecular weight (0.0001) to low-resolution instruments that only require unit resolution (28 versus 29 atomic mass units; amu). Resolution values for commonly available instruments can range from 250 to 500 000.

Before introducing the various types of mass analyzers, remember our current location of the mass analyzer in the overall ICP–MS system. The sample has been introduced to the nebulizer, atomized and ionized by the plasma, accelerated and manipulated by various lenses, sent through a CRC, and finally enters the mass analyzer. Now the packet of cations needs to be separated based on their momentum, kinetic energy, or mass-to-charge ratio (m/z). Often the terms mass filter and mass analyzer are used interchangeably, as is done in this textbook. But, first a controversy in the literature needed to be addressed with respect to how a mass filter actually separates ions.

Some resources state that all mass analyzers separate ions with respect to their mass to charge ratio, while others are more specific and contend that only quadrupoles separate ions by mass to charge ratios. The disagreement in textbooks lies in what components of the MS are being discussed. If one is discussing the effect of the accelerator plates and the mass filter, then all mass filters separate based on mass to charge ratios. This occurs because the charge of an ion will be a factor that determines the velocity a particle of a given mass has after interacting with the accelerator plate in the electronic, magnetic sector, and TOF mass analyzers. But after the ion has been accelerated, a magnetic section mass filter actually separates different ion-based momentums and kinetic energies, while the time of flight instrument separates different ions based on ion velocities (arrival times at the detector after traveling a fixed length). In the other case, no matter what the momentum or velocity of an ion, the quadrupole mass analyzer separates different ions based solely on mass to charge ratios (or the ability of the ion to establish a stable path in an oscillating electrical field). These differences may seem semantic but some users insist on this clarification. For the discussions below, in most cases, mass to charge will be used for all mass analyzers.

7.2.6.1 Quadrupole Mass Filter

Quadrupole mass filters have become the most common type of mass filters used today due to their relatively small size, lightweight, low cost, and rapid scan times (<100 ms). We will cover quadrupoles here to complete the story of MS interfaces. Other types of mass spectrometers will be covered later in Chapter 13 after chromatography and capillary electrophoresis are covered. The quadrupole mass filter is most commonly used in conjunction with ICP systems because they are able to operate at a relatively high pressure (5×10^{-5} Torr) as compared to lower pressures required in other mass filters. The quadrupole has also gained widespread use in tandem MS applications (a series of MS analyzers).

Despite the fact that quadrupoles produce the majority of mass spectra today as mentioned earlier, they are not true mass spectrometers. Actual mass spectrometers produce a distribution of ions either through time (TOFMS) or space (magnetic sector mass spectrometer). The quadrupole's mass resolving properties are instead a result of the ion's stability/trajectory within the oscillating electrical field.

A quadrupole system consists of four rods that are arranged at an equal distance from each other in a parallel manner. Paul and Steinwegen theorized in 1953 that hyperbolic cross sections were necessary. In practice, it was found that circular cross sections are both effective and easier to manufacture. Each rod is less than a cm in diameter and usually less than 15 cm long. Ions are accelerated by a negative voltage plate before they enter the quadrupole and travel down the center of the rods (in the z-direction). The ions' trajectory in the z-direction is not altered by the quadrupole's electric field. Today, due to advances in gold-coated glass shapes, hyperbolic cross sections are used.

The various ions are separated by applying a time-independent DC potential as well as a time-dependent alternating current (AC) potential. The four rods are divided up into pairs where the diagonal rods have an identical potential. The positive DC potential is applied to the rods in the X–Z plane and the negative DC potential is applied to the rods in the Y–Z plane. The subsequent AC potential is applied to both pairs of rods, but the potential on one pair is the opposite sign of the other and is commonly referred to as being 180° out of phase (Figure 7.6).

Mathematically, the potentials that ions are subjected to are described by the following equations:

$$\Phi_{X-Z} = +(U + V\cos\omega t) \tag{7.2}$$

and

$$\Phi_{Y-Z} = -(U + V\cos\omega t) \tag{7.3}$$

where Φ is the potential applied to the X–Z and Y–Z rods, respectively, ω is the angular frequency (in rad/s) and is equal to $2\pi v$ where v is the radiofrequency of the field, U is the DC potential, and V is the zero-to-peak amplitude of the radiofrequency voltage (ac potential). The positive and negative signs in the two equations reflect the change in polarity of the opposing rods (electrodes). The values of U range from 500 to 2000 V and V ranges from 0 to 3000 V.

To understand the function of each pair, consider the rods in the X–Z plane in isolation. For now, imagine that only an AC potential is applied to the rods. Half the time when the potential was positive, ions (cations) would be repelled by the rod's charge and would consequently move toward the center of the rods. Likewise, when the potential was negative, ions would accelerate toward the rods in response to an attractive force. If during the negative AC potential, an ion comes into contact with the rod, it is neutralized and is removed by the vacuum. The factors that influence whether or not a particle strikes the rod during the negative cycle include the magnitude of the potential (its amplitude), the duration of time the ions are accelerated toward the rod (the frequency of the AC potential), the mass of the particular ion, the charge of the ion, and its position within the quadrupole.

Now imagine that a positive DC potential (at a fraction of the magnitude of the AC potential) is applied to the rod in the X–Z plane. This positive DC potential alone would focus all of the ions toward the center of the pair of rods. When the AC and DC potentials are applied at the same time to the pair of rods in the X–Z plane, ions of different masses respond differently to the resulting potential. *Heavy ions are largely unaffected by the AC and, as a result, respond to the average potential of the rods. This results in heavy ions being focused toward the center of the rods. Light ions, on the*

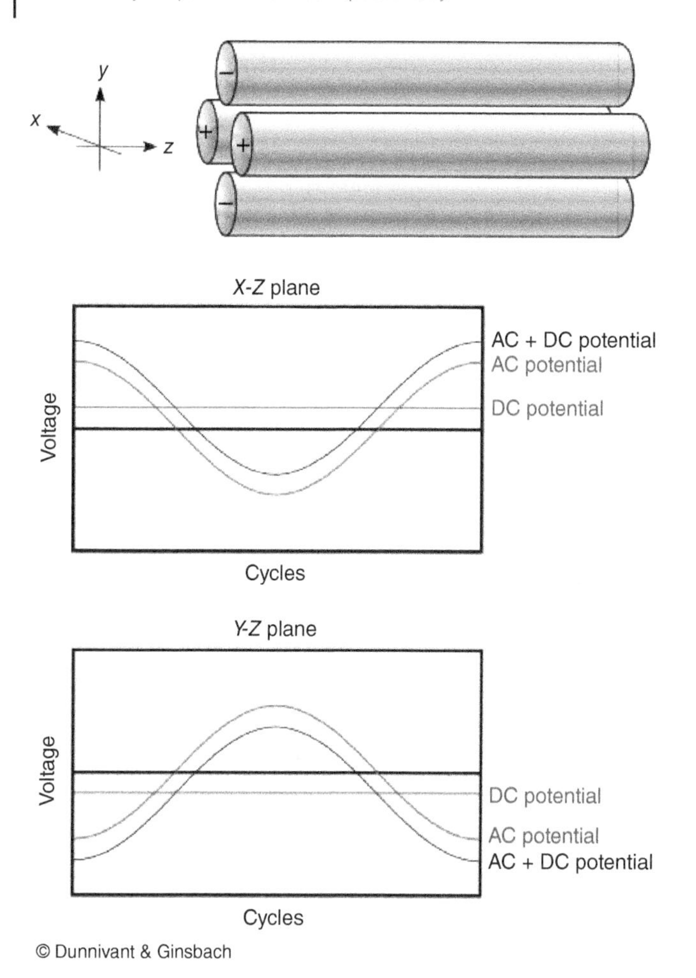

Figure 7.6 AC and DC Potentials in the Quadrupole MS. Source: Dunnivant and Ginsbach.

other hand, will respond more readily to the alternating AC. Ions that are sufficiently light will have an unstable trajectory in the X–Z plane and will not reach the detector. Only ions heavier than a selected mass will not be filtered out by the X–Z electrodes. As a result, the X–Z plane electrodes only filter light ions and form a high pass mass filter (Figure 7.7a).

Now look at the other pair of rods or the converse of the AC/DC potential. The rods in the *Y–Z* plane have a negative DC voltage and the AC potential is the same magnitude but the opposite sign as the potential applied to the *X–Z* plane. Heavy ions are still mostly affected by the DC potential, but since it is negative, they strike the electrode and are unable to reach the detector. The lighter ions respond to the AC potential and are focused toward the center of the quadrupole. The AC potential can be thought of as correcting the trajectories of the lighter ions, preventing them from striking the electrodes in the *Y–Z* plane. These electrical parameters result in the construction of a low-pass mass filter (Figure 7.7b).

When both the electrodes are combined into the same system, they are able to selectively allow a single mass to charge ratio to have a stable trajectory through the quadrupole. *Altering the magnitude of the AC and DC potential changes the mass to charge ratio that has a stable trajectory resulting in the construction of mass spectra*. Different ions possess a stable trajectory at different magnitudes

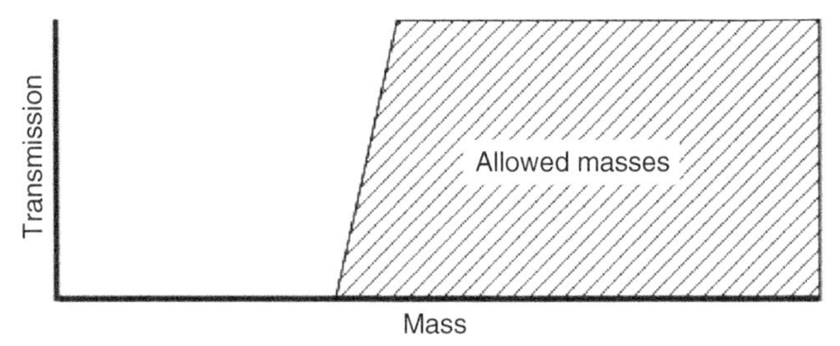

(a) The high-pass mass filter in the X-Z plane allows heavy ions to be transmitted through the quadrupole and reach the detector

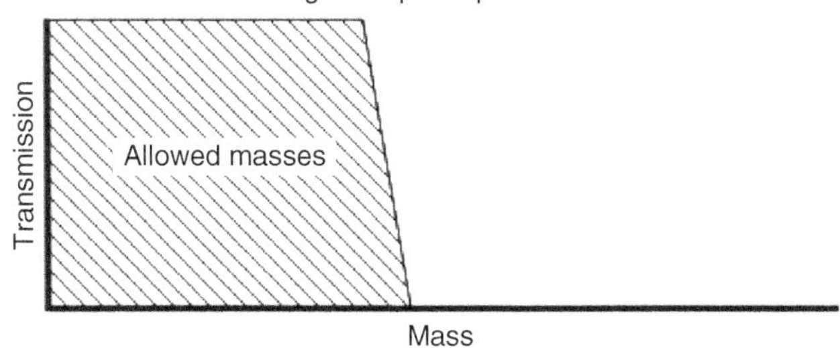

(b) The low-pass mass filter in the Y-Z plane allows light ions to be transmitted through the quadrupole and reach the detector

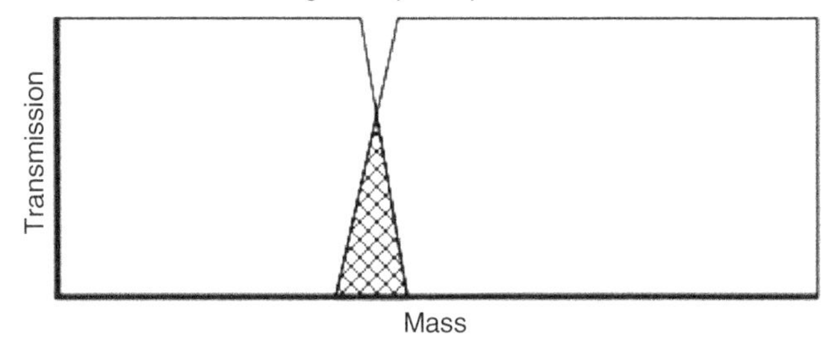

(c) The combined effect of the DC and oscilating AC potential results in an area stability for a specific mass to charge ratio

© Dunnivant & Ginsbach

Figure 7.7 A "Conceptual" Stability Diagram. Source: Dunnivant and Ginsbach.

and reach the detector at different times during a sweep of the AC/DC magnitude range. The graph of the combined effect, shown in Figure 7.7c, is actually a simplification of the actual stability diagram.

One way to generate an actual stability diagram is to perform a series of experiments where a single mass ion is introduced into the quadrupole. The DC and AC voltages are allowed to vary and the stability of the ion is mapped. After performing a great number of experiments, the resulting plot would look like Figure 7.8. The shaded area under the curve represents values of AC and DC voltages where the ion has a stable trajectory through the potential and would reach the detector. The white space outside the stability diagram indicates AC and DC voltages where the ion would not reach the detector.

While any AC and DC voltages that fall inside the stability diagram could be utilized, in practice, quadrupoles keep the ratio of the DC to AC potential constant (a constant slot of the scan line), while the scan is performed by changing the magnitude of the AC and DC potential. The result of this is illustrated as the mass scan line intersecting the stability diagram in Figure 7.8. The graphs below the stability diagram correspond to specific points along the scan and help to illustrate the

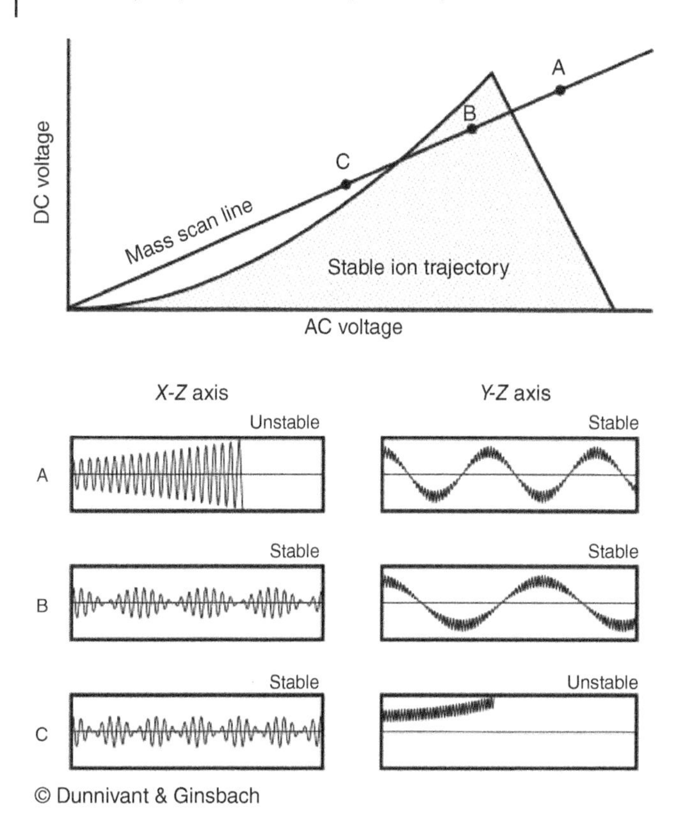

Figure 7.8 Stability Diagram for a Single Ion Mass. Used with permission from the Journal of Chemical Education, Vol. 75, No. 8, 1998, p. 1051. Source: Adapted from Steel and Henchman (1998).

ions' trajectories in the X–Z and Y–Z planes (Figure 7.8). While the mass to charge ratio of the ion remains constant in each pair of horizontal figures, the magnitude of the applied voltages changes, while their ratio stays constant. As a result, examining points along the mass scan line in Figure 7.8 is equivalent to shifting the position of the high and low pass mass filters with respect to the x-axis illustrated in Figure 7.7. Even though the mass is not changing for the stability diagram discussed here, the mass that has a stable trajectory is altered.

In Figure 7.7, the graph corresponding to point A indicates that the ion is too light to pass through the X–Z plane because of the high magnitude of the AC and DC potentials. As a result, its oscillation is unstable, and it eventually impacts the electrode/rod. The motion of the particle in the Y axis is stable because the combination of the AC potential as well as the negative DC potential yields a stable trajectory. This is the graphical representation of the AC potential correcting the trajectory of the light ions in the Y–Z plane. At point B, the magnitude of voltages has been altered so the trajectories of the ion in both the X–Z and Y–Z planes are stable and the ion successfully reaches the detector. At point C, the ion has been eliminated by the low-mass pass filter. In this case, the AC potential is too low to allow the heavy ion to pass through the detector and it strikes the rod. This is caused by the ion's increased response to the negative DC potential in the Y–Z plane. The trajectory in the X–Z-axis is stable since the DC potential focusing the ion toward the center of the poles overwhelms the AC potential.

Until this point, the stability diagram shown above is only applicable to a single mass. If a similar experiment were to be performed using a different mass, the positions of the AC and DC potential on the x- and y-axes would be altered but the overall shape of the curve would remain the

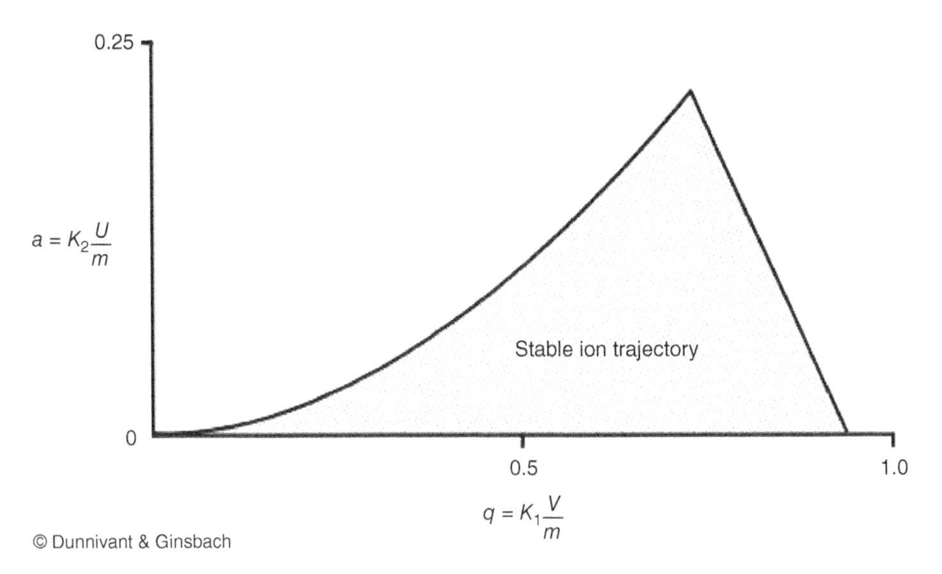

Figure 7.9 The General Stability Diagram. Source: Adapted from Steel and Henchman (1998).

same. Fortunately, there is a less time-consuming way to generate the general stability diagram for a quadrupole mass filter using a force balance approach. This derivation requires a complex understanding of differential equations and is beyond the scope of an introductory text, but the solution can be explained graphically (Figure 7.9). The parameters in the axes are explained below the figure.

While this derivation is particularly complex, the physical interpretation of the result helps explain how a quadrupole is able to perform a scan. The final solution is dependent on six variables, but the simplified two-variable problem is shown in Figure 7.9. Utilizing the reduced parameters, a and q, the problem becomes a more manageable two-dimensional problem. While the complete derivation allows researchers to perform scans in multiple ways, this discussion will focus only on the basic mode that makes up the majority of mass spectrometers. For the majority of commercially available mass spectrometers, *the magnitude of the AC potential (V) and the DC potential (U) are the only parameters that are altered during run time and allow a sweep of the mass to charge ranges.* The rest of the parameters that describe K_1 and K_2 are held constant. The values for K_1 and K_2 in the general stability diagram can be attributed to the following equations:

$$K_1 = \frac{2e}{r^2\omega^2} \tag{7.4}$$

$$K_2 = \frac{4e}{r^2\omega^2} \tag{7.5}$$

The parameters that make up K_1 and K_2 are exactly what we predicted when listing the variables earlier that would affect the point charge. Both K terms depend upon the charge of the ion e, its position within the quadrupole r, and the frequency of the AC oscillation ω. These parameters can be altered, but for the majority of applications, remain constant. The charge of the ion (e) can be assumed to be equivalent to positive one, +1, for almost all cases. The distance from the center of the quadrupole (r) is carefully controlled by the manufacturing process and an electronic lens that focuses the ions into the center of the quadrupole and is also a constant. Also the angular frequency (ω) of the applied AC waveform can be assumed to be a constant for the purposes of most spectrometers and for this discussion.

The first important note for the general stability diagram is the relationship between potential and mass. The general stability diagram (Figure 7.8) is illustrated where there is an inverse relationship between the two. Figures 7.9 and 7.10 show the lighter ions ($m - 1$) are higher on the mass scan line and the heavy ions ($m + 1$) are lower on the line. This is why in Figure 7.8, at point A, the molecule was too light for the selected frequencies, and it was too heavy at point C.

From the general stability diagram, it is also possible to explain how an instrument's resolution can be altered. The resolution is improved when the mass scan line intersects the smallest area at the top of the stability diagram (Figure 7.10). The resolution can be improved when the slope of the mass line is increased and the slope is directly related to the ratio of U and V. The resolution will subsequently increase until the line no longer intersects the stability diagram. While it would be best for the line to intersect at the apex of the stability diagram, this is impractical due to fluctuations in the AC (V) and DC (U) voltages. As a result, the line intersects a little below this point allowing the quadrupole to obtain unit resolution (plus or minus one amu).

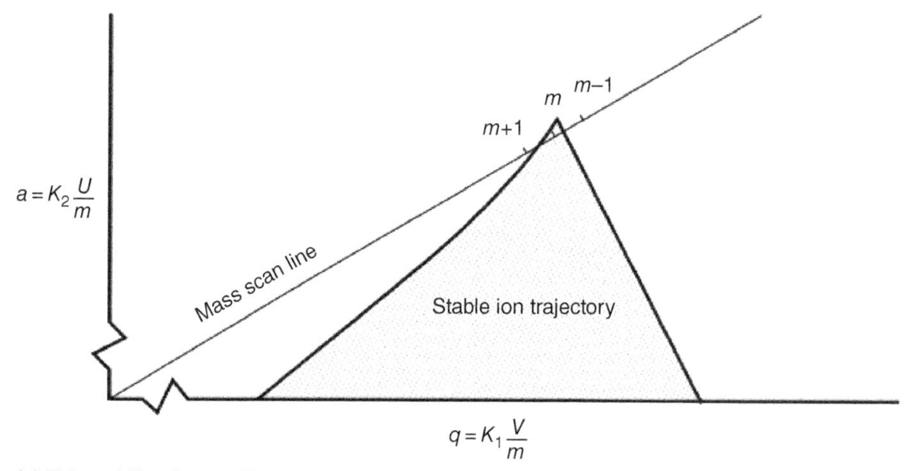

(a) This stability diagram illustrates a single value for both U and V where only particles of mass m are allowed to reach the detector.

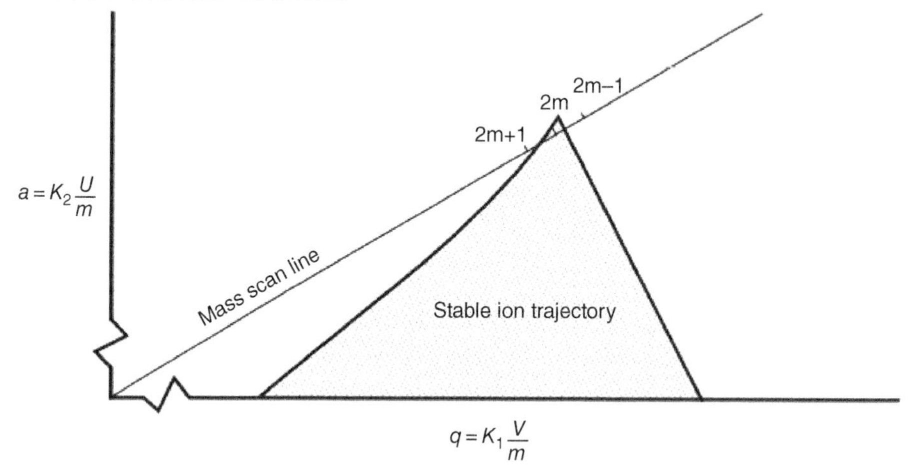

(b) This stability diagram illustrates a single value for both U and V that is double the value of figure (a). As a result, the particles corresponding to a mass of 2m are able to reach the detector

Figure 7.10 Quadrupole Mass Scan. Used with permission from the Journal of Chemical Education, Vol. 63, No. 7, 1998, p. 621. Source: Adapted from Miller and Denton (1986).

Once the resolution has been determined, the ratio of the AC to DC potential is left unchanged throughout the scan process. To perform a scan, the magnitude of the AC and DC voltages is altered, while their ratio remains constant. This places a different mass to charge inside the stability diagram. For example, if the AC and DC voltages are doubled, the mass to charge ratio of the selected ion would also be doubled as illustrated in the second part of Figure 7.10. By scanning throughout a voltage range, the quadrupole is able to create the majority of mass spectra produced in today's chemical laboratories. But it should be noted that quadrupole mass filters have an upper limit of approximately 650 charge to mass ratios. This is of no consequence in inorganic analysis since all isotopes are well below this limit.

Now that we have given a detailed description of the factors influencing the movement of a charged particle through the quadrupole, it is advantageous to summarize the entire process as a physicist would do in the form of a force balance. This is the origin of the governing equation where the French scientist E. Mathieu balanced the equations for the motion of ionized particles to the potential forces (electrical potentials) encountered in a quadrupole mass analyzer. This six-parameter differential equation, known as the Mathieu equation, is represented by

$$\frac{d^2u}{d\xi^2} + [a_u + 2qu\cos\xi]u = 0 \tag{7.6}$$

where

$$a = \frac{4eU}{\omega^2 r_0^2 m} \tag{7.7}$$

and

$$q = \frac{2eV}{\omega^2 r_0^2 m} \tag{7.8}$$

and where u is either the x or y directional coordinate, ξ is the redefining of time $(t/2)$, e is the charge of the ion, U is the magnitude of the DC potential, ω is the angular frequency $(2\pi f)$ of the applied AC waveform, r_0 is the distance from the center axis (the z-axis) to the surface of any electrode (rod), m is the mass of the ion, and V is the magnitude of the applied AC or radiofrequency waveform. By using the reduced terms, a and q, the six-parameter equation (e, ω, r_0, m, U, and V) can be simplified to a two-parameter equation (involving a and q). Thus, when the two opposing forces are balanced, the movement of a charged particle in an electrical field, the particle will pass through the quadrupole and strike the detector.

View Video S.7.3 for an illustration of how the trajectory of ions of different masses is changed during a mass scan.

7.2.6.2 Other Mass Spectrometers

Several other types of mass spectrometers are available for coupling to ICP. These include magnetic sector filters, TOF mass filters, quadrupole ion trap mass filters, double focusing systems, and combinations of filters used in Tandem Mass Spectroscopy. The theory and operation of these mass spectrometers will be covered later in Chapter 13.

7.2.7 Detectors

Once the analytes have been ionized, accelerated, and separated in the mass filter, they must be detected. This is most commonly performed with an EM, much like the photon multiplier tubes used in emission/optical spectroscopy. In MS systems, the EM is insensitive to ion charge, ion mass, or chemical nature of the ion (as a photomultiplier is relatively insensitive to the wavelength of a photon). The EMs in ICP–MS systems are usually discrete dynode EMs since these can be easily modified to extend their dynamic range.

A continuous EM is typically horn-shaped and is made of glass that is heavily doped with lead oxide. When a potential is placed along the length of the horn, electrons are ejected as ions strike the surface. Ions usually strike at the entrance of the horn and the resulting electrons are directed inward (by the shape of the horn), colliding sequentially with the walls and generating more and more electrons with each collision. Electrical potentials across the horn can range from high hundreds of volts to 3000 V. Signal amplifications are in the 10 000-fold range with nanosecond response times. Video S.7.4 illustrates the response of a continuous EM as polyatomic ions (as in Chapter 13), separated in a mass filter, strike its surface.

A standard discrete EM as shown in Video S.7.4 is actually a connected series of phototubes. In a discrete system, each dynode is held at a +90 V potential as compared to immediately adjacent dynodes. As a cation hits the first cathode, one or more electrons are ejected and pulled toward the next cathode. These electrons eject more and more electrons as they go forward producing tens to hundreds of thousands of electrons and amplifying the signal by a factor of 10^6–10^8. This allows for extremely low detection limits in the parts per billion (ppb) to parts per trillion (ppt) ranges. Such an EM is shown in Video S.7.5.

In order to extend the dynamic range of an EM to cover relatively high analyte concentrations (in the ppm range), some manufacturers have incorporated two EM detectors in one by including a switch that allows high signals to be counted in a digital manner in order to prevent the overload of signal and use analogue counting to analyze low-concentration samples.

The form of MS detector placed at the end of the EM is a Faraday Cup that counts each ion entering the detector zone.

7.3 Summary

MS detection greatly expands the applications of ICP system in the analysis of metals. Not only can more elements be analyzed, as compared to flame atomic absorption spectroscopy (FAAS), flame atomic emission spectroscopy (FAES), and ICP–AES, but isotopic data can also be collected. A variety of sample introduction techniques and mass filters make the ICP–MS a diverse instrument. But increased capabilities and lower detection limits come with a relatively high price tag. For example, a basic ICP–AES costs in the range of $80–$100 thousand, while a basic ICP–MS with reaction cell-quadrupole technology costs around $180 thousand; high-resolution instruments can cost as high as $600 thousand or more.

MS is in a constant and rapid state of development. In this chapter, we have focused on the most basic mass analyzers, such as the quadrupole, while other filters will be presented in Chapter 13. Recent technological advances have allowed for the development of two upcoming instruments. While these instruments carry a high price tag, they greatly increase the normal capabilities of the instrument.

Additional recent breakthroughs in MS include the drastic lowering of detection limits. A new technique referred to as nanostructured initiator MS (NIMS) is being used in research-grade instruments to measure biological metabolites. Utilization of these systems with laser-based systems produces detection limits is easily at the attomole (10^{-18}) amounts. These systems make the ppb and ppt detection limits discussed in this textbook seem trivial. It is likely that similar detection limits will soon be achieved for ICP-based instruments.

A summary of resolution and price for commercially available instruments is given in Table 7.2.

Table 7.2 Summary of mass filter features.

Type of Mass Filter	Resolution	Detection Limit	Approximate Instrument Price
Routine mass filters coupled with ICP			
Single quadrupole with a collision/reaction cell (CRC)	250–500	Low ppb–high ppt	\$180 000–\$200 000
Ion trap (available varies)	1000–10 000	ppb	\$400 000–\$500 000
Time of flight	3000–10 000	High ppt	\$400 000–\$500 000
Double focusing	10 000–20 000	Mid- to high ppt	\$750 000–\$1 000 000
Fourier transform	200 000–1 000 000	ppb	>\$2 000 000
New mass filters			
Magnetic sector/Multicollector with the Mattauch–Herzog geometry	~500	High ppb	\$400 000–\$500 000
Orbital trap (Electrostatic ion trap)	150 000–200 000	ppb	\$600 000

Source: Various instrument manufacturers.

Case Study: Analysis of Anthropogenic Lead in Roadside Soils by Isotopic Composition by Elizabeth W. Werner This study reports on the relationship between lead concentration in roadside soils and distance from the road, depth of the soil, traffic volume, and site construction history. In addition, this study seeks to distinguish between anthropogenic lead and lithogenic lead via isotopic composition analysis, specifically by using the value of the ratio of ^{206}Pb to ^{207}Pb as an indicator of having anthropogenic origin. Ten sampling locations, with a total of 73 soil samples, were chosen along Interstate 5 in between the cities of Centralia and Tacoma, Washington. The samples were prepared and digested according to Environmental Protection Agency (EPA) Method 3050B—Acid Digestion of Sediments, Sludges, and Soils. Samples were analyzed for lead isotope concentration on an Agilent 7500ce inductively coupled plasma mass spectrometer. The highest ^{206}Pb + ^{207}Pb concentration found for an individual sample was 5.22×10^2 ppb. The lowest ^{206}Pb + ^{207}Pb concentration found was 2.51×10^5 ppb. The average ^{206}Pb + ^{207}Pb concentration for all samples was 7.27×10^4 ppb. It was found that lead concentration in soil increases with increasing traffic volume. The isotopic composition of the majority of samples expressed an anthropogenic origin – 93% of samples had a ^{206}Pb/^{207}Pb ratio within the range that was chosen to signify anthropogenic origin (1.115–1.250). The slope of the line in Figure 7.11 (1.170) corresponds to the ^{206}Pb/^{207}Pb ratio.

7.4 Questions

1 Draw a diagram of a common ICP quadrupole MS system.

2 Why are most mass filters maintained at a low pressure?

3 List the common ways samples are introduced into an MS system.

4 What is the purpose of the cones in the sample interface?

5 Draw and explain an Einzel lens.

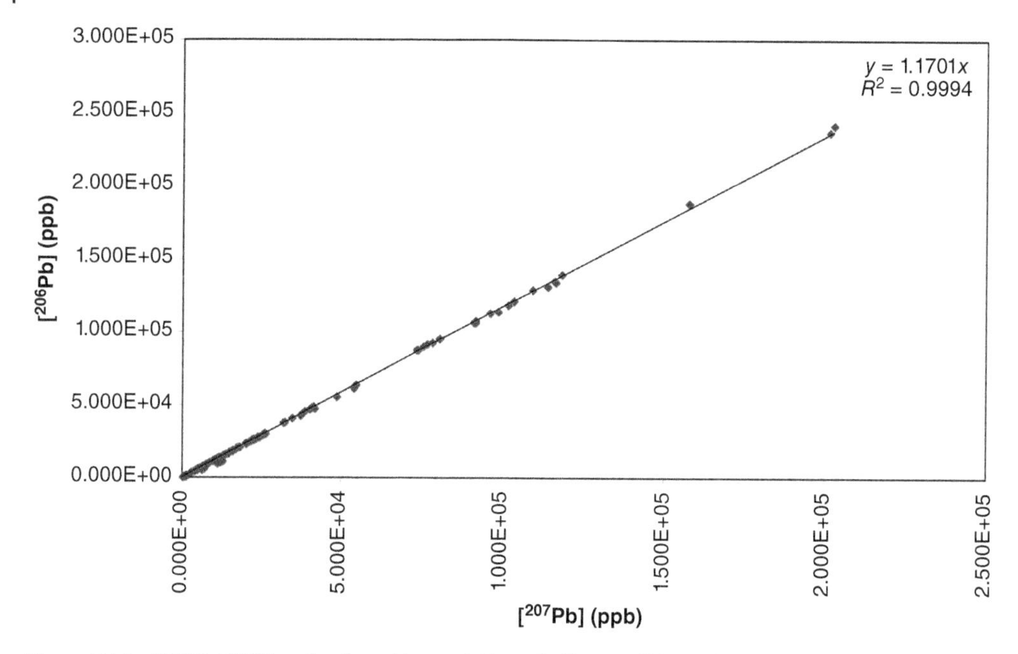

Figure 7.11 206Pb/207Pb ratios found in roadside soil. (Source: Elizabeth Werner).

6 Explain how a Bessel Box removes neutral particles and photons from the ion beam in an ICP–MS system.

7 Explain how an Omega lens removes neutral particles and photons from the ion beam in an ICP–MS system.

8 Explain polyatomic interferences and how they can be removed.

9 Explain how interference equations are used to eliminate the effects of mass interferences in ICP–MS.

10 What is the purpose of the cool plasma technique?

11 Explain how a collision cell works. What elements work best with this technique?

12 Explain how a reaction cell works.

13 How does a high-resolution mass filter work to remove virtually all interferences?

14 Give an example of a charge transfer reaction in a chemical reaction cell.

15 Give an example of an atom addition reaction in a chemical reaction cell.

16 Explain resolution with respect to mass filters.

17 Draw and explain how a magnetic sector mass filter works.

18 Draw and explain (in detail) how a quadrupole mass filter works.

19 The governing equation of the quadrupole mass filter consists of a six-parameter differential equation. Which two parameters are used to control the mass filter?

20 What is the purpose of the DC voltage in the quadrupole MS?

21 What is the purpose of the AC cycle in the quadrupole MS?

22 How do the low-mass and high-mass filters work to create a stable cation region in the quadrupole MS?

23 Explain the mass scan line in the quadrupole MS figures.

24 What is the purpose of sweeping the DC–AC voltages?

25 Extend the concepts of a linear quadrupole mass filter, explained above, to explain how the quadrupole ion trap mass filter works.

Supporting Information

Additional supporting information may be found online in the Supporting Information section in the HTML rendition of this article.

Reference

Koppenaal, D.W., Eiden, G.C., and Barinaga, C.J. (2004). Collision and reaction cells in atomic mass spectrometry: development, status, and applications. *Journal of Analytical Atomic Spectrometry* 19(5): 561–570.

Further Reading

de Hoffmann, E. and Stroobant, V. (2002). *Mass Spectrometry Principles and Applications*, 2e. New York: John Wiley & Sons.

Blakely, C.R. and Vestal, M.L. (1983). Thermospray interface for liquid chromatography/mass spectrometry. *Analytical Chemistry* 55(4): 750

Golhke, R.S., McLafferty, F., Wiley, B., and Harrington, D. (1956). *First Demonstration of GC/MS*, Bendix Corporation.

Jonscher, K. and Yates, J. (1997). The quadrupole ion trap mass spectrometer – a small solution to a big challenge. *Analytical Biochemistry* 244: 1–15.

March, R. (1997). An introduction to quadrupole ion trap mass spectrometry. *Journal of Mass Spectrometry* 32: 351–369.

Miller, P. and Denton, M. (1986). The quadrupole mass filter: basic operating concepts. *Journal of Chemical Education* 63(7): 617–622.

Skoog, D., Holler, F., and Nieman, T. (1992). *Principles of Instrumental Analysis*, 5e. Philadelphia: Saunders College Publishing.

Steel, C. and Henchman, M. (1998). Understanding the quadrupole mass filter through computer simulation. *Journal of Chemical Education* 75(8): 1049–1054.

Wong, P. and Cooks, R. (1997). Ion trap mass spectrometry. *Current Seperations.com and Drug Development* 16(3).

Yamashita M., Fenn, J.B. (1984). Electrospray ion source. Another variation on the free-jet theme. *Journal of Physical Chemistry* 88(20): 4451.

8

Contrasts and Comparisons of Instrumentation

8.1 Introduction

A number of techniques and instruments for the measurement of metal concentrations in a variety of matrices have been presented in this book. The purpose of this chapter is to condense some of the more important information into a few tables for comparison.

8.2 Figures of Merit

There are a number of ways to determine whether a technique is *best suited* for a particular situation. These include the detection limit that is required, the chemical and labor cost associated with the analysis, the cost of the instrument required to perform the analysis, and secondary advantages that may or may not be important. These are summarized and discussed below.

8.2.1 Detection Limits

One of the first decisions an analyst must make when determining the appropriate technique or instrument is the required detection limits. Table 8.1 gives a comparison of selected metals for a variety of techniques and instruments. In general, the detection limits decrease in concentration from FAAS to ICP–MS (left to right) in the table, but better (lower) detection limits also cost more money. Detection limits generally range from ppm to ppb levels. And as noted in the relevant chapters, as an instrument achieves better detection limits, higher quality (purer) dilution water and digestion acids are required.

8.2.2 Cost of Analysis

Table 8.2 provides a summary of the price range an analysis would cost when performed at an industrial laboratory per metal and per sample. *The key to reducing analysis costs for every technique is to automate the technique by using automatic samplers and data collection and to analyze as many samples as possible with one set of standards and quality assurance/quality control (QA/QC) samples.* QA/QC requirements vary from discipline to discipline, but it is common to recalibrate after every 20 samples. The information in Table 8.2 indicates that flame-based techniques that can only analyze one element at a time (FAAS and FAES) and labor-intensive techniques that are more difficult to automate (cold vapor and hydride generation) have slightly fallen out of favor due to

Essential Methods of Instrumental Analysis, First Edition. Frank M. Dunnivant and Jake W. Ginsbach.

Companion Website: www.wiley.com/go/essentialmethodsofinstrumentalanalysis1e

Table 8.1 A comparison of detection limits for a variety of techniques and instruments (in µg/L or ppb).

Element	FAAS	Hg/Hydride	Graphite Furnace	ICP–AES	ICP–MS
As	150	0.03	0.05	2	0.0006
Ca	1.5	NA	0.01	0.05	0.0002
Cd	0.8	NA	0.002	0.1	0.00009
Cr	3	NA	0.004	0.2	0.0002
Cu	1.5	NA	0.014	0.4	0.0002
Fe	5	NA	0.06	0.1	0.0003
Hg	300	0.009	0.6	1	0.016
K	3	NA	0.005	1	0.0002
Mg	0.15	NA	0.004	0.04	0.0003
Mn	1.5	NA	0.005	0.1	0.00007
Na	0.3	NA	0.005	0.5	0.0003
Ni	6	NA	0.07	0.5	0.0004
P	75 000	NA	130	4	0.1
Pb	15	NA	0.05	1	0.00004
Sb	45	0.15	0.05	2	0.0009
Se	100	0.03	0.05	4	0.0007
Sn	150	NA	0.1	2	0.0005
Sr	3	NA	0.025	0.05	0.00002
U	15 000	NA	NA	10	0.0001
Zn	1.5	NA	0.02	0.2	0.0003

(Source: Data from Perkin Elmer)

Table 8.2 The cost of analysis of an element for each technique/instrument for drinking and wastewater samples.

Technique/Instrument	Price per Element per Sample (as of 2008)	Comment(s)
Cost of digestion for nondrinking water samples	$10 per digestion; an additional charge may be required for complex matrices	Required when solid matter is present in the water or for sediment and tissue samples
FAAS/FAES	$10	No longer commonly used in commercial service labs since only one element can be analyzed at a time
Hg cold vapor/Hydride	$20–30	~Not commonly used by most labs due to labor intensive operations; being replaced by ICP–MS

Table 8.2 (Continued)

Technique/Instrument	Price per Element per Sample (as of 2008)	Comment(s)
Graphite furnace	\$20–30	Not commonly used by most labs due to labor-intensive operations; being replaced by ICP–MS
ICP–AES	\$10	Most labs offer a price reduction for specific sets of metal analyses
ICP–MS	\$10	Most labs offer a price reduction for specific sets of metal analyses

Prices as based on the analysis of a batch of samples (not an individual sample) and include the costs of calibration of the instrument and all QA/QC samples.
(Source: Confidential data from a Commercial Consulting Environmental Laboratory (2015 cost estimates))

Table 8.3 Approximate cost of instruments used in measuring metal concentrations.

Instrument	Approximate Price (as of 2020)
FAAS/FAES with automatic sampler	\$55 000+
FAAS with graphite furnace with automatic sampler	\$105 000
Flame fluorescence for Hg	\$35 000+
ICP–AES with automatic sampler	\$70 000–110 000
ICP–MS (with collision/reaction cell, automatic sampler, and quadrupole mass analyzer)	\$180 000–200 000

increasing labor costs but achieve excellent detection limits. These techniques are still commonly used in situations where only a few elements need to be analyzed infrequently. Otherwise, it rapidly becomes more economical to purchase a more automated system that will provide better detection limits and one that will analyze multiple elements at the same time.

8.2.3 Costs of Instruments

The cost of the instrument is a major factor determining whether analyses will be completed "in house" or whether samples will be sent to a consulting or contract laboratory. Table 8.3 shows the approximate costs of common instruments and those discussed in the book as of 2020. Three major manufacturers (Agilent, Perkin Elmer, and Thermo Scientific) were consulted and prices were averaged.

8.2.4 Summary of Advantages and Disadvantages

Several factors can go into the decision process for determining which instrument to use in an analysis. A summary of these is shown in Table 8.4.

Table 8.4 Advantages and disadvantages of each instrument.

Criterion	FAAS/FAES	Graphite Furnace	ICP–AES	ICP–MS (Quadrupole and CRC)
Detection limits	High ppb–ppm	Sub-ppb	Sub-ppb–ppm	Sub-ppt
Capability	Single element	Single element	Multielement	Multielement and isotope
Precision	1–2%	0.5–5%	0.1–2%	0.5–2%
Interferences:				
Spectral	Few	Very few	Some	Few
Chemical	Many	Many	Very few	Some
Physical	Some	Very few	Some	Some
Number of elements that can be analyzed	>35	>40	>70	>80
Sample volume requirements	4–8 mL/min	0.2–1 mL	1–2 mL/min	0.02–2 mL/min
Ease of use	Very easy	More difficult	Easy	More difficult
Capable of independent operation	No	Yes	Yes	Yes
Instrument costs	Low	Medium	High	Very high
Operating costs	Low	High	Medium	Very high

(Source: Misc. instrument vendors)

8.3 Questions

1. What are “figures of merit” with respect to analytical measurements?

2. Contrast detection limits for FAAS, graphite furnace, ICP–AES, and ICP–MS.

3. Give approximate instrument costs for the techniques discussed in this book.

4. Give approximate detection limits for each instrument/technique discussed in this book.

9

Chromatography Introduction, Chromatography Theory, and Instrument Calibration

9.1 Introduction

Analytical chemists have few tools as powerful as chromatography to measure distinct analytes in complex sample matrices. The power of chromatography comes from its ability to separate a mixture of compounds, or "analytes," and determine their respective identity (chemical structure) and concentration. Chromatography can be divided into three basic types that include gas, liquid, and supercritical fluid chromatography. Liquid chromatography (LC) can further be divided into ion exchange, separations based on size, and even extended to gel-based electrophoretic techniques. This book will provide a basic introduction to different types of liquid and gas chromatography (GC). The relationship between each type of chromatography is illustrated in Figure 9.1.

In general, each type of chromatography comprises two distinct steps: chromatography (or separation of individual compounds in distinct elution bands) and identification (detection of each elution band). GC is the process of taking a sample and injecting it into the instrument, turning the solvent and analytes into gaseous form, and separating the mixture of compounds into individual peaks (and preferably individual compounds). LC completes the same process except that the separations occur in a liquid phase. Individual bands or peaks exit the column and identification occurs by a relatively universal detector. One particularly common detector for both gas and LC is mass spectrometry (MS), which transforms each analyte from a chemically neutral species into a positive cation, usually breaking various bonds in the process. Detecting the mass of the individual pieces (referred to as fragments) allows for conclusive identification of the chemical structure of the analyte. Principles of GC will be covered in Chapter 10, LC in Chapter 11, capillary electrophoresis (CE) in Chapter 12, and MS in Chapter 13.

In MS, the combination of compound separation and ion fragment identification (the subject of Chapter 14) yields an extremely powerful analysis that is said to be confirmatory. Confirmatory analysis means the analyst is absolutely sure of the identity of the analyte. In contrast, many other individual techniques and detectors are only suggestive, meaning the analyst thinks they know the identity of an analyte. This is especially true with most universal GC and LC detectors since these detectors respond similarly to many compounds. The only identifying factor in these chromatographic systems is their elution time from the column. In order to obtain confirmatory analysis, the sample would need to be analyzed by at least two or more techniques (for example, different separation columns) that yield the same results. MS and nuclear magnetic resonance (NMR) are two confirmatory techniques in chemistry.

At this point, it is important to understand the different applications GC–MS and LC–MS offer for two different types of chemists: analytical and synthetic organic chemists. Organic chemists

Essential Methods of Instrumental Analysis, First Edition. Frank M. Dunnivant and Jake W. Ginsbach.

Companion Website: www.wiley.com/go/essentialmethodsofinstrumentalanalysis1e

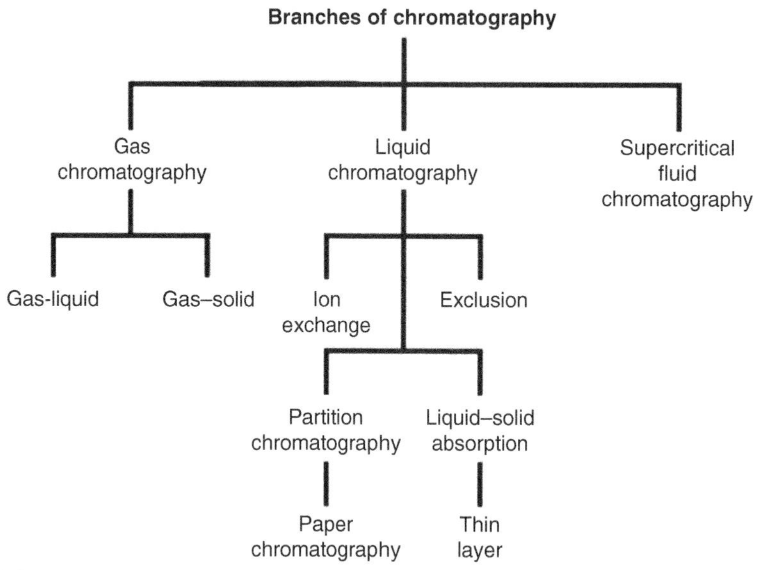

Figure 9.1 Categories of chromatography and their relationship to each other (Source: Dunnivant and Ginsbach).

attempt to create a desired chemical structure by transforming functional groups and intentionally breaking or creating bonds; in their resulting identification procedures, they already have a relatively good idea of the chemical structure. To characterize the resulting product, the chemist will use infrared (IR) spectroscopy to observe functional groups, MS to obtain the compound's molecular weight, and NMR spectroscopy to determine the molecular structure. Information from all three techniques is used to conclusively identify the synthesized product.

Analytical chemists are forced to approach identification in a different way because they have no a priori knowledge of the chemical structure and because the analyte is usually present at low concentrations where IR and NMR are inaccurate and of little to no use. Often, analysis is performed to look for a desired compound by comparing the sample analysis, again at low concentrations, to that of a known (reference) compound. The reference is used to identify the unknown compound by matching retention time (in chromatography) and ion fragmentation pattern (in MS). With today's computer mass spectral libraries that contain ion fractionation patterns for numerous chemicals, the analyst has the option of not using a reference standard. This is especially valuable if a reference compound is not available or is expensive. In some cases, especially with low analyte concentration, this approach may only result in a tentative identification.

This book will focus on GC–MS and LC–MS applications from an analytical chemistry perspective even though many synthetic chemists will also find much of this information useful for their applications.

9.2 Chromatographic Theory

All chromatographic systems have a *mobile phase* that transports the analytes through the column and a *stationary phase* coated onto the column or on the resin beads in the column. The stationary phase loosely interacts with each analyte based on its chemical structure, resulting in the separation of each analyte as a function of time spent in the separation column. The less analytes interact with the stationary phase, the faster they are transported through the system. The reverse is true for less mobile analytes that have stronger interactions. Thus, the many analytes in a sample are identified by retention time in the system for a given set of conditions. In GC, these conditions include the gas (mobile phase) pressure, flow rate, linear velocity, and temperature of the separation column. In high-performance liquid chromatography (HPLC), the mobile phase (liquid) pressure, flow rate, linear velocity, and the polarity of the mobile phase all affect a compound's retention time. An illustration of retention time is shown in Figure 9.2. The equation at the top of the figure will be discussed later during our mathematic development of chromatography theory.

In Figure 9.2, the minimum time that a nonretained chemical species will remain in the system is t_M. All compounds will reside in the injector, column, and detector for at least this long. Any affinity for the stationary phase results in the compound being retained in the column, causing it to elute from the column at a time greater than t_M. This is represented by the two larger peaks that appear to the right in Figure 9.2, with retention times t_{RA} and t_{RB}. Compound B has more affinity for the stationary phase than compound A because it exited the column last. A net retention ($t_{R'A}$ and $t_{R'B}$) time can be calculated by subtracting the retention time of the mobile phase (t_M) from the peaks' retention time (t_{RA} and t_{RB}).

Figure 9.2 also illustrates how peak shape is related to retention time. The widening of peak B is caused by longitudinal diffusion (diffusion of the analyte as the peak moves down the length of the column). This relationship is usually the reason why integration by area, and not height, is utilized. However, compounds eluting at similar retention times will have near identical peak shapes and widths.

A summary of these concepts and data-handling techniques is shown in Video S.9.1.

Chromatographic columns adhere to the old adage "like dissolves like" to achieve the separation of a complex mixture of chemicals. Columns are coated with a variety of stationary phases or chemical coatings on the column wall in capillary columns or on the inert column packing in packed columns. When selecting a column's stationary phase, it is important to select a phase possessing similar intermolecular bonding forces to those characteristic of the analyte. For example, for the

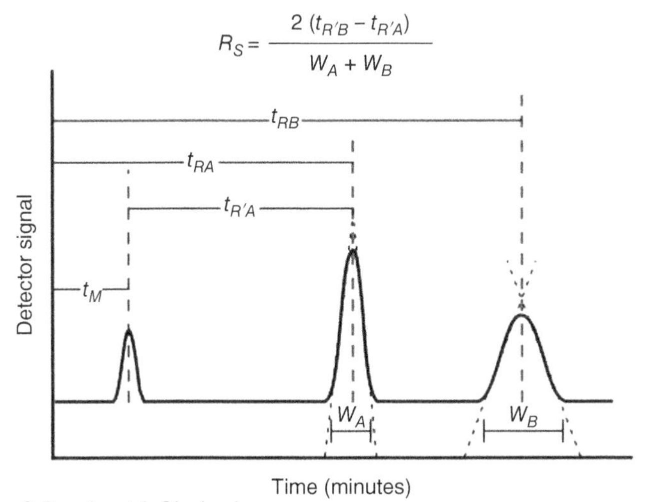

Figure 9.2 Identification of analytes by retention time (Source: Dunnivant and Ginsbach).

separation of a series of alcohols, the stationary should be able to undergo hydrogen bonding with the alcohols. When attempting to separate a mixture of nonpolar chemicals such as aliphatic or aromatic hydrocarbons, the column phase should be nonpolar (interacting with the analyte via van der Waals forces). Selection of a similar phase with similar intermolecular forces will allow more interaction between the separation column and structurally similar analytes and increase their retention time in the column. This results in a better separation of structurally similar analytes. Specific stationary phases for GC and HPLC will be discussed later in Chapters 10 and 11, respectively.

9.2.1 Derivation of Governing Equations

The development of chromatography theory is a long-established science and almost all instrumental texts give nearly exactly the same set of symbols, equations, and derivations. The derivation below follows the same trends that can be found in early texts such as Karger et al. (1973) and Willard et al. (1981), as well as the most recent text by Skoog et al. (2007). *The reader should keep three points in mind as they read the following discussion. First, the derived equations establish a relatively simple mathematical basis for the interactions of an analyte between the mobile phase (gas or liquid) and the stationary phase (the coating on a column wall or resin bead). Second, while each equation serves a purpose individually, the relatively long derivation that follows has the ultimate goal of yielding an equation that describes a way to optimize the chromatographic conditions in order to yield maximum separation of a complex mixture of analytes. Third, do not become lost in all of the following equations; all chromatography calculations can be performed only using retention time data from a chromatogram, as illustrated in an example at the end of this section.*

To begin, we need to develop several equations relating the movement of a solute through a system to properties of the column, properties of the solute(s) of interest, and mobile phase flow rates. These equations will allow us to predict (1) how long the analyte (the solute) will be in the system (retention time), (2) how well multiple analytes will be separated, and (3) what system parameters can be changed to enhance separation of similar analytes. The first parameters to be mathematically defined are flow rate (F) and retention time (t_m). Note that "F" has units of cubic volume per time. Retention behavior reflects the distribution of a solute between the mobile and stationary phases. We can easily calculate the volume of stationary phase. In order to calculate the mobile phase flow rate needed to move a solute through the system, we must first calculate the flow rate.

$$F = (pr_c^2)\, e \left(\frac{L}{t_m}\right) \tag{9.1}$$

$$F = p\left(\frac{d_c}{2}\right)^2 e \left(\frac{L}{t_m}\right) \tag{9.2}$$

$$F = (pd_c/4)e \left(\frac{L}{t_m}\right) \tag{9.3}$$

where $(pd_c/4)$ is the cross-sectional area of the column, e is the porosity of the column packing, and $\left(\frac{L}{t_m}\right)$ is the average linear velocity of the mobile phase.

In the equations above, r_c is the internal column radius, d_c is the internal column diameter, L is the total length of the column, and t_m is the retention time of a nonretained analyte (one which does not have any interaction with the stationary phase). Porosity (e) for solid spheres (the ratio of the volume of empty pore space to total particle volume) ranges from 0.34 to 0.45, for porous materials ranges from 0.70 to 0.90, and for capillary columns is 1.00. The average linear velocity, a later factor of these, is represented by $\overline{\mu}$.

The most common parameter measured or reported in chromatography is the retention time of particular analytes and, in the end, will be the basis of all the calculations that follow. For a nonretained analyte, we can use the retention time (t_M) to calculate the volume of mobile phase that was needed to carry the analyte through the system. This quantity is designated as V_m,

$$V_m = t_M F \tag{9.4}$$

where V_m is in mL, t_M is in minutes, F is in mL/minute, and V_m is called the dead volume. For a retained solute, we calculate the volume of mobile phase needed to move the analyte through the system by

$$V_r = t_R F \tag{9.5}$$

where V_R is in mL, t_R is the retention time of the analyte in minutes, and F is in mL/minute.

In actual practice today, the analyst does not calculate the volume of the column but measures the flow rate and the retention time of nonretained and retained analytes. When this is done, note that the retention time not only is the transport time through the detector but also includes the time spent in the injector and detector! Therefore

$$V_m = V_{column} + V_{injector} + V_{detector} \tag{9.6}$$

The net volume of mobile phase (V'_R) required to move a retained analyte through the system is

$$V'_R = V_R - V_M \tag{9.7}$$

where V_R is the volume for the retained analyte and V_M is the volume for a nonretained (mobile) analyte.

This can be expanded to

$$t'_R F = t_R F - t_M F \tag{9.8}$$

and dividing by F, yields

$$t'_R = t_R - t_M \tag{9.9}$$

Equation 9.9 is important since it gives the net time required to move a retained analyte through the system (illustrated in Figure 9.2).

Note, for GC (as opposed to LC), the analyst has to be concerned with the compressibility of the gas (mobile phase), which is done by using a compressibility factor, j

$$j = \frac{3\left[\left(\frac{P_i}{P_o}\right)^2 - 1\right]}{2\left[\left(\frac{P_i}{P_o}\right)^2 - 1\right]} \tag{9.10}$$

where P_i is the gas pressure at the inlet of the column and P_o is the gas pressure at the outlet. The net retention volume (V_N) is

$$V_N = jV'_R \tag{9.11}$$

The next concept that must be developed is the partition coefficient (K), which describes the spatial distribution of the analyte molecules between the mobile and stationary phases. When an analyte enters the column, it immediately distributes itself between the stationary and mobile phases. To understand this process, the reader needs to look at an instant in time without any flow of the mobile phase. In this "snapshot of time," one can calculate the concentration of the analyte in each phase. The ratio of these concentrations is called the equilibrium partition coefficient,

$$K = C_S/C_M \tag{9.12}$$

where C_s is the analyte concentration in the solid phase and C_M is the solute concentration in the mobile phase. If the chromatography system is used over analyte concentration ranges where the "K" relationship holds true, then this coefficient governs the distribution of analytes anywhere in the system. For example, a K equal to 1.00 means that the analyte is equally distributed between the mobile and stationary phases. The analyte is actually spread over a zone of the column (discussed later) and the magnitude of K determines the migration rate (and t_R) for each analyte (since K describes the interaction with the stationary phase).

$$k' = \frac{t_r - t_m}{t_m} = \frac{V_s - V_m}{V_m} \tag{9.13}$$

Equation 9.7 ($V'_R = V_R - V_M$) relates the mobile phase volume of a nonretained analyte to the volume required to move a retained analyte through the column. K can also be used to describe this difference. As an analyte peak exits the end of the column, half of the analyte is in the mobile phase and half is in the stationary phase. Thus, by definition

$$V_R C_M = V_M C_M + V_S C_S \tag{9.14}$$

Rearranging and dividing by C_M yields

$$V_R = V_M + KV_S \quad \text{or} \quad V_R - V_M = KV_S \tag{9.15}$$

Now three ways to quantify the net movement of a retained analyte in the column have been derived, Equations 9.4, 9.9, and 9.15.

Now we need to develop the solute partition coefficient ratio, k' (also known as the capacity factor), which relates the equilibrium distribution coefficient (K) of an analyte within the column to the thermodynamic properties of the column (and to temperature in GC and mobile phase composition in LC, discussed later). For the entire column, we calculate the ratio of total analyte mass in the stationary phase ($C_S V_S$) as compared to the total mass in the mobile phase ($C_M V_M$), or

$$k' = \frac{C_s V_s}{C_m V_m} = K \frac{V_s}{V_m} \tag{9.16}$$

V_S/V_M is sometimes referred to as b, the volumetric phase ratio.

Stated in more practical terms, k' is the additional time (or volume) an analyte band takes to elute as compared to an unretained analyte divided by the elution time (or volume) of an unretained band, or rearranged, gives

$$k' = \frac{t_r - t_m}{t_m} = \frac{V_s - V_m}{V_m} \tag{9.17}$$

$$t_r = t_m(1 + k') = \frac{L}{u}(1 + k') \tag{9.18}$$

where μ is the linear gas velocity and the parameters in Equation 9.18 were defined earlier. So, the retention time of an analyte is related to the partition ratio (k'). Optimal k' values range from ~1 to ~5 in traditional packed column chromatography, but the analysts can use higher values in capillary column chromatography.

9.2.2 Multiple Analytes

The previous discussions and derivations were concerned with only one analyte and its migration through a chromatographic system. Now we need to describe the relative migration rates of analytes in the column; this is referred to as the selectivity factor, α. Notice Figure 9.2 had two analytes in the sample and *the goal of chromatography is to separate chemically similar compounds*. This is possible when their distribution coefficients (K_s) are different. We define the selectivity factor as

$$\alpha = \frac{K_B}{K_A} = \frac{k'_B}{k'_A} \tag{9.19}$$

where subscripts A and B represent the values for two different analytes and solute B is more strongly retained. By this definition, α is always greater than 1. Also, if one works through the math, you will note that

$$\alpha = \frac{V'_B}{V'_A} = \frac{t_{R,B} - t_m}{t_{R,A} - t_m} = \frac{t'_{R,B}}{t'_{R,A}} \tag{9.20}$$

The relative retention time, α, depends on two conditions: (1) the nature of the stationary phase and (2) the column temperature in GC or the solvent gradient in LC. With respect to these, the analyst should always first try to select a stationary phase that has significantly different K values for the analytes. If the compounds still give similar retention times, you can adjust the column temperature ramp in GC or the solvent gradient in LC; this is the general elution problem that will be discussed later.

Appropriate values of α should range from 1.05 to 2.0 for traditional packed columns but today's GC capillary column systems may have greater values.

Now, we finally reach one of our goals of these derivations, an equation that combines the system conditions to define analyte separation in terms of column properties such as column efficiency (H) and the number of separation units (plates, N) in the column (both of these terms will be defined later). As analyte peaks are transported through a column, an individual molecule will undergo many thousands of transfers between each phase. As a result, packets of analytes and the resulting chromatographic peaks will broaden due to physical processes discussed later. This broadening may interfere with "resolution" (the complete separation of adjacent peaks) if their K (or k') values are close (this will result in a value close to 1.0). Thus, the analyst needs a way to quantify a column's ability to separate these adjacent peaks.

First, we will start off with an individual peak and develop a concept called the theoretical plate height, H, which is related to the width of a solute peak at the detector. Referring to Figure 9.2, one can see that chromatographic peaks are Gaussian in shape, and can be described by

$$H = \frac{\sigma^2}{L} \tag{9.21}$$

where H is the theoretical plate height (related to the width of a peak as it travels through the column), σ is one standard deviation of the bell-shaped peak, and L is the column length. Equation 9.15 is a basic statistical way of using standard deviation to mathematically describe a bell-shaped peak. One standard deviation on each side of the peak contains ~68% of the peak area and it is useful to define the band broadening in terms of the variance, σ^2 (the square of the standard deviation, σ). Chromatographers use two standard deviations that are measured in time units (t) based on the baseline width of the peak, such that

$$t = \frac{\sigma}{Lt_R} \tag{9.22}$$

Here, L is given in cm and t_R in seconds. Note that in Figure 9.2, that the triangulation techniques for determining the base width in time units (t) result in 96% if the area or ±2 standard deviations, or

$$W = 4t = 4\frac{\sigma}{Lt_R} \text{ or } \sigma = \frac{WL}{4_R} \tag{9.23}$$

Substitution of Equation 9.22 into Equation 9.23 yields

$$H = \frac{LW^2}{16t^2{}_R} \tag{9.24}$$

H is always given in units of distance and is a measure of the efficiency of the column and the dispersion of a solute in the column. Thus, the lower the H value, the better the column in terms of separations (one wants the analyte peak to be as compact as possible with respect to time or distance in the column). Column efficiency is often stated as the number of theoretical plates in a column of known length, or

$$N = \frac{L}{H} = 16\left(\frac{t_R}{W}\right)^2 \tag{9.25}$$

This concept of H, theoretical plates, comes from the petroleum distillation industry as explained in Video S.9.2.

To summarize Video S.9.2 with respect to gas and LC, a theoretical plate is distance in a column needed to achieve baseline separation; the number of theoretical plates is a way of quantifying how well a column will perform.

We now have the basic set of equations for describing analyte movement in chromatography, but it still needs to be expanded to more practical applications where two or more analytes are separated. Such an example is illustrated in Video S.9.3 for a traditional packed column.

Separation of two chemically similar analytes is characterized mathematically by resolution (R_s), the difference in retention times of these analytes. This equation, shown earlier in Figure 9.2, is

$$R_s = \frac{2(t_{R'B} - t_{R'A})}{W_A + W_B} \tag{9.26}$$

where $t_{R'B}$ is the corrected retention time of peak B, $t_{R'A}$ is the corrected retention time of peak A, W_A is the peak width of peak A in time units, and W_B is the width of peak B. Since $W_A = W_B = W$, Equation 9.26 reduces to

$$R_s = \frac{t_{R,B} - t_{R,A}}{W} \tag{9.27}$$

Equation 9.27 expressed W in terms of N and t_R,

$$R_s = \left(\frac{t_{R,B} - t_{R,A}}{t_{R,B}}\right)\frac{\sqrt{N}}{4} \tag{9.28}$$

which, upon rearrangement, yields

$$R_s = \left(\frac{k'_B - k'_A}{1 + k'_B}\right)\frac{\sqrt{N}}{4} \tag{9.29}$$

Recall that we are trying to develop an equation that relates resolution to respective peak separations and although k' values do this, it is more useful to express the equation in terms of α, where $\alpha = k'_B/k'_A$. Substitution of α into Equation 9.29, with rearrangement, yields

$$R_s = \frac{\alpha - 1}{\alpha}\frac{k'_B}{1 + k'_B}\frac{\sqrt{N}}{4} \tag{9.30}$$

or the analyst can determine the number of plates required for a given separation by

$$N = 16R_{s^2}\left(\frac{\alpha}{\alpha - 1}\right)^2\left(\frac{1 + k'_B}{k'_B}\right)^2 \tag{9.31}$$

Thus, the number of plates present in a column can be determined by direct inspection of time units in a chromatogram, where R_s is determined from Equation 9.27, k'_A and k'_B are determined using Equation 9.17, and α is determined using Equation 9.19.

$$R_s = \frac{t_{R,B} - t_{R,A}}{W} \tag{9.32}$$

$$k' = \frac{t_r - t_m}{t_m} = \frac{V_s - V_m}{V_m} \tag{9.33}$$

$$\alpha = \frac{K_B}{K_A} = \frac{k'_B}{k'_A} \tag{9.34}$$

Another use of Equation 9.31 is that it can be used to explain improved separation with temperature programming of the column in GC and gradient programming in HPLC. Recall that poorer separation will result as peaks broaden as they stay for extended times in the column and several factors contribute to this process. N can be changed by changing the length of the column to increase resolution but this will further increase band broadening. H can be decreased by altering the mobile phase flow rate, the particle size of the packing, the mobile phase viscosity (and thus the diffusion coefficients), and the thickness of the stationary phase film.

9.3 Case Study

To better understand the application of the equations derived above, a useful exercise is to calculate all of the column quantification parameters for a specific analysis. First the output from an HPLC for the analysis of triazines (Figure 9.3 and Table 9.1). The next chromatogram and data table (Figure 9.4 and Table 9.2) were obtained from a capillary column GC with a flame ionization detector. Separations of hydrocarbons commonly found in auto petroleum were made on a 30-m long, 0.52-mm diameter DB-1 capillary column.

LC data:

Column length: 10.0 cm

$t_m = 1.31$.

Calculate: t'_A, t'_B (where A is 2.328 minutes peak and B is 2.922 peak), k'_A, k'_B, α, R_s (short and long equation), H, and N (short and long equation).

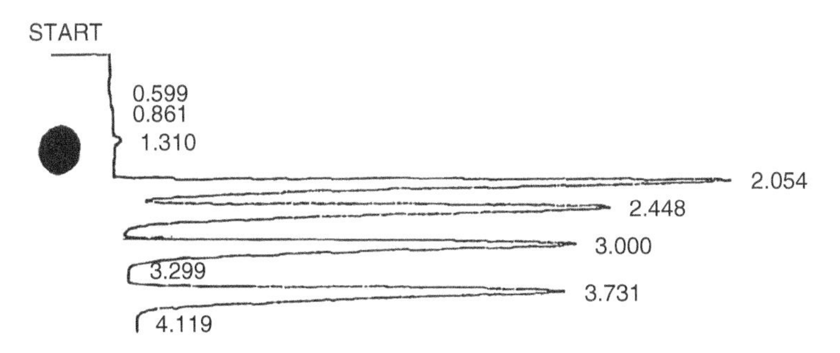

Figure 9.3 LC output from an HPLC-UV-vis C-8 column for triazines (Source: Dunnivant and Ginsbach).

Table 9.1 Integrator output from an HPLC chromatogram.

Retention Time (min)	Peak Width (min)	Area (Not Used Here)
2.054	0.191	1753345
2.448	0.206	1521755
3.000	0.206	1505381
3.731	0.198	1476639

LC integrator output (chromatogram above).

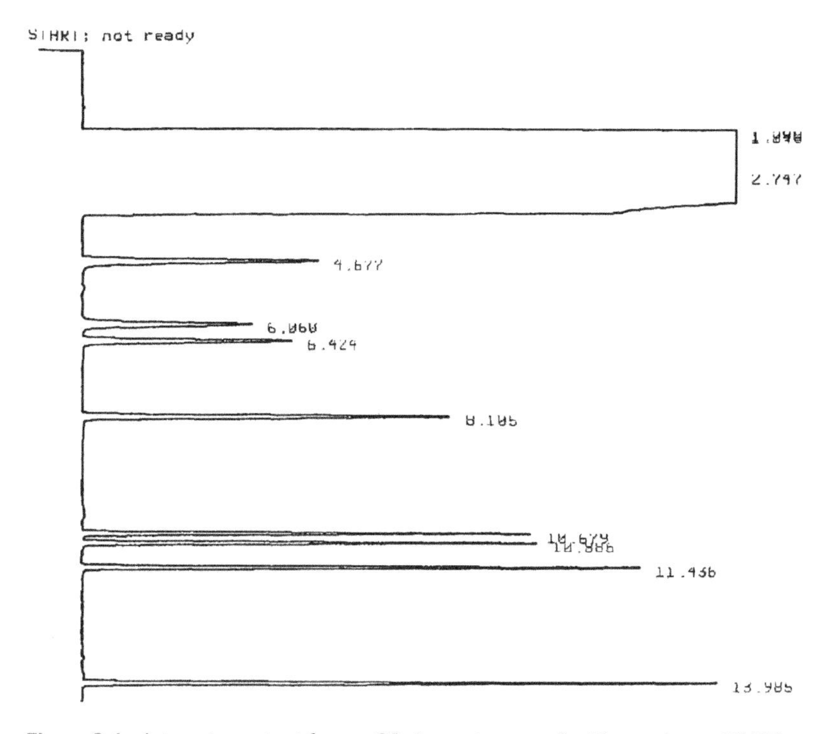

Figure 9.4 Integrator output from a GC chromatogram. Capillary column GC-FID output for gasoline hydrocarbons on a 30-M DB-1 column (Source: Dunnivant and Ginsbach).

Table 9.2 Integrator summary from a GC chromatogram.

Retention Time (min)	Peak Width (min)	Area (Not Used Here)
4.677	0.102	141 811
6.060	0.109	109 128
6.424	0.086	106 660
8.105	0.069	150 569
10.679	0.058	153 564
10.886	0.058	154 676

GC data:
Column length: 30.0 m.
$t_m = 1.846$.
GC integrator output (no chromatogram).

Table 9.3 Calculation results for an LC and GC chromatogram.

Parameter	LC	GC
k'_a	1.29	4.79
k'_b	1.85	4.90
α	1.43	1.02
R_s	3.62	Short eqn: 3.57 (Long eqn: 3.50)
H	18.3 μm	53 μm
N	Short eqn: 5682 (Long eqn: 5458)	Short eqn: 542 000 (Long eqn: 583 000)

Calculate: t'_A, t'_B (where A is 10.679 minutes peak and B is 10.886 peak), k'_A, k'_B, α, R_s (short and long equation), H, and N (short and long equation).

Results (Table 9.3):

Note that the plate height diameter is similar for HPCL and GC. But the number of plates in each column is much greater for the GC due to the length of the column. The length of HPLC column is limited by the higher viscosity of liquid and the unrealistic pressure needed to pass the liquid through a packed column.

9.4 Optimization of Chromatographic Conditions

Now we will review and summarize this lengthy derivation and these complicated concepts. Optimization of the conditions of the chromatography system (mobile phase flow rate, stationary phase selection, and column temperature or solvent gradient) is performed to achieve baseline resolution for the most difficult separation in the entire analysis (two adjacent peaks). This process results in symmetrically shaped peaks that the computer can integrate to obtain a peak (analyte) area or peak height. A series of known reference standards are used to generate a linear calibration line (correlating peak area or height to analyte concentration) for each compound. This line, in turn, is used to estimate the concentration of analyte in unknown samples based on peak area or height.

An instrument's resolution can be altered by changing the theoretical plate height and the number of theoretical plates in a column. The plate height, as explained in the animation shown later, is the distance a compound must travel in a column needed to separate two similar analytes. The number of theoretical plates in a column is a normalized measure of how well a column will separate similar analytes.

Now it is necessary to extend the concept of theoretical plate height (H) a bit further to understand its use in chromatography. Since gas and LC are dynamic systems (mobile flow through the column), it is necessary to relate a fixed length of the column (the theoretical plate height) to flow rate in the column. Flow rate is measured in terms of linear velocity, or how many centimeters a mobile analyte or carrier gas will travel in a given time (cm/s). The optimization of the relationship between H and linear velocity (μ), referred to as a van Deemter plot, is illustrated in Figure 9.5 for GC.

It is desirable to have the smallest plate height possible, so the maximum number of plates can be "contained" in a column of a given length. Three factors contribute to the effective plate height, H, in the separation column. The first is the longitudinal diffusion, B (represented by the blue line in Figure 9.5) of the analytes that are directly related to the time an analyte spends in the column.

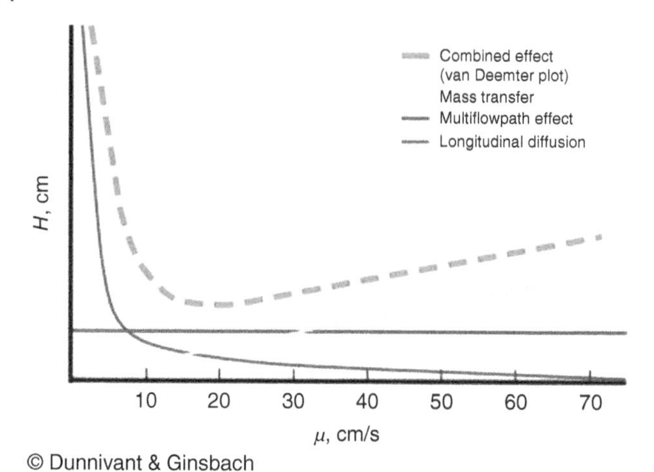

Figure 9.5 A theoretical van Deemter plot for a capillary column showing the relationship between theoretical plate height and linear velocity (Source: Dunnivant and Ginsbach).

When the linear velocity (μ) is high, the analyte will only spend a short time in the column and the resulting plate height will be small. As linear velocity slows, more longitudinal diffusion will cause more peak broadening resulting in less resolution. The second factor is the multiflow path effect represented by the red line in Figure 9.5. This was a factor in packed columns but was effectively eliminated when open tubular columns (capillary columns) became the industry standard. Third are the limitations of mass transfer between and within the gas and stationary phases, C_u (the yellow line in Figure 9.5) defined by

$$\text{Mass transfer to and from the stationary phase}: C_s u = \frac{f(k)d_f^2}{D_s}\mu \tag{9.35}$$

$$\text{Mass transfer in the mobile phase}: C_M u = \frac{f'(k)d_p^2}{D_M}\mu \tag{9.36}$$

where μ is the mobile phase linear velocity, D_s and D_m are diffusion coefficients in the stationary and mobile phases, respectively, d_f and d_p are the diameter of the packing particles and the thickness of liquid coating on the stationary phase particles, respectively, k is the unitless retention or capacity factor, and $f(k)$ and $f'(k)$ are mathematical functions of k.

If the linear velocity of the mobile phase is too high, the entire "packet" of a given analyte will not have time to completely transfer between the mobile and stationary phase or have time to completely move throughout a given phase (phases are coated on the column walls and therefore have a finite thickness). This lack of complete equilibrium of the analyte molecules will result in peak broadening for each peak or skewing of the Gaussian shape. This, in turn, will increase H and decrease resolution.

The green line in Figure 9.5 represents the van Deemter curve, the combined result of the three individual phenomena. Since the optimum operating conditions have the smallest plate height; the flow rate of the GC should be set to the minimum of the van Deemter curve. For GC, this occurs around a linear velocity of 15–20 cm/s. However, in older systems, as the oven and column were temperature programmed, the velocity of the gas changed, which, in turn, changed the mobile phase flow rate and the linear velocity. This has been overcome in modern systems with mass flow regulators, instead of pressure regulators, that hold the linear velocity constant.

These concepts are reviewed in Video S.9.4.

Now that the theoretical basis for understanding chromatographic separation has been established, it is necessary to extend these ideas one step further. Remember, the power of chromatography is the separation of complex mixtures of chemicals; not just for two chemicals as illustrated previously. In most cases, separating mixtures of many compounds is required. This requires that the resolution, R_S, be constantly optimized by maintaining H at its minimum value in the van Deemter curve.

This optimization is accomplished by systematically altering the column temperature in GC or the solvent composition in HPLC. Analytes in the separation column spend their time either "dissolved" in the stationary phase or vaporized in the mobile phase. When analytes are in the stationary phase, they are not moving through the system and are present in a narrow band in the length of the column or resin coating. As the oven temperature is increased, each unique analyte has a point where it enters the mobile phase and starts to move down the column. In GC, analytes with low boiling points will move down the column at lower temperatures, exit the system, and be quantified. As the temperature slowly increases, more and more analytes (with higher boiling points) likewise exit the system. In reverse-phase HPLC, analytes with more polarity will travel fastest and less polar analytes will begin to move as the polarity of the mobile phase is decreased. Thus, the true power of GC separation is achieved by increasing the oven/column temperature (referred to as ramping) while in LC, the separation power is in gradient programming (composition of the mobile phase). This is "the general elution problem" that is solved by optimizing the mobile phase, linear velocity, and the type of stationary phase. As noted, temperature programming is used to achieve separation of large numbers of analytes in GC. An example of the effects of temperature programming on resolution is illustrated in Figure 9.6. In Figure 9.6a, a low isothermal temperature is used to separate a mixture of six analytes with limited success as some peaks contain more than one analyte. A higher isothermal temperature, shown in Figure 9.6b, is more successful for analytes with higher boiling points but causes a loss of resolution for peaks that were resolved at the lower isothermal temperature. The temperature program used to produce Figure 9.6c achieves adequate separation and good peak shape for a complex solution.

9.5 Calibration of an Instrument/Detector

We now have a basis for understanding separation science with respect to chromatography. All chromatography systems rely on these principles. But how does the analyst relate instrument output to analyte concentration in a sample? Instruments yield signals (also referred to as responses) that are specific to the type of detector being used. Most GC detectors result in electrical currents while most LC detectors yield absorbance values. MS units can be attached to both GC and LC systems and yield counts of ions per time. But before actual samples are analyzed, each instrument detector must be calibrated. Two common forms of calibration are internal and external calibration.

Detector response yields two useful means of quantification in chromatography: peak area and peak height. In the "old days," these measurements were made manually; a strip chart recording was obtained by passing a strip of paper consisting of uniform weight past a pen that moved relative to the detector signal. The shape of the peak was drawn on the paper and the peak height was measured with a ruler or the peak area was measured either by triangulation or by actually weighing a cutout of the paper containing the peak! Fortunately for us, these archaic methods are no longer required. The major disadvantage of these techniques is that the range of detector responses was limited by the height of the paper. Today, peak area and height measurements are calculated by electronic integrators or computers, and most systems are automated such that peak

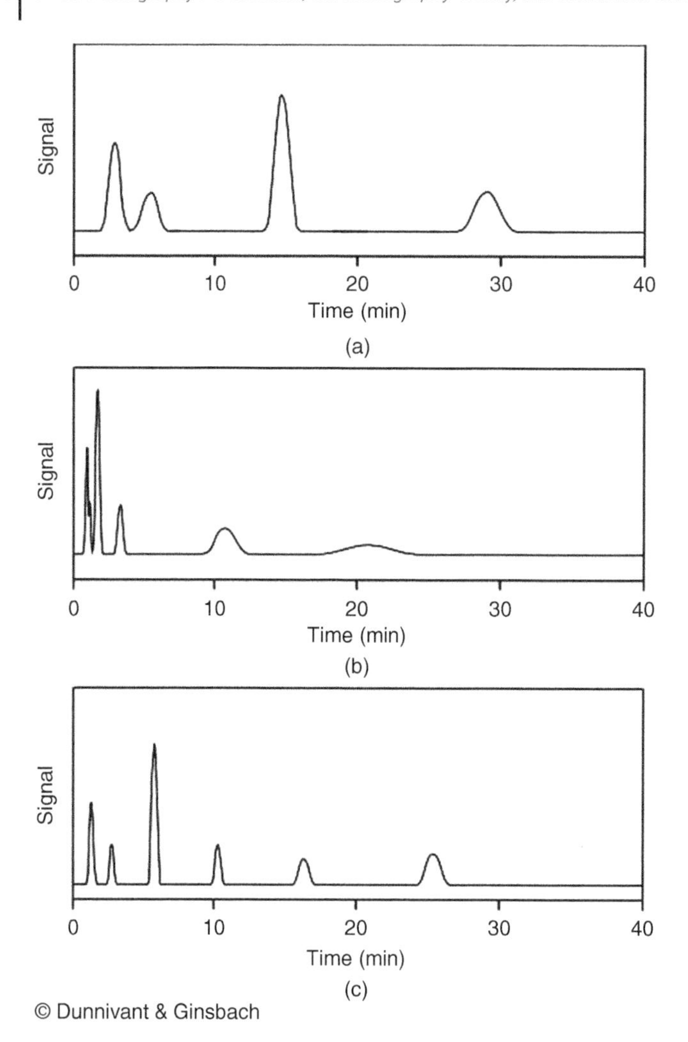

Figure 9.6 Temperature programming: the solution to the general elution problem for GC applications. (a) Isothermal temperature at 40 °C, (b) isothermal temperature at 200 °C, and (c) temperature program from 40 to 200 °C (Source: Dunnivant and Ginsbach).

area/height are directly correlated between standards and samples. Most systems use peak area to generate calibration lines, which are usually linear relationships between the detector response and the concentration or mass of analyte injected into the instrument. Such a plot is shown for an external calibration method in Figure 9.7.

A summary of integration concepts is illustrated in Video S.9.5.

After an instrument has been calibrated, a sample extract is analyzed under the same conditions as the standards. The calculated area for the sample is then analyzed by a linear regression of the standard line and a mass or concentration of the analyte in the sample is calculated. Usually, a dilution factor adjustment is made and the concentration of analyte in the original sample is then calculated.

A special type of additional calibration is used in capillary column GC because of analyte losses during sample injection and due to the possibility of inconsistent injections when manual injections are performed. This method is referred to as an "internal standard" where every sample and

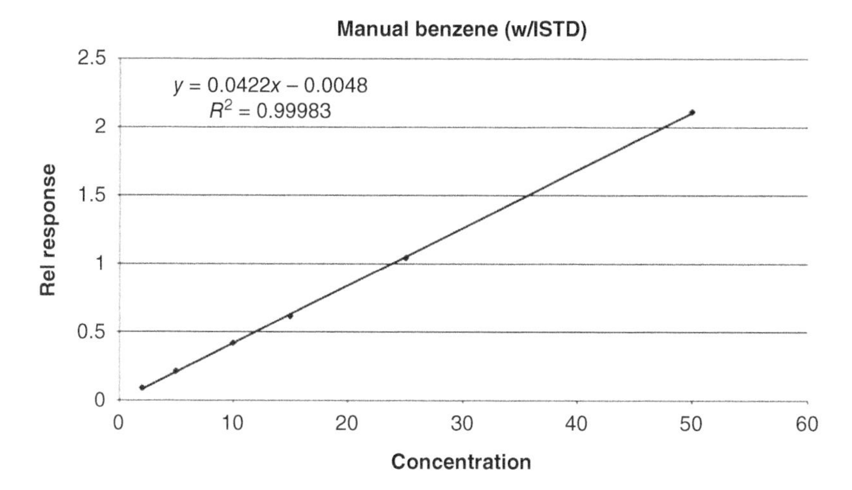

Figure 9.7 External calibration of benzene on a capillary column GC (Source: Dunnivant and Ginsbach).

standard injected into the instrument contain an identical concentration of a compound with a chemical structure similar to the analyte but one that has a unique retention time in the column. The instrument is set to measure a constant concentration (and therefore measured area) of the internal standard and adjusts all injections to that constant value. For example, if a sample is found to only contain 90% of the internal standard, then it is assumed that 10% of the injection was lost and all analyte concentrations are increased by 10%. Similarly, adjustments can be made by over-injecting a sample.

9.6 Evolution of Peak Integration

Calibration curves or lines in chromatography are based on correlating peak area to analyte concentration. Our ability to more accurately quantify peak area has evolved over several decades. Starting prior to the 1960s and 1970s with the tracing of chromatographic peaks on strip charts, where the chart paper was then cut out and weighted for correlation to mass injected, to strip chart integrators, to crude electronic integrators, to stand-alone combination strip chart–integrator units, and finally to computer-based control of the instrument. We have finally reached the plateau of our quantitation evolution with computers.

The control of our instruments by computers is divided into two programming components: the *sequence* and the *method*. First, a *method* on the computer is created that controls most instrument functions such as mobile flow rates, temperature zones and settings, temperature or elution gradient programming, and most of the electronic components of the instrument. The method is directly linked to the *sequence* that controls the data collection, including the auto-sampler and which sample vial is in which slot, what slots in the auto-sampler contain blanks, samples, and standards, and where the data are stored on the computer for each instrument run. The combined instrument–computer systems are slightly complicated to learn but greatly save time, money, and effort in the long run. Most computer-controlled systems can also take a three-to-five-day instrument run of 100 samples, create a calibration line, and report all sample results back to the original sample concentration.

The next chapters of this book will focus on the components of GC, LC, CE, and MS with an additional chapter on interpretation of MS fragmentation patterns. Both GC and LC rely on the chromatography theory discussed in this chapter, and all instruments rely on some form of calibration if quantitative results are required.

Supporting Information

Additional supporting information may be found online in the Supporting Information section in the HTML rendition of this article.

References

Karger, B.L., Synder, L.R., and Horvath, C. (1973). *An Introduction to Separation Science*, 1e, New York, USA: John Wiley and Sons.

Skoog, D.A., Holler, F.J., and Crouch, S.R. (2007). *Principles of Instrumental Analysis*, 6e. USA: Thomson Publishing.

Willard, H.W., Merritt, Jr. L.L., Dean, J.A., and Seattle, Jr. F.A. (1981). *Instrumental Methods of Analysis*, 6e. Belmont, CA, USA: Wadsworth Publishing Company.

10

Gas Chromatography

10.1 Introduction and History

This chapter will focus on the components and operation of basic gas chromatography (GC). The general field of chromatography dates back to 1903 with the work of Russian scientist Mikhail Tswett, who separated plant pigments using liquid chromatography. Fritz Prior, as part of his graduate work, developed solid-state GC in 1947. Modern GC is generally considered to have been invented by Martin and James in 1952. A review of the history of GC can be found in Bartle and Myers (2002). Since 1952, GC has advanced from using solid spheres to act as the stationary phase (gas–solid chromatography) to using liquid-coated resins as the stationary phase, and finally to using covalently bonded stationary phases attached to wall of a capillary column (gas–liquid chromatography). Components of the actual chromatograph have also advanced from many manual parts such as rotary gas flow regulators being updated to electronic flow or mass flow controllers, resin-packed columns have been replaced with fused silica capillary columns, and manual injection and control of the instrument have been replaced with automated injection and computer control. Most notable is the diversity of detectors utilized today with GC, especially the ability to connect capillary column GCs with mass spectrometers for confirmatory analysis. Additionally, in the past, analyses of a set of samples could take days to complete and required the constant attention of an analyst, but today with the help of computers, a set of samples and standards can be started on the instrument and the scientist can return later with all of the samples analyzed and the data processed. Typical automatic sampler units can hold up to 100 samples. These improvements have greatly increased the capabilities of laboratories and advanced scientific endeavors but, in many cases, have decreased the analyst's knowledge of the chromatographic system. But such is the price of advancement and economics. In this chapter, we will discuss the types of samples and analytes that can be analyzed by GC, the components of the GC and their operation, the variety of detectors available today for use with GC, and examples of specific analyses.

10.2 Types of Samples and Sample Introduction

A basic GC of reasonable quality costs from \$40 000 to 60 000 US today (2022) depending on the detectors that are purchased with the GC, although more inexpensive models can be purchased

☆ Several instrument and column manufacturers were consulted while researching this chapter. Manufacturers have excellent websites for researching their products, talented sales staff, and very helpful technical assistance.

Essential Methods of Instrumental Analysis, First Edition. Frank M. Dunnivant and Jake W. Ginsbach.

Companion Website: www.wiley.com/go/essentialmethodsofinstrumentalanalysis1e

for limited routine analysis. A capillary column gas chromatography–mass spectrometer (GC–MS, quadrupole) will cost slightly less than $120 000. With this relatively high price tag, students sometimes trust the results as unquestionably accurate. Reality could not be farther from this belief. Every step, including extraction of the analytes from the sample matrix, conducting serial dilutions, injection into the GC, and identification of the fragmentation pattern in MS is prone to errors. From experience, sample extraction can be the most difficult and is the source of considerable error. Samples come in a variety of forms: gaseous, liquid, solid, and biological. In order for the reader to fully appreciate sample preparation for GC analysis, the following discussion will present several sample collection, extraction, and sample preparation techniques.

10.2.1 Gaseous Samples

Gaseous samples are the easiest samples to analyze. For on-site analysis, a gaseous sample can simply be drawn into a syringe and the sample injected into a sampling valve/loop. Sampling loops are necessary in GC analysis in order to inject a consistent volume of a compressible sample. When a gas sample is taken at atmospheric pressure and injected into a GC inlet, the pressure in the GC will compress the gas in the syringe and not allow all of the sample volume to be injected. A sampling valve and loop consist of a four- or six-port valve that allows the sample to be injected into a fixed-volume (loop of tubing that is at atmospheric pressure). A valve is then turned that transfers all of the gas contained in the sample loop into the GC injection port. For field gaseous samples that need to be transported to the laboratory for analysis, a variety of sampling containers are available including Teflon bags and sealable metal cylinders (referred to as bombs) that can be filled with the sample gas. It should be noted that when these containers are analyzed, they should be adjusted back to their field temperature in order to avoid condensation of some gaseous analytes to liquids; this is especially true when industry smoke stack or process gases are being sampled and analyzed. Another possibility for sampling gaseous analytes is a resin tube. To collect a sample, a known volume of gas is passed through a glass or metal tube containing a resin that has a strong affinity for the analytes. The analytes adsorb to the resin and after a sufficient volume of gas has passed through the system, each end of the resin tube is capped and transported back to the lab. In the lab, the resin is extracted with a solvent specific to the analysis and the solvent/analyte solution is injected into the GC. A relatively simple calculation yields the concentration of each analyte in the original gas volume. The obvious advantage of this method is concentration of the gaseous analytes and the improvement of detection limits, as opposed to analyzing the gas by direct injection. The resin tube method is commonly used in the monitoring of solvents in the workplace where an industry worker will wear a portable personal pump that takes in atmospheric gases at the same rate as a human would breathe under working conditions. At the end of the day, the tube is extracted and analyzed to determine if the worker was exposed to chemicals in excess of workplace limits according to Occupational Safety and Health Association (OSHA) standards.

10.2.2 Liquid Samples

Liquid samples are the next easiest to analyze by GC since they are already in an injectable matrix. Samples from organic synthesis procedures usually have products (analytes) present at high concentrations and are analyzed by direct injection. Unfortunately, relatively few products fit the requirements of GC—that analytes be *volatile and thermally stable*—so most products are analyzed by HPLC, the subject of Chapter 11. Analytes in aqueous samples are also easy to analyze by GC. Some GC detectors and columns allow the direct injection of aqueous samples if the concentration of the analyte is sufficiently high. The aqueous sample is frequently extracted into an organic solvent using a standard separatory funnel when there is a low concentration of analyte present in a sample or when water could harm the GC column or detector. Usually, the aqueous sample is extracted three times with a relatively small volume of organic solvent, the organic extracts are combined, the volume is reduced by evaporation, and the resulting organic extract is injected into the GC. One disadvantage of organic extraction is the need to purchase expensive and very pure organic solvents (typically priced at approximately $150 for 4 L) and the expensive disposal costs of the resulting organic waste solvents. A more automated version of the separatory funnel is a liquid–liquid extractor, but the glassware is relatively expensive and typical extraction times run from 8 to 24 hours. In this extraction setup, the organic solvent is evaporated, condensed, and passed through a water vessel where the nonvolatile hydrophobic analytes partition into the organic solvent that is constantly recycled into the boiling vessel. The recycled solvent is re-evaporated again, leaving the analytes in the boiling flask, and passes again into the condenser and water column. Figure 10.1 shows a typical liquid–liquid extraction system.

A relatively easy way to avoid the need for expensive glassware is to use resin packs (solid phase extraction—SPE; Figure 10.2) that are available from a variety of vendors. In this technique, the water sample is passed through a resin pack (a solvent-resistant tube usually one to a few centimeters in diameter and slightly taller in height). Again, the resin has a high affinity for the analytes. After the passage of the aqueous sample through the packet, the resin is dried by passing ultrapure gas through it and the adsorbed analytes are removed by passing a small volume of organic solvent

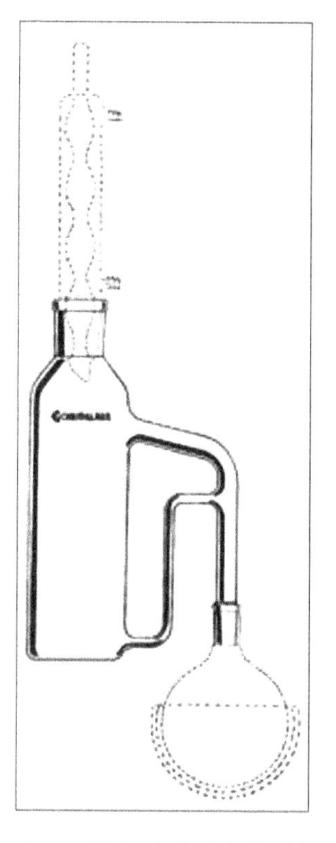

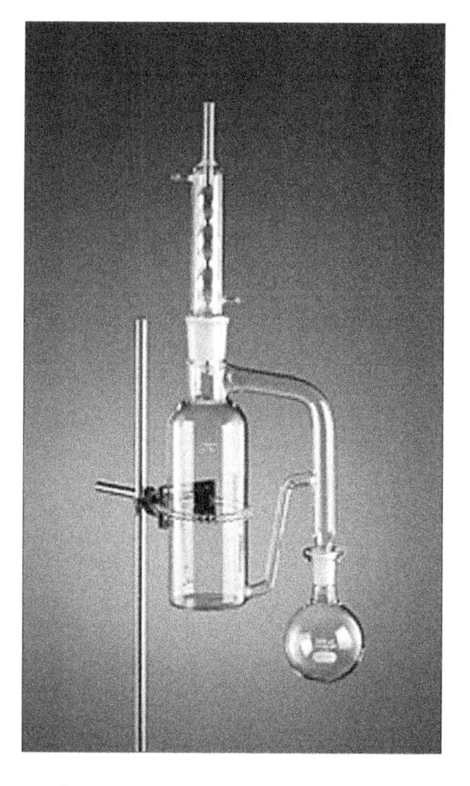

Figure 10.1 A liquid–liquid extractor for extracting analytes from water samples (Source: with permission of ChemGlass Life Sciences).

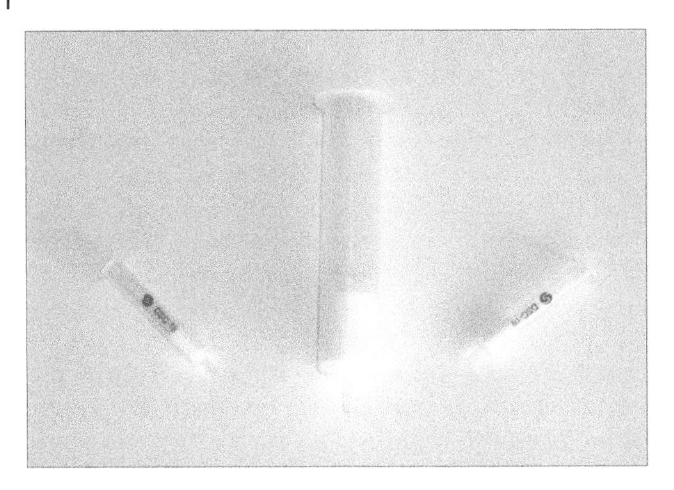

Figure 10.2 Three resin packets for extracting analytes from aqueous samples (Source: Dunnivant and Ginsbach (Book Authors)).

(usually a few mL) through the packet. The solvent volume is adjusted to a known volume and injected into the GC.

An even more novel way of extracting analytes from water samples is to use solid-phase microextractors (SPMEs) that consist of a syringe containing a fused silica capillary fiber coated with a chromatography stationary phase with a high affinity for the analytes of interest. The fiber is housed in a metal needle where it can be extended for collecting analytes or for desorption in a GC injection inlet. The SPME needle and fiber are passed through a septum in the sample bottle, either into the gaseous space above the water or directly into the water, the fiber is exposed through the end of the needle and allowed to equilibrate (adsorb the analytes) for typically 10–30 minutes while the sample is mixed with a stir bar. After this time, most or all of the analytes are transferred to the SPME fiber. The fiber is drawn into the metal needle; the needle is withdrawn from the sample bottle and placed directly into the GC injector. Following are the advantages of this technique: (1) no need for expensive organic extraction solvents, (2) relatively rapid analysis, (3) possibly improved extraction recovery, and (4) significant concentration of the analytes and improvement of detection limits (up to 10 000–1 000 000-fold concentrations). Fibers can be reused from 50 to 100 times. The minor disadvantage of the SPME technique is the cost of the apparatus (approximately $600 for three fibers and a holder/injector). The SPME syringe and fiber are shown in Figure 10.3.

Volatile analytes present an additional problem since considerable quantities of the analyte can be lost during sample preparation. Analyses of volatile analytes are best performed with some type of commercial head-space analyzer or "purge and trap" device where the actual water sample (with no gaseous headspace) is attached to a sample processing unit, a gas is used to transfer the volatile analytes to a resin trap or directly into the GC injector, and after the required purge time, the transferred analytes are analyzed by GC.

10.2.3 Soil/Sediment Samples

Soil and sediment samples present considerable difficulties in sample preparation since the analytes must be extracted and transferred to a liquid phase before introduction into the GC. Early techniques focused on simply washing the air-dried solid matrix with organic solvent, but these methods proved to yield low extraction efficiencies (analyte recoveries were considerably less than

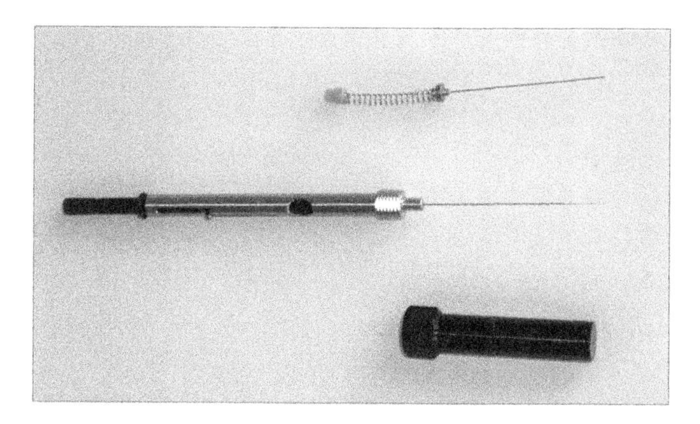

Figure 10.3 An SPME device with the microfiber exposed (middle item). Extra needles are available (top item) since the injector syringe can be reused indefinitely. Needles can be reused from 50 to 100 times depending on the composition of the sample. The bottom item is the needle protection guard and GC injection guide (Source: Dunnivant and Ginsbach (Book Authors)).

100%). The gold standard for the extraction of analytes from soil and sediment matrices is the Soxhlet extraction technique (Figure 10.4). The Soxhlet is a glass distillation setup that repeatedly passes pure solvent through the soil/sediment matrix over a period of 24–48 hours. After this time, the solvent is collected and the volume is reduced and analyzed by GC. Laboratory studies have recovered approximately 100% of analytes with this method, but Soxhlet glassware is expensive (each setup costs at least $300), uses expensive organic solvents (approximately $150 for 4 L), and is very labor- and time-intensive. Alternatives to the Soxhlet technique include relatively rapid sonication procedures and automated heated solvent extraction systems.

10.2.4 Biological Samples

Biological tissue samples are undoubtedly the most difficult to extract and analyze. During the extraction process, the analytes need to be effectively transferred from the outside and inside of cellular matter to the solvent phase. The approaches used are as diverse as the high number of sample tissue types. Common approaches include (1) drying the tissue, followed by grinding, and Soxhlet extraction and (2) a combination of grinding and sonication, followed by liquid extraction. Whichever method is used, extensive sample cleanup (the removal of interfering substances and analytes) is necessary since the analyst should not inject nonvolatile biological material into a GC.

An additional point should be made here. GC is only used for analytes with boiling points below approximately 300 °C, and this limits the utility of GC analysis for both the organic and analytical chemist (HPLC was developed for most other nonvolatile compounds). However, some analytes can be reacted with derivatizing agents to remove functional groups that tend to make them nonvolatile. A common derivatizing agent (also referred to as a silylating agent) is N,O-bis (trimethylsily) acetamide which converts groups such as –OH, –COOH, $-NH_2$, =NH, and –SH to a $-O-Si(CH_3)_3$ group that renders the compound volatile. It should be noted that derivatizing agents are very hazardous and usually carcinogenic.

10.2.5 Analyte Recovery

Now that we have presented some of the common extraction techniques, another problem must be pointed out. How is it possible to know all of the analytes were extracted from the sample (i.e.

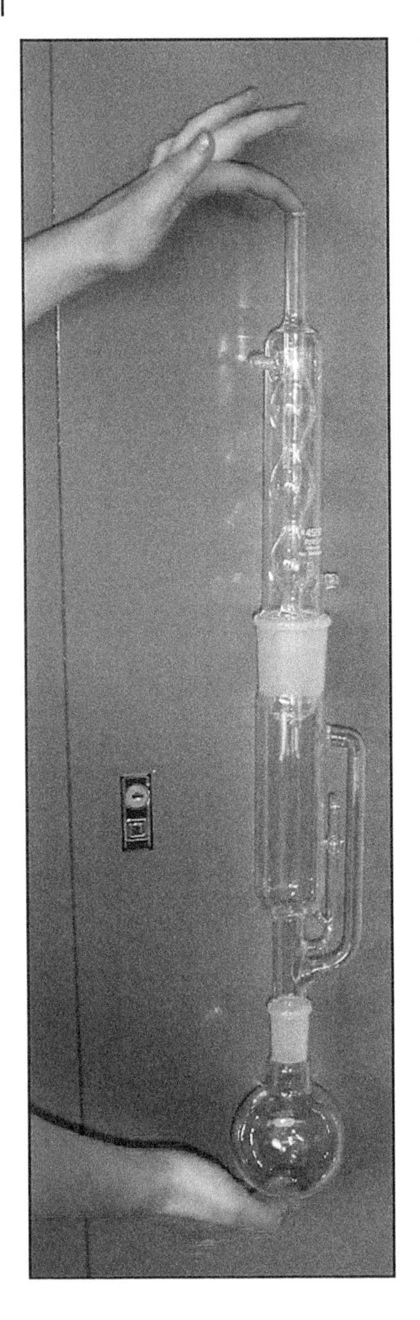

Figure 10.4 Soxhlet extraction glassware (Source: Dunnivant and Ginsbach (Book Authors)).

water, urine, soil, and fish)? This question becomes more difficult to answer as the sample matrix becomes more complex. For example, how does the chemist quantitatively recover all of the analyte from lake sediment or from food items? These sample matrices can have analytes contained within every clay particle or biological cell and require the development and testing of rigorous extraction procedures. Fortunately, many of these procedures have been developed and are published by governmental agencies, industry, or research scientists. As a result, incorporation of these procedures into the laboratory is relatively easy. As an aid to determining how well your extraction procedure works, relatively expensive "reference" samples that contain a known amount of analytes can be obtained for a variety of sample matrices (i.e. fish, sediment, and manufactured goods). A procedure can be validated if the results from your method are statistically equivalent to the known concentration. *For many procedures, it is not necessary to have a high recovery (i.e. 98%) but it is necessary to have a known and consistent recovery, even if it is low.*

In addition to the potential human errors present in an analysis, instrument detectors can also contain errors due to nonoptimum instrumental setting, out-of-date tuning or calibration, and when peaks elute from the column with more than one analyte or in mass spectrometry when more than one reference spectra are identified in the computer search library. This latter situation is common with low concentrations of analytes.

Now that the basic problems and common errors associated with sampling handling and instrumentation have been identified, we will move on to distinctions between gas and liquid chromatography. Gas and liquid chromatography were originally developed due to the existence of two basic different types of analytes: (1) those that are thermally stable (do not degrade at temperatures up to 300 °C) and are volatile at relatively low temperatures (below 300 °C) and (2) for analytes that are not volatile and/or thermally degrade at temperature above room temperature. GC is used for thermally stable and volatile chemicals while HPLC is used for both nonvolatile compounds and ones that degrade at high temperatures. Recent advances in the stationary phases on separation columns and mobile phase selection (solvent gradient in HPLC) allow many analytes that were exclusively analyzed by GC to be analyzed by HPLC. For example, GC was the exclusive technique for analyzing mixtures of volatile organic solvents. Yet today, by changing HPLC to a reverse-phase system (where the separation column is the nonpolar phase and the solvent is the more polar phase), it can now analyze components of organic solvents. HPLC will be discussed in depth in the next chapter.

GC analysis can also have special concerns. Impurities introduced during sample preparation can result in contamination that may interfere with the analysis of a desired analyte or introduce additional peaks into the chromatogram (the output of a chromatograph). A notable case is a class of compounds known as phthalates that are found in plastics that interfere with the analysis of chlorinated pesticides such as DDT and PCBs in GC analysis with an electron capture detector (ECD). Even with detection by mass spectrometry, the analysis may conclude that these compounds were present in the original samples when, in fact, they are laboratory contaminants. As a result, contact with plastics must be avoided regardless of the detector used. It is also important to purchase GC-grade solvents that are certified to contain an extremely low amount of impurities when trace analyses are being conducted.

Some functional groups of analytes, such as in the analysis of Bisphenol A, a known endocrine disruptor present in some plastic bottles may react with or irreversibly adsorb to the glass surfaces in the GC injector liner and result in the analyst reporting the absence of Bisphenol A in a sample when, in fact, it was present but lost during the analysis. This can be overcome by deactivating the surfaces with a silanization agent that coats the glass with a nonreactive trimethylsilane group, and allows the analyte(s) to pass through the system to the detector. What and when to worry about these problems, and many others, come with experience and knowledge of the literature.

10.3 Gas Chromatography

The main purpose of chromatography is to separate a complex mixture of compounds into discrete chromatographic peaks containing only one analyte. Today's capillary column chromatographic systems are ideal for this task and interface well with detection by mass spectrometry due to the

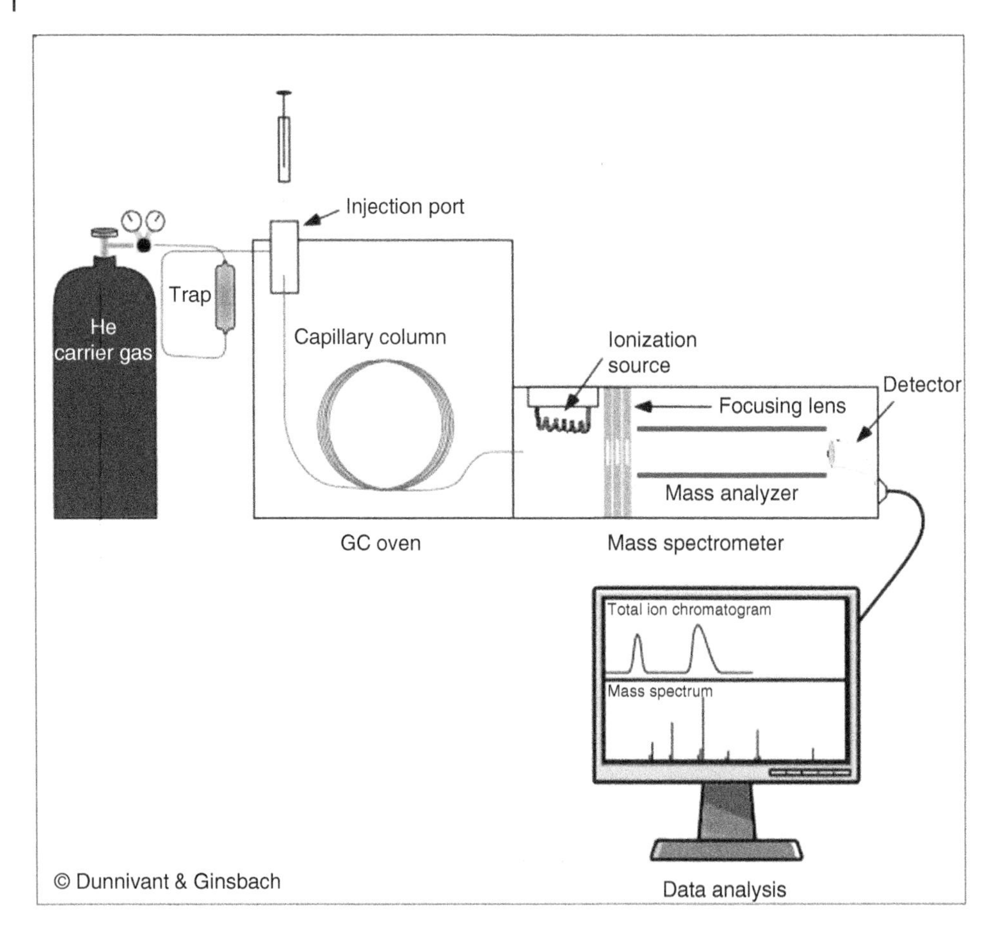

Figure 10.5 A basic GC–MS system (Source: Dunnivant and Ginsbach).

low volume of carrier gas used in capillary columns (1–5 mL/min as opposed to 60–100 mL/min in packed column GC used prior to the 1980s). Figure 10.5 illustrates the major components of a modern capillary column gas chromatography–mass spectrometry (GC–MS) system.

10.3.1 Carrier Gases

The first important component is the carrier gas or mobile phase. For a basic GC system, extremely pure helium is usually used, and hydrogen is less frequently encountered due to its explosive nature. Helium is used due to its inertness, nonreactive nature, and the shape of its van Deemter curve (Section 9.4), which allows for a relatively wide range of optimum mobile phase linear velocities. The common grade of helium used is referred to as "five-nine gas," meaning that it is 99.999% pure. But this level of purity is still not sufficient for most systems when trace (parts per million, ppm or parts per billion, ppb) analyses are being conducted. Before entering the GC, the 2500 psi (18 000 kPa), pressure in the gas cylinder is reduced to approximately 60 psi (400 kPa) with a two-stage regulator before entry into the GC. But first, the helium gas is passed through at least one resin trap to further remove hydrocarbons, oxygen, trace analytes, and/or water vapor that could interfere with analysis, degrade the column, or interfere with the detector.

10.3.2 Injectors

After passing through the purification traps, helium enters the injector where it acts as the mobile phase and helps "push" the analytes through the analytical (separation) column. A variety of injectors are used in GC, but this text is concerned with the most common, a split-splitless injector. This type of injector can be used in two modes. For solutions containing extremely concentrated levels of analytes (in the parts per thousand, ppt or percent level as is encountered in synthesis operations), the injector is operated in the split mode. In this mode, only a small fraction of the 0.2–1 μL of solution injected actually enters the separation column and the majority of the sample is vented to the atmosphere. The high concentration of analytes in the solvent allows for adequate identification and quantification. For solutions containing lower levels of analytes (ppm and ppb), the injector is operated in a dual or splitless-split mode. Upon injection of a sample, the injector is operated in a splitless mode where all of the injected volume is being pushed onto the column. But if this mode of operation is allowed to continue throughout the chromatographic run, the peaks will be nonsymmetrical (they will tail or be skewed), which will interfere with peak integration because of a continual addition of solvent molecules entering from the injection port. To avoid this problem, the split mode is switched on approximately 30–60 seconds after injection. This splitless–split mode allows the majority of the sample to "load" onto the column while clearing out the remainder of the sample from the injector to allow for a "clean," well-shaped chromatographic peak. A typical split-splitless injection is shown in Video S.10.1.

A few more points need to be made about the injector inlet. The interface between the atmosphere and the injector is separated by a septum such as the ones shown in Figure 10.6. Septa are manufactured out of various materials, all of which must be inert to leaching organic constituents into the GC or are coated on the GC side of the septum with Teflon. Septa are inexpensive and are routinely replaced, usually daily, after 30–50 injections.

Samples are injected through the septa and enter a glass liner in the injection port. The purpose of the glass inserts (liners; Figure 10.7) is to avoid exposure of the analytes to reactive hot metal

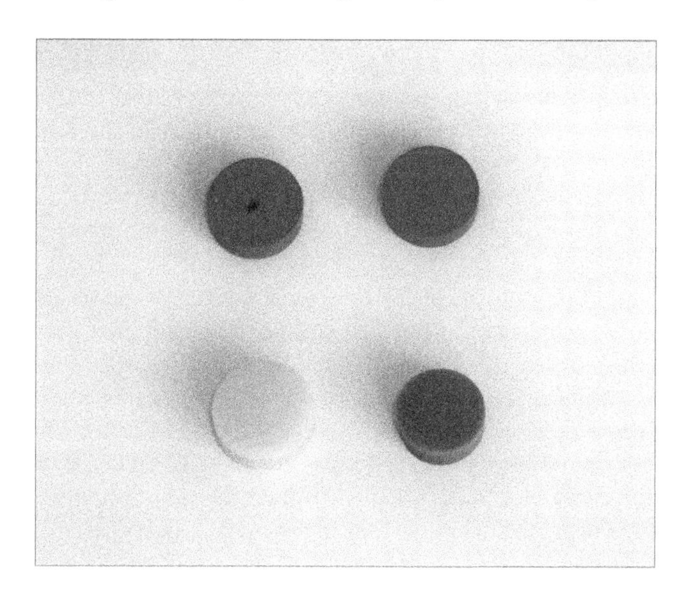

Figure 10.6 GC septa. Note the puncture holes in the top left septum. The lower left septum contains a Teflon coating (yellow). The septa shown here are about 1 cm in diameter (Source: Dunnivant and Ginsbach (Book Authors)).

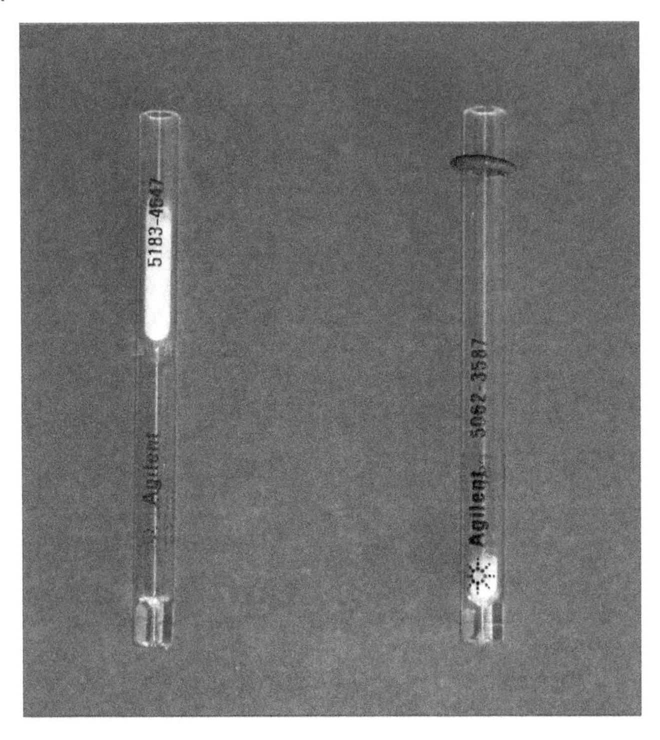

Figure 10.7 Injection glass inserts. The insert on the right shows the o-ring that seals the injection chamber and forces carrier gas through the column (Source: Dunnivant and Ginsbach (Book Authors)).

surfaces such as those contained in an unlined injector. Inserts come in a variety of forms. All liners contain a hole in the top to allow entry of the injection needle, a wider middle space for the expansion of liquid solvents into the vapor phase, and a hole in the bottom for insertion of the capillary column. Glass wool is usually present in the glass insert to keep pieces of the septum from blocking the inlet of the capillary column and to trap nonvolatile components of the sample. Figure 10.7 shows two common injector inserts, one with the glass wool in the middle and one with the glass wool at the end.

As noted earlier, samples are typically introduced into the GC with a glass syringe with a metal needle. Samples can be injected manually or with an automatic sampler. Standard 10-μL syringes are shown in Figure 10.8.

All connections in GCs, from the carrier gas cylinder to the detector, are made with Swedge Lock (aka Swagelok) fittings that seal the connections at high gas pressure. These fittings consist of a threaded nut, back ferrule, and front ferrule, all placed around a piece of tubing (refer to Figure 10.9). Fittings come in Teflon, stainless steel, and copper and in a variety of sizes ranging from smaller sizes for capillary columns as small as 0.2–6.0 mm packed columns. Ferrules are also available in graphite and in a variety of advanced materials such as Vespel, a composite of graphite and ceramic. A gas-tight fitting is achieved by tightening the nut and compressing the ferrule around the metal tube. A selection of ferrules and how they fit around a piece of tubing is shown in Figure 10.10.

In addition to the split-splitless injector, other types of injector systems are available including an on-column injector and a cryogenic focusing injector. In the past, when packed columns were used, most injections were made directly onto a section of the column that did not contain any

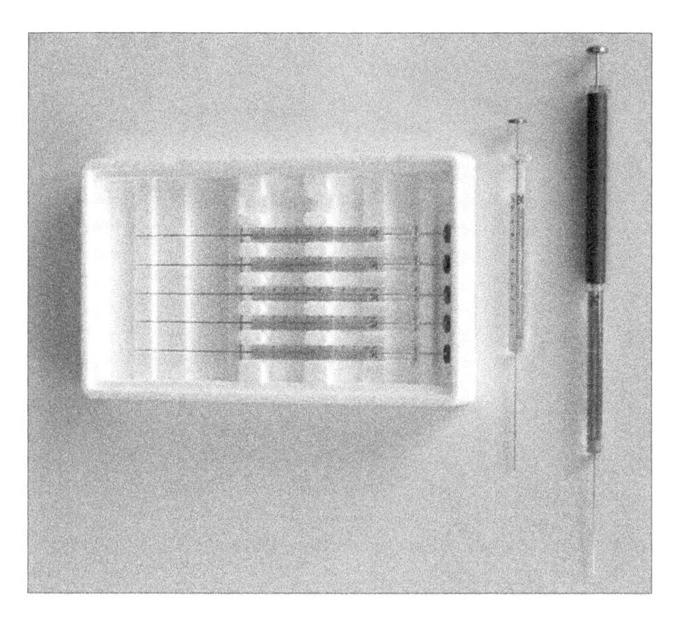

Figure 10.8 GC injection syringes (Source: Dunnivant and Ginsbach (Book Authors)).

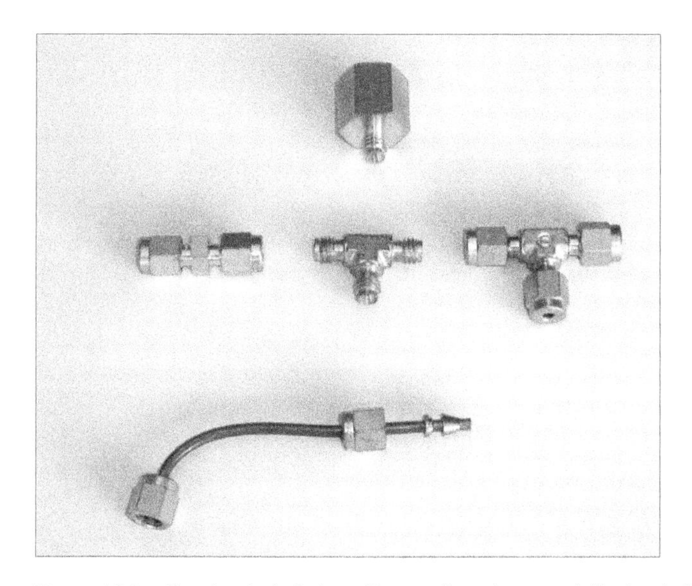

Figure 10.9 Swedge lock fittings (Source: Dunnivant and Ginsbach (Book Authors)).

stationary phase resin. This concept has been extended to capillary column technology by using a wide-bore column (typically 0.5 mm in diameter or greater). A syringe with a very narrow capillary column needle (0.2 mm in diameter) is placed through a special port at the head of the injector and liquid samples are placed directly onto the column. The needle is withdrawn, the injector port is sealed, and the chromatographic run is started. On-column injection avoids exposing the analytes to any reactive surfaces.

Highly volatile analytes, normally not separated by standard GC conditions, can be analyzed using a cryogenic focusing injector. Here, gas or liquid samples are injected through a septum, but

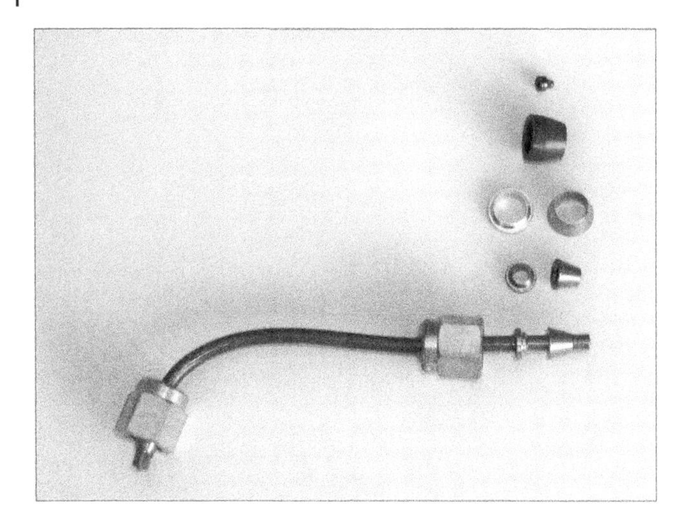

Figure 10.10 Several types of ferrules (Source: Dunnivant and Ginsbach (Book Authors)).

the bottom of the injector contains a cryogenic fluid (liquid N_2) around the head of the capillary column. The cryogenic fluid cools the column and causes the analytes to condense at the head of the column. After injection, the cryogenic fluid is removed, the column oven is slowly heated, and the volatile analytes are analyzed.

With today's extremely small injection volumes (0.2–1 μL), reproducibility of sample injection can be a problem. It is necessary to inject exactly the same volume of sample (to within three significant figures) to avoid introducing considerable error in the results. Two solutions have been devised for this problem: (1) automatic samplers/injectors and (2) internal standards. Mechanical automatic samplers can accurately and consistently reproduce small-volume injections and save considerable labor costs (and time if you happen to be a graduate student or on a slim budget). A typical automatic sampler today can hold up to 100 samples and be programmed to run the samples in any order. This is more convenient than manually injecting a sample and waiting for the GC run to end, which can range anywhere from five minutes to over an hour. The automatic sampler allows the user to simply return hours to days later to find the samples analyzed and the data stored and ready for processing on the computer-controlled station. Automatic samplers are common features today even in graduate schools since they run hour after hour, day after day, with minimal oversight or maintenance. The typical cost of an automatic sampler is only $12 000 and they rapidly pay for their self in high-sample volume work environments.

The second option for overcoming injection errors, an internal standard, is also common in capillary GC and is usually used in conjunction with automatic samplers. An internal standard is a chemical (analyte) that is not originally present in any sample but structurally similar to the analyte (typically based on boiling points). Equal amounts of the internal standard are added to every sample and reference standard. The computer program in the control station can then be programmed to correct injection errors (low or high introduction of the internal standard and analyte to the GC). One way to use this technique is to average the peak area for the internal standard that is measured in every external standard and compare the area to that observed for each sample. If the internal standard measured in a sample falls below or above this average due to a poor injection, the computer will automatically adjust the areas of every peak in the chromatogram accordingly. By using a combination of automatic injectors and internal standards, highly accurate and consistent results can be obtained.

10.3.3 Columns/Ovens

Separation columns are the heart of the GC and are housed in a temperature-controlled and temperature-programmable oven that can control and reproduce temperatures to within 0.5 °C. Considerable advances were made in GC in the 1980s, especially with regard to columns. Prior to the advent of capillary columns, chromatographic systems used packed columns. Packed columns are 1/8 to 1/4 in. metal, Teflon, or glass tubes filled with an inert resin coated with the stationary phase. Early stationary phases were highly viscous, nonvolatile liquids that interacted with the analytes to achieve separation or were molecular sieves that separated the analytes by molecular size and diffusion. In the early 1980s, packed columns were mostly replaced with capillary columns, which are open tubular columns (internal diameters in the tenths of millimeters) with the stationary phase placed on the column walls. Initially, stationary phases were simply applied to the walls as nonvolatile liquids. However, today most phases are covalently bonded to the fused silica wall, which yields more thermal stability (less loss of the stationary phase at high temperatures). Capillary columns have dramatically more theoretical plates than packed columns and greatly improve resolution. The reader should recall this relationship from Example 9.1 in Section 9.3. For example, capillary columns can have as many as 1000 times more theoretical plates as compared to a packed column. Figure 10.11 shows several GC columns. Figure 10.12 shows a typical packed column resin.

A summary of the most common stationary phases is given in Table 10.1. The selection of a particular phase depends on the intermolecular interactions expected for the analyte of interest. As discussed in Chapter 9, the selection of the phase follows the adage "like dissolves like" or in this case, "like stationary phases attach to like chemicals." The uses of each resin are also shown in Table 10.1. Additional stationary phases are available for a variety of analyte applications. Even some chiral compounds can be separated with specialty stationary phases.

Today most columns are fused silica capillary columns with typical internal diameters ranging from 0.25 to 0.53 mm. Column lengths range from 5 to 100 m. As noted in Chapter 9, the longer the column, the more theoretical plates it will contain and therefore long columns are capable of separating almost any mixture of compounds. Film thicknesses range from 0.25 to 3.00 μm with thicker films usually providing more resolution (and longer analysis times). Cross-linking of films is also common and provides more thermal stability and less "column bleed" of the stationary phase at high oven temperatures. Lower column bleed provides for a more stable detector baseline and these columns are preferred in mass spectrometer applications.

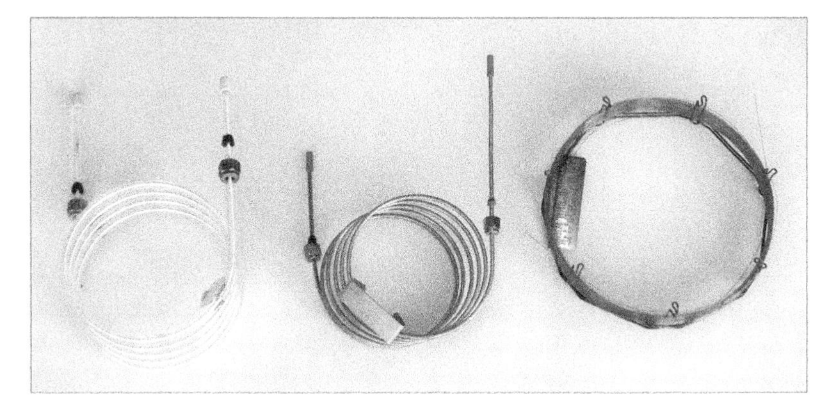

Figure 10.11 A 1/4-in. glass-packed GC column, a 1/8-in. stainless steel GC, and a fused silica capillary column (from left to right) (Source: Dunnivant and Ginsbach (Book Authors)).

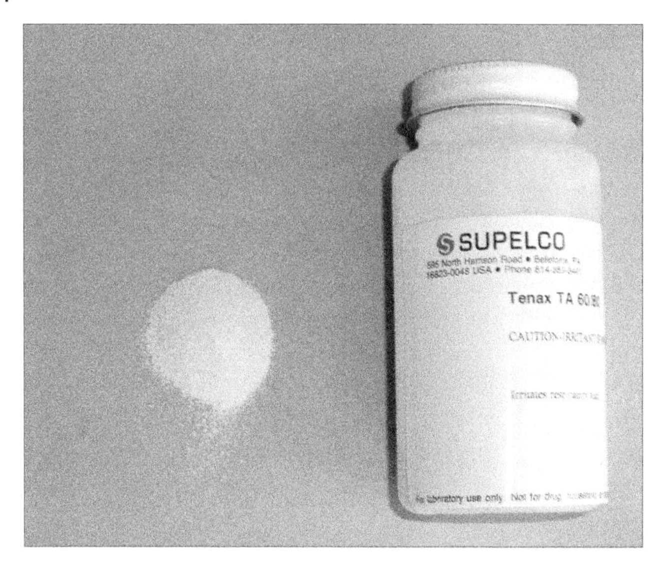

Figure 10.12 Resin for GC-packed column (Source: Dunnivant and Ginsbach (Book Authors)).

Table 10.1 Common stationary phases and their primary use.

Stationary Phase	Applications
Polydimethyl siloxane (Trade names are DB-1, HP-1, OV-1, and SE-30)	This is a general-purpose nonpolar phase for separating hydrocarbons, polynuclear aromatics, nonpolar drugs, chlorinated pesticides, and PCBs
Poly(phenylmethyldimethyl) siloxane (5–10% phenyl) (Trade names are DB-5, HP-5)	Still mostly nonpolar but with some polarity. Used to separate fatty acid methyl esters, alkaloids, drugs, and halogenated chemicals
Poly(phenylmethyl) siloxane (50% phenyl) (Trade names are DB-15 and OV-17)	Slightly more polar. Used to separate more polar drugs, pesticides, and glycols
Poly(tritluoropropyldimethyl) siloxane (Trade names are DB-210 and OV-210)	More polar. Used to separate chlorinated aromatics nitroaromatics, and alkyl-substituted benzenes
Polyethylene glycol (Trade names are DB-WAX and Carbowax)	The glycol functional group makes this phase considerably polar. Used to separate free acid, alcohols, ethers, essential oils, and glycols
Poly(dicyanoallyldimethyl) siloxane (Trade names are DB-1701 and OV-275)	The most polar phase shown here. Used to separate polyunsaturated fatty acids, free acids, and alcohols

10.3.4 Detectors

While the general focus of this book is utilizing mass spectrometry (Chapter 13) as the detector, it is informative to note that a variety of detection systems are available for GC. The most common and commercially available ones are listed in Table 10.2 with information on their detection limits and analytes of interest.

Table 10.2 Commercially available GC detectors.

Detector	General Type	Analytes It Is Used to Measure	Typical Detection Limits
Flame ionization detector (view the FID animation below)	Selective	Any chemical that will burn in an H_2/O_2 flame	Parts per million
Thermoconductivity detector (view the thermoconductivity animation below)	Universal	Any chemical with a thermal conductivity (~specific heat) different from helium	Parts per thousand or hundred
Electron capture (view the ECD animation below)	Selective	Electrophores such as halogenated hydrocarbons	Parts per billion or less
Flame photometric	Specific	P- and S-containing compounds	Parts per million or less
Fourier transform infrared	Specific	Chemicals with specific molecular vibrations	Parts per thousand or hundred
Mass spectrometry	Universal	Any chemical species	Parts per million or less

Three of the most common GC detectors will be discussed in detail here, while types of mass spectrometers will be presented in Chapter 13. One of the earliest GC detectors was the thermal conductivity detector (TCD). The basis for this detector is that most analytes have a thermal conductivity lower than that of helium. Helium is used as the carrier gas and as it passes a wire with a current applied to it, the wire heats up via electrical resistance. Helium molecules remove the maximum amount of heat and the wire reaches thermal equilibrium and a constant current reading. As an analyte with a different thermal conductivity enters the detector, the wire heats up with increasing electrical resistance and the measured current decreases. Since the analytes pass through the detector as a "chromatographic plug," a bell-shaped current reading results known as a chromatographic peak. After the analyte has passed through the detector, the current returns to the original baseline reading as the helium recools the wire. Usually, two matched columns and detectors are used, where only helium is passed through one setup and samples are injected into the other. While this detector responds to any chemical with a thermal conductivity different than helium (which includes almost every other compound), these detectors suffer from relatively poor detection limits (ppth to parts per hundred, pph). Video S.10.2 illustrates the operation of a TCD.

The next most common detector in GC is the flame ionization detector (FID). This detector is based on the fact that most chemicals will burn in an H_2-air flame and current can be passed through the path of ions produced in the flame. Helium is again used as the carrier gas and analytes are injected in the standard split-splitless injector. As individual packets of analytes are separated in the column and enter the detector, they burn in the flame. As illustrated in Video S.10.3, a potential is placed across the flame jet and an electron collector plate is placed above the flame. As ions are produced, electrons are passed through the ion cloud, and a current is measured that is proportional to the mass of analytes produced in the flame. The FID is also considered a universal detector, although not every chemical will burn in an H_2-air flame. FIDs are relatively sensitive with detection limits of 1 ppm for most chemicals.

The ECD is perhaps the most sensitive detector for a GC and was developed primarily to detect chlorinated hydrocarbons in the environment. It relies on the electrophilic nature of halogens contained in an organic chemical but can be used to detect other electrophilic-contained elements such as oxygen. The detector is a sealed unit and contains a radioactive isotope of nickel, ^{63}Ni. This isotope gives off a steady supply of beta particles that are essentially high-speed electrons. These high-speed electrons collide with trace amounts of methane carrier gas that enters the column after the column effluent and produces slower-speed electrons (thermal electrons). These thermal electrons are captured by the anode in the middle of the detector and provide a constant current in the absence of any electrophilic analytes. As electrophilic analytes enter the detector they attract the thermal electrons and carry them out of the detector. This removal process results in a lowering of the current measured by the detector and the change in current is measured as an inverse chromatographic peak. ECDs are extremely sensitive and yield detection limits of pg or sub-ppb concentrations in the injection solvent. The operation of an ECD is illustrated in Video S.10.4.

10.4 Advanced GC Systems

Most commercial GC systems come with ports for at least two injectors, and therefore two columns and two detectors. This allows for versatility and some inventive operational designs. With a special ferrule, two columns with different stationary phases can be inserted into one injection port allowing two analyses per injection and confirmatory analysis.

Confirmatory analysis, with respect to chromatography, is usually restricted to mass spectrometry detection, but when a sample is analyzed on two columns with different stationary phases, the likelihood of two different compounds yielding the same retention time on both columns and the same response on identical detectors is highly unlikely. Thus, dual-column analysis usually produces confirmatory identification.

More enhanced arrangements include detectors aligned in series where the effluent of one detector is passed into another detector. This arrangement does not necessarily provide confirmatory analysis but does allow considerably more information to be collected about the analyte. Note that the first detector cannot be a destructive detector since the chemical integrity of the analyte must be intact for the operation of the second detector.

10.5 Applications/Case Studies

An almost endless variety of chromatographic separations are achievable today due to the diversity of analytical columns. Major column manufacturers and distributors provide very useful Internet sites that contain chromatograms for common analytes that can be used to help select the appropriate column for your needs. In addition, technical help is available from professional chromatographers at these companies for more complex separations. The chromatograms below were selected from the hundreds available from Agilent Technologies. There are many excellent additional examples from chemical and chromatography suppliers.

One of the most commonly published lists of GC applications is for the analysis of environmental pollutants. The chromatographs below are for a variety of chemicals, analytical columns, and detectors. As you review these, note the correlation between the intermolecular forces available to the analytes and the stationary phases used to separate them. Also note the GC detectors used for each type of analyte.

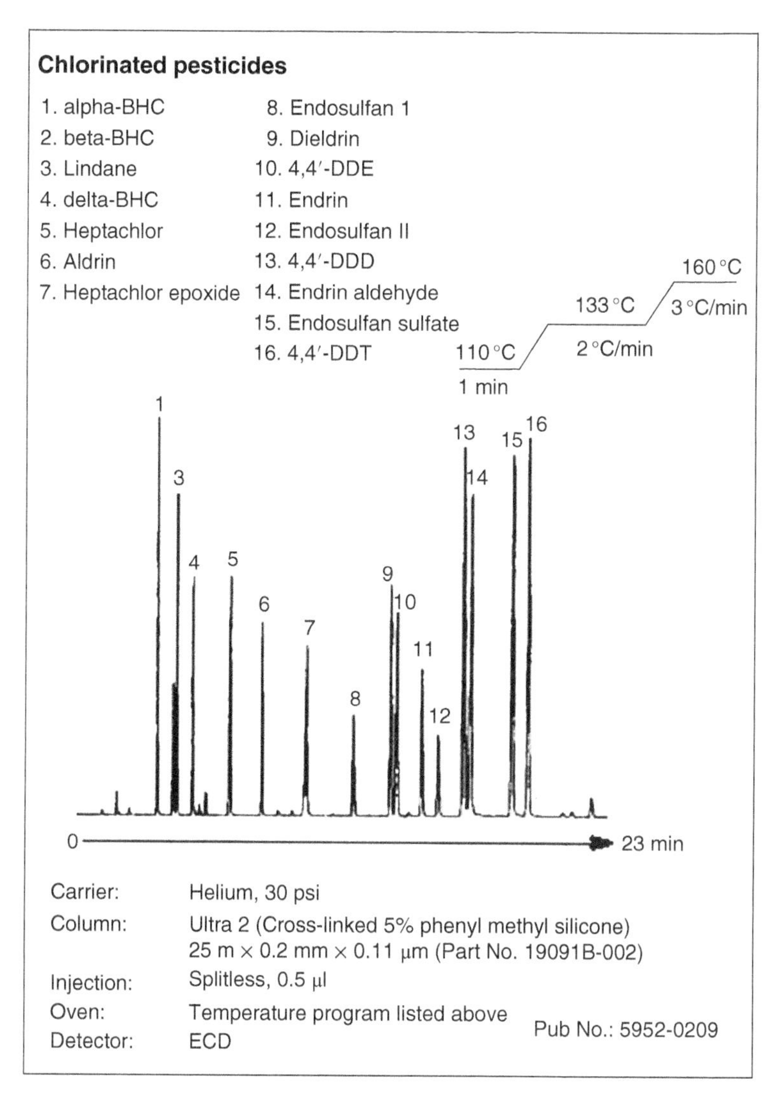

Figure 10.13 Analysis of chlorinated pesticides by GC–ECD using an ultra 2 column (5% cross-linked phenyl methyl silicone) (Source: With permission of Agilent Technologies, Inc., 2006).

The pharmaceutical industry also heavily uses GC and HPLC to determine the purity of reagents, the identity of synthesis products, and the identity of medicines and illicit drugs. A few examples are shown below (Figures 10.13–10.20).

GC can also be used to determine the identity of natural products containing complex mixtures of similar compounds. For example, the geographic source of crude oil or natural gas can be determined by the "fingerprint," or relative distribution of major and trace compounds in each oil. Natural produce oils, such as food products or fragrances, can be identified by GC–FID or GC–MS. A few examples of the separation of these complex mixtures are shown below (Figures 10.13–10.20).

The purity of solutions, from relatively pure solvents such as xylene, to liquors such as scotch can also be determined by GC. Two examples are shown in Figures 10.19 and 10.20.

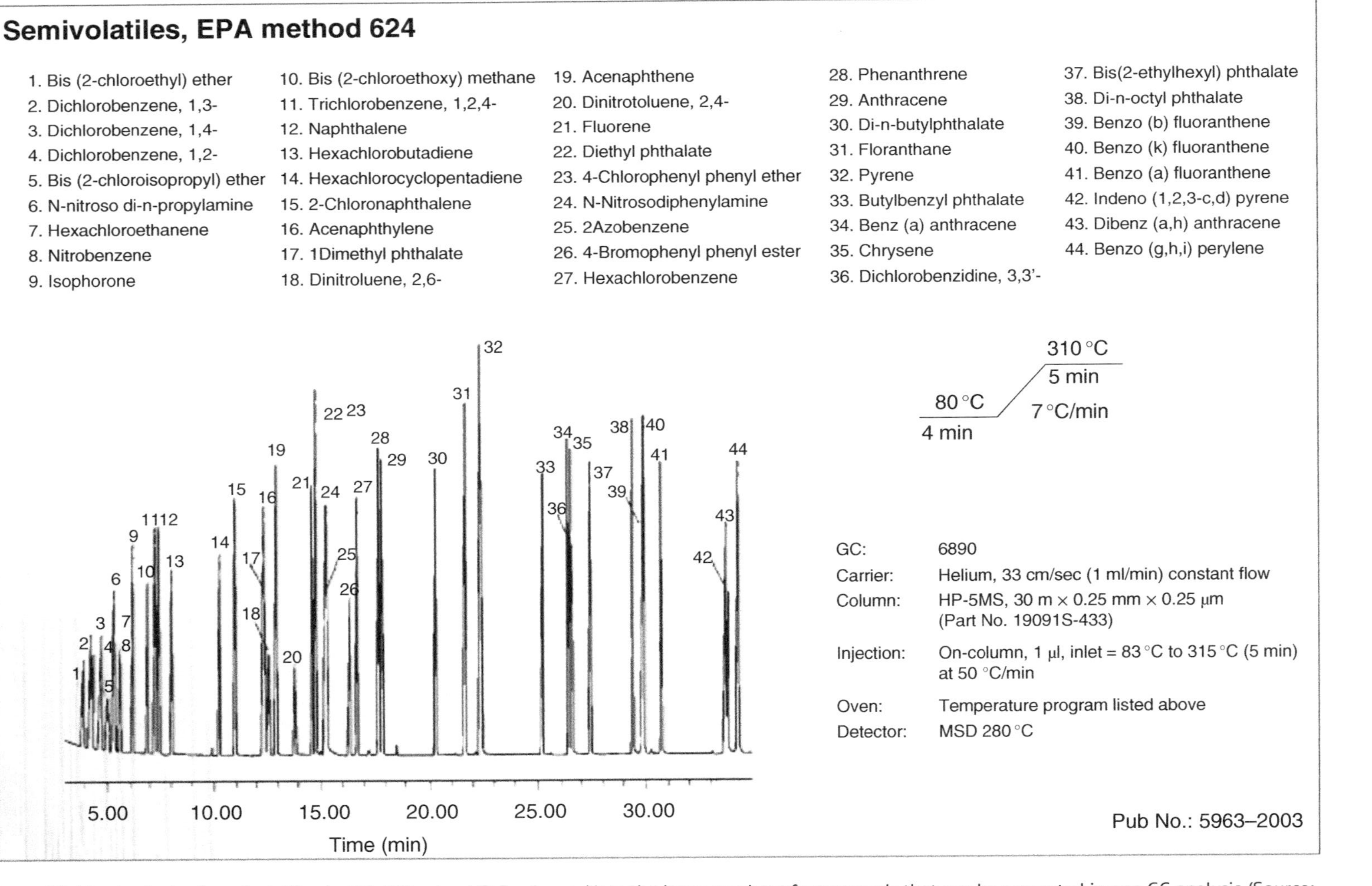

Figure 10.14 Analysis of semivolatiles by GC–MS using HP-5 column. Note the large number of compounds that can be separated in one GC analysis (Source: With permission of Agilent Technologies, Inc., 2006).

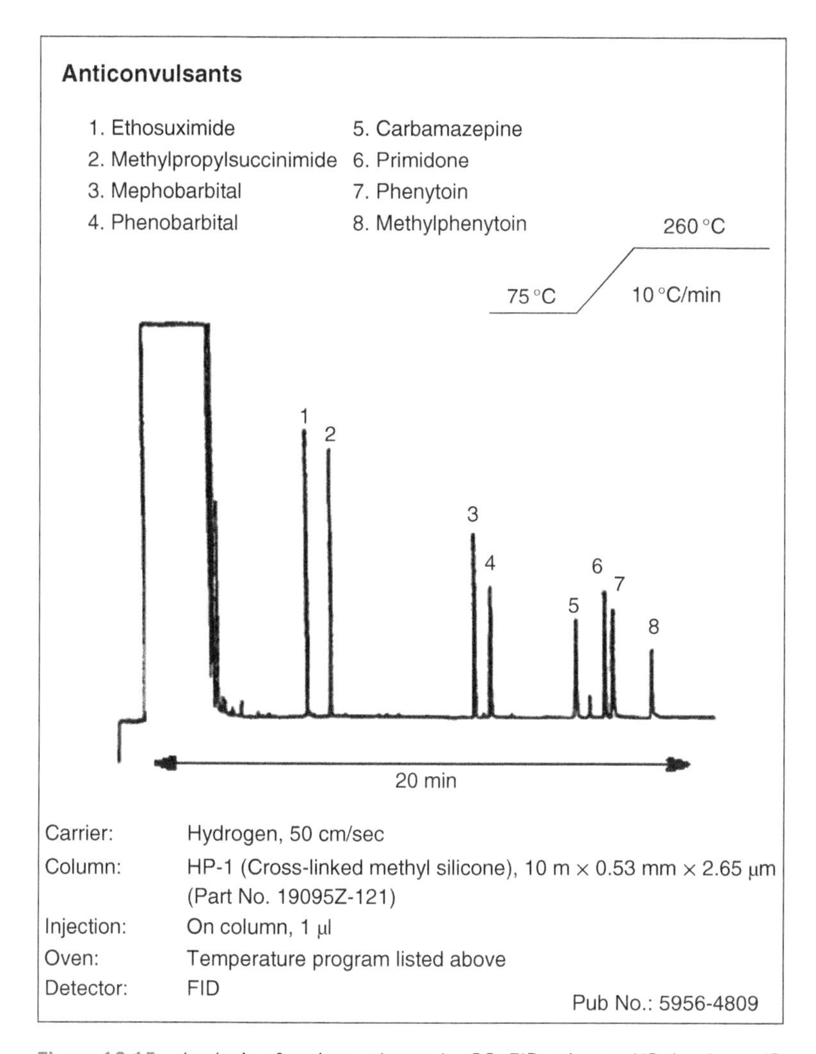

Figure 10.15 Analysis of anticonvulsants by GC–FID using an HP-1 column (Source: With permission of Agilent Technologies, Inc., 2006).

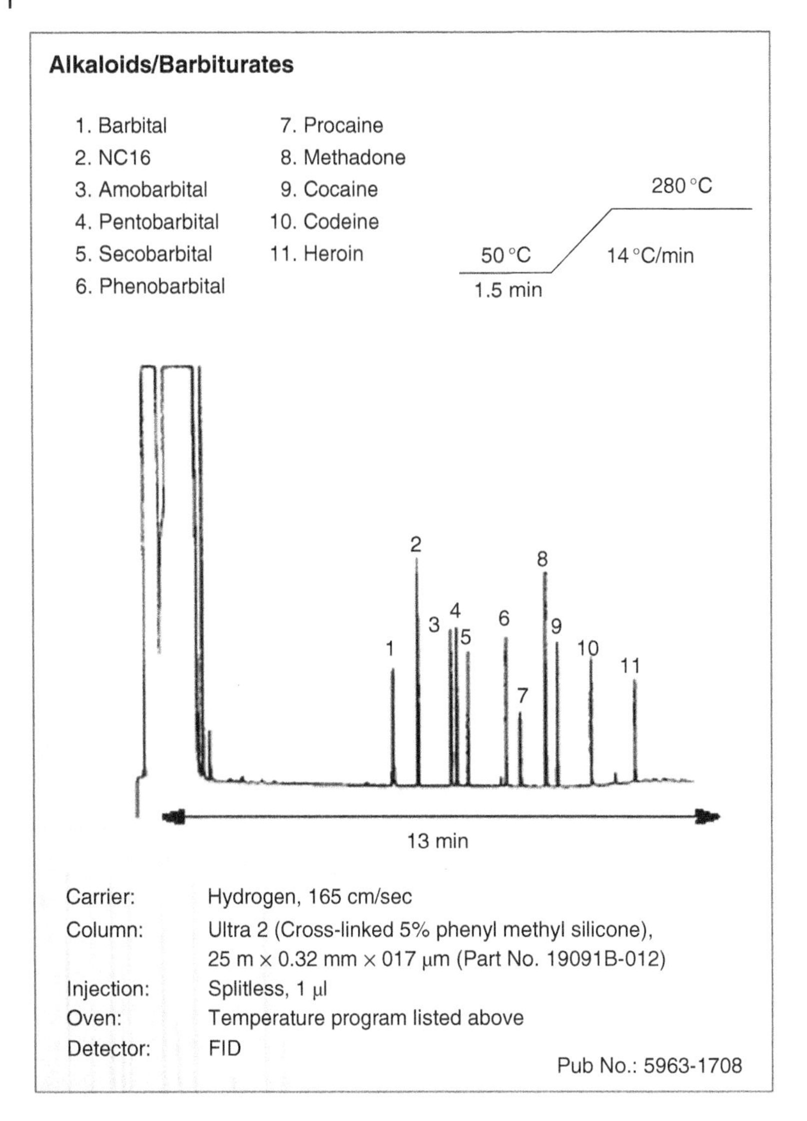

Figure 10.16 Analysis of alkaloids and barbiturates by GC–FID with an ultra-2 column (Source: With permission of Agilent Technologies, Inc., 2006).

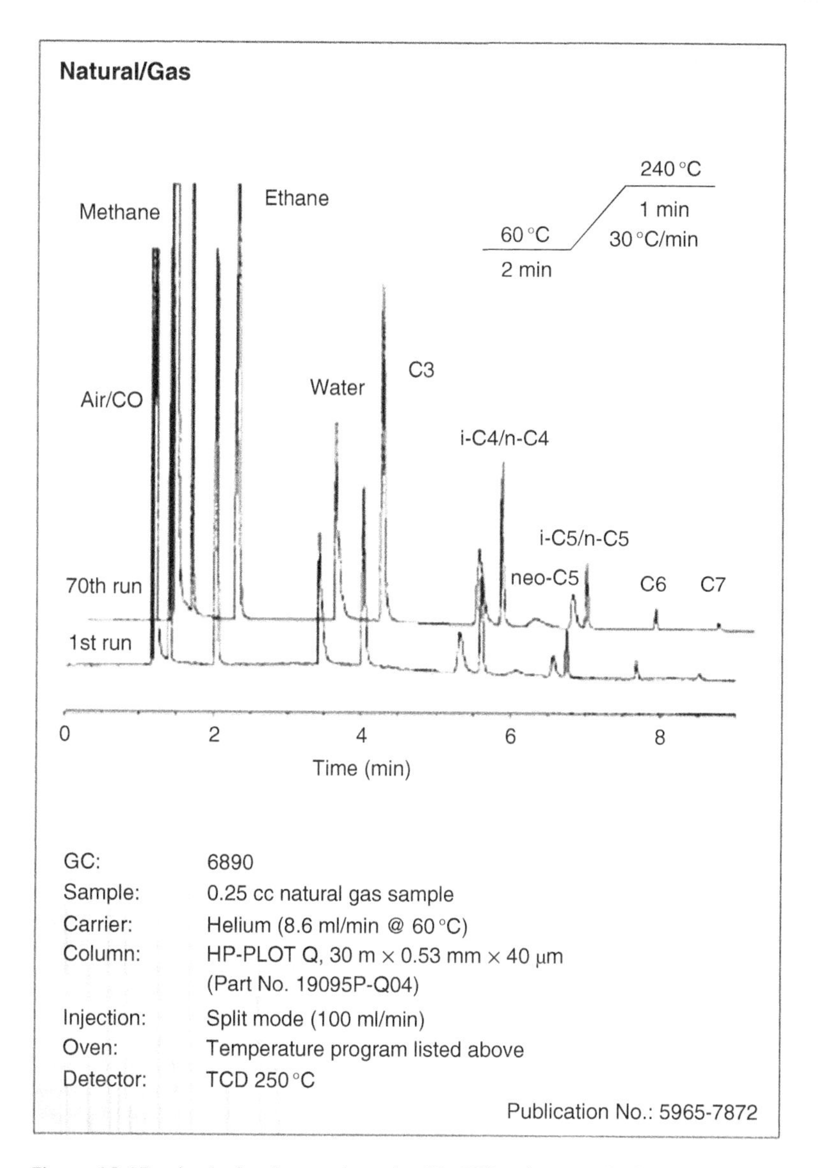

Figure 10.17 Analysis of natural gas by GC–TCD using an HP–PLOT Q column (Source: With permission of Agilent Technologies, Inc., 2006).

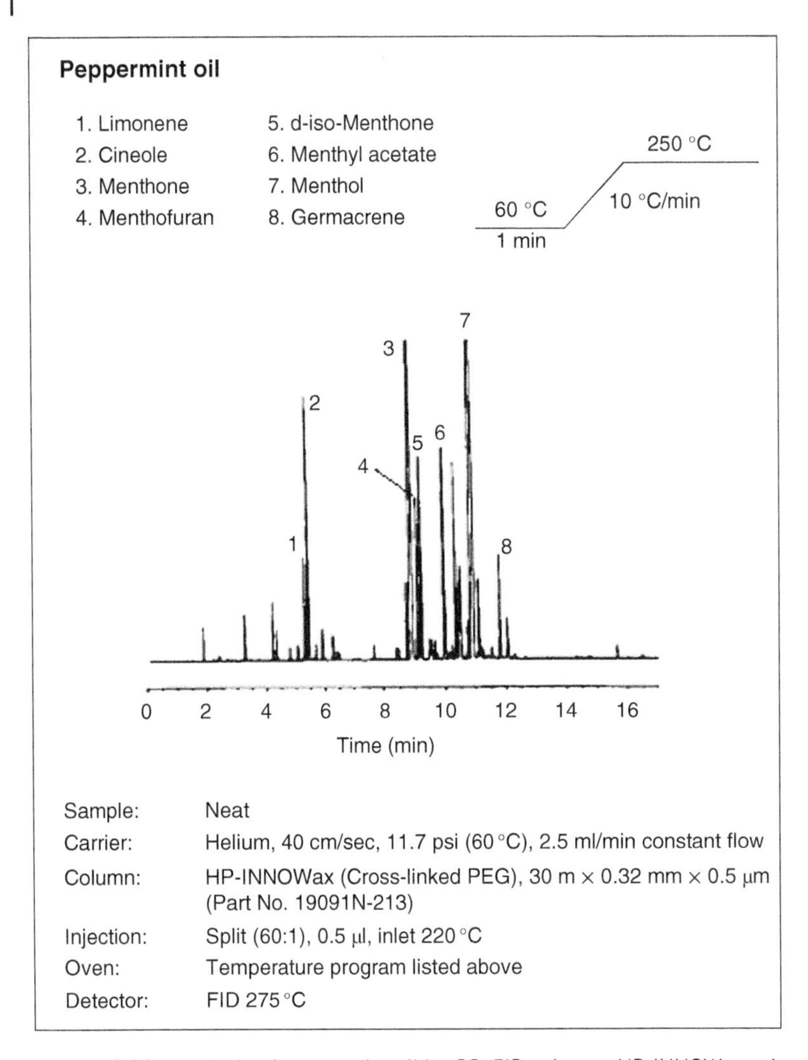

Figure 10.18 Analysis of peppermint oil by GC–FID using an HP-INNOWax column (Source: With permission of Agilent Technologies, Inc., 2006).

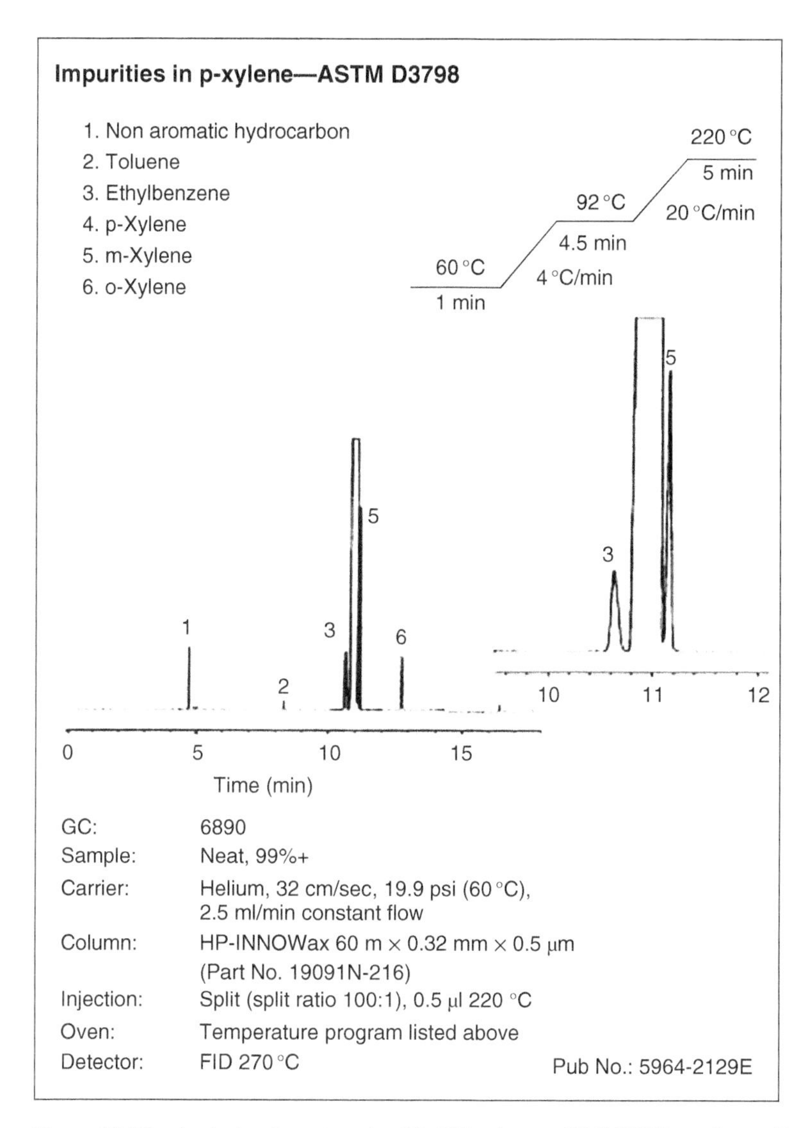

Figure 10.19 Analysis of *p*-xylene by GC–FID using an HP-INNOWax column (Source: With permission of Agilent Technologies, Inc., 2006).

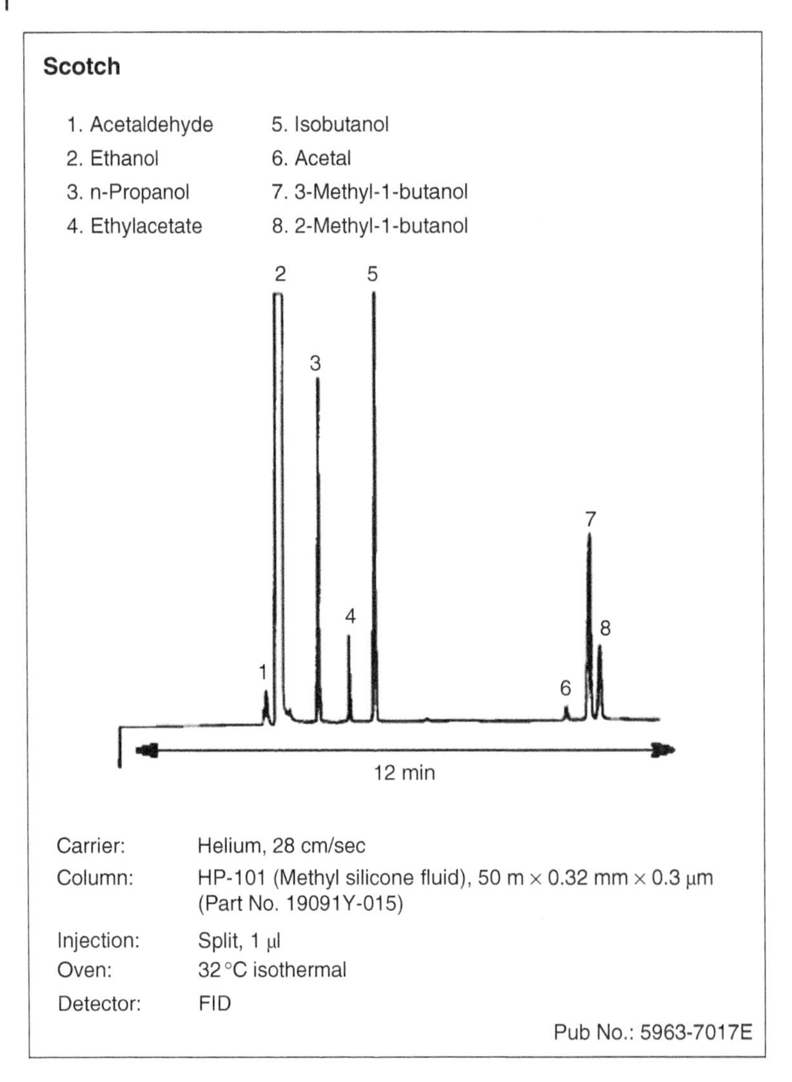

Figure 10.20 Analysis of minor components of scotch by GC–FID using an HP-101 column (Source: With permission of Agilent Technologies, Inc., 2006).

10.6 Summary

This chapter focused on the use of GC. A variety of separation columns and detectors allow the analysis of a diverse set of chemical structures and the separation of complex mixtures of chemicals. Advances in technology have increased the utility of GC analysis and automated instrument controls have greatly decreased the cost of analysis by decreasing labor costs. GC allows the relatively rapid analysis of analytes present in high concentrations, such as in product quality assurance/quality control and in product identification, as well as in the analysis of trace analysis such as the identity of pollutants in environmental media and confirmation of medicines or illicit drugs in human samples. As noted several times in this chapter, GC is used to analyze volatile, thermally stable compounds, and, in general, cannot be used to analyze the more extensive and diverse compounds found in biological systems. These compounds are typically analyzed by HPLC, the subject of the Chapter 11. Later, in Chapter 13, GC, HPLC, and CE will be coupled with MS, the ultimate detector since it allows immediate confirmatory identification. Identification by fragmentation pattern in GC analysis is the subject of Chapter 14.

10.7 Questions

1. In approximately what decade was chromatography first discovered?
2. What is the purchase price of a basic GC?
3. What are the four basic types of sample matrices?
4. List and describe the common ways gaseous samples are collected for GC analysis.
5. List and describe the common ways liquid (aqueous) samples are prepared for GC analysis.
6. List and describe the common ways soil and sediment samples are extracted and prepared for GC analysis.
7. How are biological samples prepared for GC analysis?
8. Why are analysts concerned with extraction efficiency when they prepare samples for chromatographic analysis?
9. What chemical characteristics must a chemical have in order to be analyzed by GC?
10. How are derivatizing agents used in GC analysis?
11. Draw a diagram showing all of the components of a basic GC.
12. How are the mobile phase flow rates different between packed column and capillary column GC?
13. What is the most common carrier gas used in capillary column GC?
14. Draw and explain how a split–splitless injector works. Why do we use a combination of splitless and split injection? When would it be advantageous or necessary to use a total splitless injection?
15. Why is it important to use a Teflon-coated septum in some GC analyses?
16. What is the purpose of the glass liner in the GC injector?
17. What are the typical sizes (diameters and column lengths) of fused silica capillary columns?
18. What are typical injection volumes for capillary column analysis?
19. Contrast the cost of an automatic sampler with the advantages of using one.

20 List the six common types of stationary phases used in GC and describe what types of analytes they can be used to analyze.

21 What are the advantages (and disadvantages) of cross-linking the stationary phase coating on a capillary column?

22 List five common GC detectors, give their acronym, list the types of chemicals they are commonly used to detect, and also their detection limits.

23 Explain, with drawings, how a thermal conductivity detector (TCD) works.

24 Explain, with drawings, how a flame ionization detector (FID) works.

25 Explain, with drawings, how an electron capture detector (ECD) works.

26 What is meant by confirmatory analysis in chromatography?

27 For each of the chromatograms shown in Section 10.2.5, identify the intermolecular force involved in the separation of each class of analytes.

28 Select a compound that can be analyzed by GC (relatively volatile and thermally stable) and use the Internet to find what GC column and temperature program is used in its analysis.

29 Say that you are analyzing a mixture of compounds by GC and that you are having trouble achieving separation of some of them (they coelute or appear as a shoulder peak). What three major things can you change about your GC to possibly improve separation?

Supporting Information

Additional supporting information may be found online in the Supporting Information section in the HTML rendition of this article.

Reference

Bartle, K.D. and Myers, P. (2002). History of gas chromatography. *Trends in Analytical Chemistry* 21 (9): 547–557.

11

High-performance Liquid Chromatography

11.1 Introduction and History

In Chapter 9, Michael Tswett was credited as the father of chromatography due to his 1903 separation of the green leaf pigments into bands of colors, a demonstration of liquid chromatography. While similar work was being conducted in the petroleum industry, Tswett is credited with coining the term "chromatography." Despite Tswett's results, chromatography did not develop quickly. The next major developments were the use of thin-layer chromatography (TLC) in 1937–1938 and the use of paper chromatography in the mid-1940s, but TLC quickly won popularity. TLC was originally developed by Nikolai Izmailov and his graduate student Maria Shraiber for pharmaceutical preparations. Early TLC was conducted with microscope slides that were coated with suspensions of calcium, magnesium, and aluminum oxides. As used today, a small spot of solution was placed on one end of the slide, the slide was dipped into a solvent, and the analytes migrated at different rates through the oxide coatings where they were later detected (today by a UV lamp or chemical stain).

TLC advanced slowly during the next few years, but a major advancement was made in 1956 by Egon Stahl when he attempted to standardize the preparation of the sorbents used to make the plates. While these advances and others, such as forced flow TLC, significantly matured TLC into an accepted (and reproducible) practice, it was still only a qualitative technique, at best. However, Izmailov and Shraiber's spot chromatography, commonly known today as TLC, is the workhorse of undergraduate organic synthesis labs where synthesis reactions are conducted and the resulting products are selected for using common glass open columns filled with silica gel (refer to Figure 11.1). Eluent from these columns is collected in fractions that are then run by TLC to identify which column fraction contains the desired product.

Since the development of TLC, liquid chromatography needed a technique comparable to gas chromatography where a complex mixture of analytes could be quantitatively separated and identified. Several attempts were made to pressurize the relatively large glass preparatory column (shown in Figure 11.1) with little success due to the fragile nature of the column. High-pressure liquid chromatography (HPLC) was later developed to meet this goal in the 1970s. The pressure was first delivered by a large syringe, but this approach limited the volume of solvent that would be passed through a column and therefore limited the analysis time. Syringes were later replaced by a single reciprocating pump, but these delivery systems experienced flow surges, between strokes of the single piston, interfering with stable detector baselines. The placement of two reciprocating pumps, operating opposite to each other with respect to flow, greatly minimized the flow fluctuations which

Essential Methods of Instrumental Analysis, First Edition. Frank M. Dunnivant and Jake W. Ginsbach.

Companion Website: www.wiley.com/go/essentialmethodsofinstrumentalanalysis1e

Figure 11.1 An atmospheric pressure open-column chromatographic column. Fractions are collected in the test tubes and later run by TLC to determine the purity of the fraction and the presence of synthesis products (Source: Dunnivant and Ginsbach (Book Authors)).

were later removed completely with a pulse damper. This form of chromatography is referred to as HPLC, where HP stands for *high performance* or *high pressure*. Some jokingly refer to the HP as meaning *high priced* since it replaced TLC plates, which only cost pennies, with \$40 000 to \$50 000 instruments. The inflation of the cost of an LC analysis is even greater when an HPLC–MS is considered, a minimum of \$200 000. But regardless, HPLC–MS is considered the technique of choice for isolating a synthetic product and is widely utilized in most synthesis laboratories.

Chromatography, as noted in Chapter 9, is divided into gas, liquid, and supercritical fluid (SCF) techniques. Liquid chromatography can be divided into a relatively large collection of techniques. Those mentioned above include atmospheric or low-pressure open-column chromatography and TLC. Pressurized liquid chromatography can be divided into ion exchange, exclusion, partition, and liquid–solid chromatography as summarized in Video S.11.1.

As noted in Video S.11.1, a variety of separation techniques are available with HPLC. Most relevant to this chapter is partition chromatography, although a few others will be discussed in later sections. The most important point here is to distinguish between normal and reverse-phase liquid chromatography. HPLC was first developed using normal phase conditions (NP–HPLC) that followed the logic of atmospheric open-column chromatography, where the stationary phase acted as the polar phase and the mobile phase was nonpolar, specifically an organic solvent. NP–HPLC focused on the separation of analytes that were readily soluble in nonpolar solvents but had slight affinities for the polar stationary phase. However, NP–HPLC required the use of relatively expensive and large volumes of organic solvents, which also led to high waste disposal costs. NP–HPLC was effectively replaced by reverse-phase HPLC (RP–HPLC) that operates with a nonpolar stationary phase and an aqueous, moderately polar, mobile phase. Gradient programming, an additional development, changes the composition of the mobile phase during a chromatographic run which greatly enhances the utility of RP–HPLC. In RP–HPLC, the gradient is initially more polar (i.e. water) and as the chromatographic run progresses, more and more less-polar solvent is added to the mixture (i.e. methanol or acetonitrile) to end the gradient program with pure organic solvents. As noted in Video S.11.2, the retention order of analytes in RP–HPLC is opposite NP–HPLC. For example, the first analyte that eludes on an RP system would elude last is separated by NP–HPLC.

Additionally, the pH of the polar solvent in RP–HPLC can play an important role in optimizing analyte separations. Ionic or ionizable analytes that would not normally be separated on an RP (nonpolar) analytical column can also be analyzed by ion-pair chromatography. In this type of chromatography, the ionic analyte is bound to another ion (usually a large organic counter-ion such as quaternary ammonium or alkyl sulfonate) to form a neutral pair that has a selective affinity for the nonpolar stationary phase. Even many chiral compounds can be separated on special chiral stationary phases or by the addition of chiral-resolving agents that selectively bind to one of the enantiomers. Additional modifications to an HPLC system such as ultrahigh-pressure LC, ion exchange, and SCF chromatography will be discussed in Sections 11.5 and 11.6.

One final point should be made with respect to HPLC. Chemists use HPLC for two completely different purposes. Organic chemists, especially in the pharmaceutical industry, use large-scale systems, referred to as preparatory HPLC or flash chromatography, to recover relatively large-scale milligram quantities of their products. The main differences of a preparatory HPLC, as opposed to an analytical HPLC, are the pump flow rates and the size of the columns. In contrast, analytical chemists use HPLC to separate and identify nanograms or smaller quantities of analytes. Whichever practice is needed, the overall chromatography is the same.

11.2 Types of Analytes, Samples, and Sample Introduction

Gas chromatography is somewhat limited in that the analyte has to be volatile below 300 °C and thermally stable with molecular weights less than 1000 Da. HPLC greatly expands the range of possible analytes with RP–HPLC and can include many of the same analytes as GC (but with less resolution as compared to capillary column GC; see Section 9.3). Analysis of compounds commonly used in HPLC increases the analyte range of molecular weights to just less than one million Daltons. Many chemicals, such as petroleum hydrocarbons, solvents, illicit drugs, and environmental chemicals, can be analyzed by both GC and HPLC. But with the ability of HPLC to analyze nonvolatile chemicals, many biomolecules can be analyzed, including sugars, amino acids, proteins, and a large variety of other nonvolatile compounds.

Samples analyzed by HPLC are always in liquid form, in either aqueous or organic solvents. Little to no sample preparation is needed, except that it is good practice to filter all samples through a 0.2 μm low-volume cartridge filter prior to or during injection. Filtration of samples and solvents avoids the buildup and eventual clogging of the in-line filter, guard column, or expensive analytical column. Samples containing high concentrations of analytes may need to be diluted in order to avoid overwhelming the capacity of the stationary phase and remaining in the linear range of the detector (governed by Beer's law in UV–vis applications). Solvent exchange of the samples may be necessary depending on the solvent gradient conditions needed for separation. Samples containing compounds with significantly different chemical structures may interfere with the detector of the analyte(s) and may require removal with a microcolumn cleanup (such as silica gel or alumina) or may require the use of a specialty cleanup cartridge (such as those used in GC for concentration of the analytes from aqueous samples described in Section 10.2). Samples containing relatively low concentrations of analytes may need to be concentrated using one of these same microresin columns.

11.3 Liquid Chromatography

11.3.1 Overview

In recent decades, the modern HPLC system has become increasingly complex. Automatic injection systems have all but replaced manual injections, manual six-port values are now pressure actuated, the instrument is usually computer controlled, and data are collected and processed via a specified computer method. It should come as no surprise that the cost of a basic HPLC system (without mass spectrometry detection) has risen from \$10 000 a few decades ago to over \$50 000 today. In this section, each component of an HPLC will be presented. Figure 11.2 illustrates an overview of a semimodern HPLC–MS system.

11.3.2 Actuator Gas

Pressurized gas in Figure 11.2 is required to degas solvents and to drive many of the mechanical functions of modern HPLCs. Solvents used as the mobile phase contain trace concentrations (ppm) of atmospheric gases (N_2 and O_2), which create bubbles in the pump (during the low-pressure intake of solvents) causing pumping problems. Dissolved gases can also evolve in the low-pressure detectors in ultra HP–LC, ion exchange, and SCF chromatography giving the appearance of a chromatographic peak depending on the specific detector being used. Today, helium (He), necessary to degas gradient solvents, has been replaced with a mechanical vacuum degasser. Pressurized gas is also needed to turn the six-port injection valve during automatic injection mode.

11.3.3 Solvents and Solvent Preparation

Solvent purity requirements depend on the type of samples being analyzed. All solvents used as the mobile phase must be filtered through 0.2 μm filters prior to entry to an HPLC system in order to avoid the scratching of the pump pistons by particles and to avoid clogging the in-line filter or guard column. Organic chemists are usually not concerned with the purity of solvents since their analytes are present in high concentrations and their samples usually contain many side-products and solvents. Hence, relatively low purity, and therefore inexpensive, solvents can be used. In contrast, the analytical chemist must purchase HPLC-grade solvents that are considerably more expensive. For example, a 20-L metal can of ACS grade methanol costs less than \$50, while a 4-L

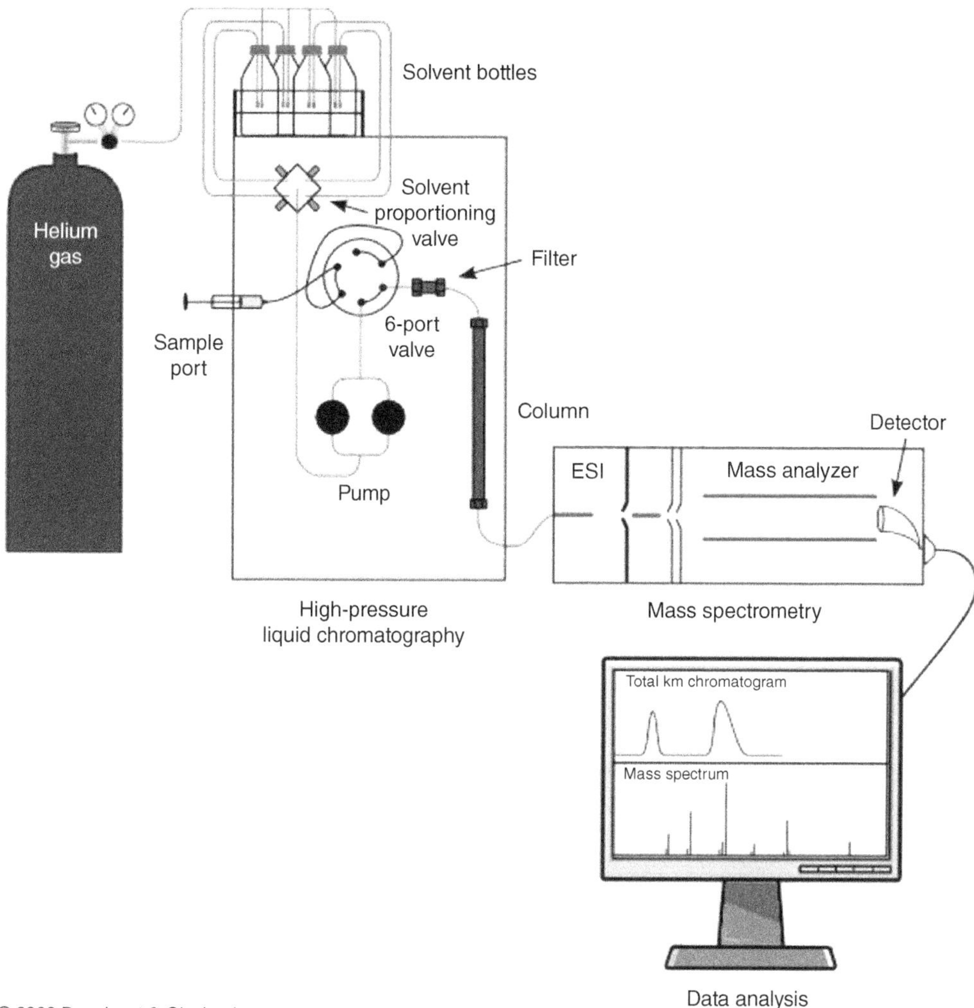

Figure 11.2 An overview of a semimodern HPLC–MS system. More modern changes to this system eliminate the helium purge gas with a vacuum degasser. Note modern modifications mentioned below (Source: Dunnivant and Ginsbach).

glass bottle of HPLC UV–vis grade bottle costs \$150. Many solvent gradients call for acetonitrile, which in recent years (2009) has risen considerably in costs due to a decrease in production, from \$100 in 2005 to \$250 in 2009 (if supplies allow you to even order it). The increase in cost has tempted many chromatographers to switch their acetonitrile-solvent gradients to one based on methanol while maintaining the same polarity index (a measure of the polarity of the solvent mixture that is used to optimize analyte separation). However, as noted by William Campbell in a recent Supelco Analytical Note (volume 27.1, p. 13), while analyte retention orders may (or may not) remain the same, the relative retention times and resolution can significantly change, in some cases, where analyte separations are no longer achievable. Solvents used for elution are contained in glass reservoirs (usually one liter in capacity) and connected to HPLC with Teflon tubing.

11.3.3.1 Tubing and Connections

Different components of an HPLC system operate under different pressures ranging from atmospheric to 4500 psi (3.10×10^7 Pa) in standard HPLC and to 20 000 psi (1.38×10^8 Pa) in ultra HPLC. Systems are currently under development that will operate at 100 000 psi. While GC fittings are usually standardized to Swagelok fittings, many HPLC manufacturers have established their own type of high-pressure fittings. All operate on the same ferrule-nut configuration as in GC, but the shape of the ferrule and receiving system is different. A few of the more common fittings are shown in Figure 11.3. Tubing sizes are much smaller in HPLC as compared to GC. Common tubing materials are stainless steel and PEEK (polyaryletheretherketone), a very strong organic, solvent-resistant polymer.

11.3.4 Proportioning Valve

Up to four separate elution solvents enter the HPLC first through a proportioning valve that adjusts the flow of each solvent to a predetermined amount. A proportioning valve is shown on the right-hand side of Figure 11.4. Prior to the development of the proportioning valve, a separate pumping system was needed for each solvent. Since the pumping system (left-hand side of Figure 11.4) is typically the most expensive component of an HPLC system, the proportioning valves help reduce the cost of an HPLC system.

The relatively inexpensive proportioning valve allows the mixing of up to four solvent systems which enter a single pumping system. The mobile phase can be delivered as a constant composition of solvents (isocratic) or as a changing composition (gradient programming). In gradient programming, the solvent composition changes from a more polar solvent mixture to a more nonpolar organic composition. Isocratic operations are used for relatively simple separations, while gradient programming is used for complex separations. In the past, the relative flow rates of multiple pumps were used to control the mobile phase composition. Today, the less expensive proportioning valves adjust the composition. The need for gradient programming was illustrated in Video S.11.2 (earlier) and is similar to temperature programming in GC and is again referred to as the general elution problem.

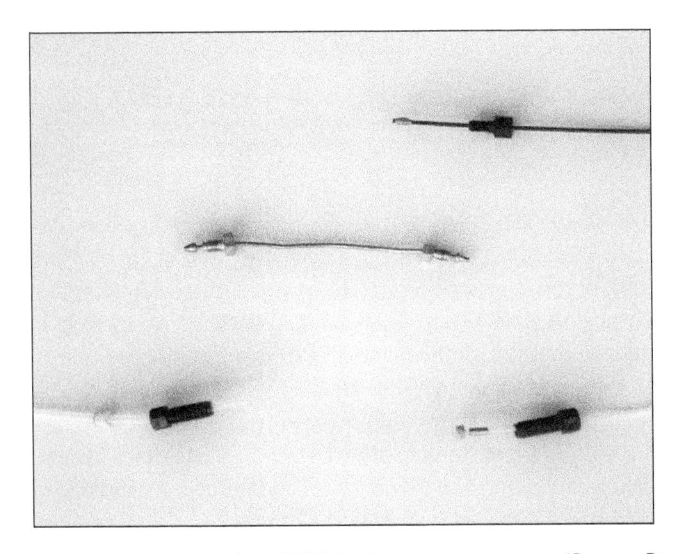

Figure 11.3 A selection of HPLC tubing and connectors (Source: Dunnivant and Ginsbach (Book Authors)).

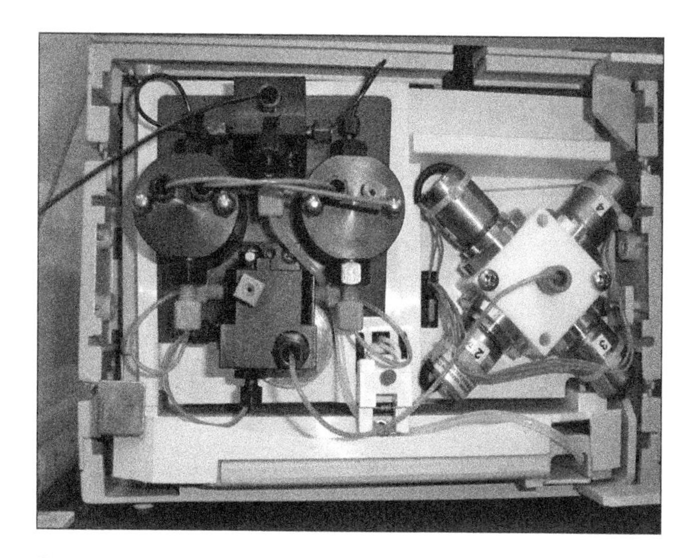

Figure 11.4 A dual reciprocating piston pump (left) and a four-way proportioning valve (right) (Source: Dunnivant and Ginsbach (Book Authors)).

11.3.5 Pump

As noted earlier, HPLC pumping systems have evolved more than any other component of LC. Gravity-fed open column systems were later pressurized with gas, which was replaced by syringe pumps in early HPLC, then by single piston pumps, and today with specially engineered dual reciprocating piston pumps. Today's pumps provide constant pressure and mobile phase flow by alternating the pumping actions between two pumping systems. When either pump is in full stroke, constant flow is easily provided but as each pump reaches the end of a stroke, flow rates may be altered. This is overcome in some systems by an oval-shaped camming device that speeds the pumping rate at the end of each piston stroke. At the end of the stroke, one pump speeds up the flow of mobile phase as the other pump head decreases the pumping rate. This combined action provides constant flow rates in modern HPLC systems. While the simultaneous operation of two pumps and their associated pistons, cams, and check valves are difficult to describe in words, an animation makes the process easy to understand. Such an animation is shown in Video S.11.3.

11.3.5.1 Pulse Damper

Early HPLC pumps produced pressure fluctuations, and therefore flow pulses, this was especially true with single piston pumps which required a pulse damper to be installed down-line from the pump. Today's systems, with advanced pump designs, produce little pressure fluctuations and either do not require a pulse damper or electronically compensate for the very small pressure fluctuations.

11.3.6 Six-port Injector

High-pressure systems require a special type of sample injection since injections with standard syringes are not possible. Four-, six-, and eight-port valves are used for high-pressure systems. These systems are equipped with a fixed volume loop of tubing that serves as the sample loop that is loaded with a standard blunt-tipped syringe. These multiple port systems allow uninterrupted flow to the column during the loading of a sample on a sample loop and during injection of a sample under high pressure. As with a dual reciprocating HPLC pump, it is easier to show how a six-port valves works through an animation. Video S.11.4 shows the alignment of the valves during the loading of a sample onto a sample loop, switching of the values, and injection of the contents of the sample loop onto an HPLC column. View the animation by clicking on the figure below.

The sampling loop of a six-port valve can be loaded by a syringe or by pumping liquid samples into the valve by an automatic sampler. Sample loop sizes range from 1 to 100 μL for analytical-scale HPLC to milliliter volumes in preparatory-scale HPLC. The advantage of a syringe (manual) injection is that less sample volume is required (since filling of the tubing in an automatic sampler is not required). The obvious advantage of an automatic sampler is that numerous samples can be automatically analyzed by a computer-controlled system.

11.3.6.1 In-line Filter

Although samples and solvents are filtered prior to injection or use with an HPLC, an in-line filter is usually present immediately after the injection valve. This filter removes any remaining particles in the mobile phase that may clog the guard or analytical column. In-line filters are one of the most commonly maintained items for HPLC systems.

11.3.7 Columns

11.3.7.1 Guard Column

The next component of an HPLC is the guard column. Guard columns are miniature versions of the analytical (separation) column and they contain the same stationary phase. The purpose of the guard column is to adsorb any permanently adsorbing chemicals that could destroy the more expensive analytical column. Guard columns typically cost one-fourth or less of the cost of an analytical column.

11.3.7.2 Analytical Column

The heart of an HPLC system is the analytical column. This is where the mixture of chemicals injected in the system is separated into individual analytes that appear as peaks in the chromatogram. Columns are available in a variety of diameters and lengths ranging from large preparatory columns (20–50 mm in diameter by 50–250 mm in length) to analytical columns (typically 4.5 mm in diameter by 12–25 mm in length), to narrow bore analytical columns for improved performance and MS applications (1–2 mm in diameter by 10 cm in length), to capillary columns for MS detectors (from 0.075 to 0.1 mm in diameter). Larger columns require a larger mobile phase volume to push the analytes through the system. A collection of HPLC columns is shown in Figure 11.5.

With the exception of relatively new capillary columns, all HPLC columns are packed with small resin beads (typically 2–5 μm in diameter or less) that contain a coating of stationary phase. As in GC, stationary phases are selected based on the expected intermolecular forces available to the analyte for attraction/bonding. The adage "like dissolves like" is also appropriate for liquid chromatography applications. Separations in HPLC are considerably more complicated than in GC. In GC, the chromatographer is only concerned with the stationary phase and the temperature program. In HPLC, one must also be concerned with the polarity of the mobile phase. Adequate analyte separations are only achieved with an appropriate match of gradient polarity and stationary phase interactions. In more advanced applications, the pH and ionic composition of the mobile

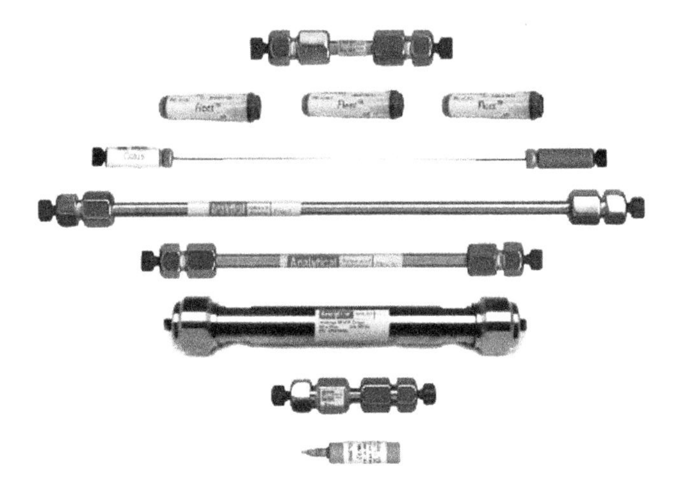

Figure 11.5 Various sizes of HPLC analytical and guard columns (Source: With permission of Analytical Sales and Services, Inc).

phase must also be controlled. The most common stationary phases used in RP–HPLC include alkylamine, octodecyl (C18), octyl (C8), butyl (C4), cyanopropyl (CN), and methyl functional groups. Additional and custom-made stationary phases can be ordered from many suppliers.

Early chromatography columns that were open to the atmosphere were easy to pack. The chromatographer filled a glass column (1–5 cm in diameter) with solvent, slowly poured in silica gel with stirring to remove air pockets, and applied the solution to be separated to the top of the column. These types of columns yielded adequate separation for organic chemists attempting to isolate their synthesis products but have extremely limited use for analytical chemists who require more theoretical plates in the column and on-line detection of column effluents. Today, analytical chromatographers almost exclusively purchase their separation columns from manufacturers because of the need for perfectly packed columns with no void spaces. The emphasis here is on perfect column packing since the presence of only a few void areas in a column will significantly affect the theoretical plate height in the column and decrease resolution. A comparison of a poorly packed column and an adequately packed column is shown in Video S.11.5.

The particle size of the stationary phase also significantly affects the operation of an HPLC by determining the shape of the van Deemter curve. Figure 11.6 shows the van Deemter curves for a variety of particle-sized stationary phases. Recall from our discussions in Chapter 9 that better separations occur at the minimum point (linear velocity) in the van Deemter curve and a more stable system operation will occur with a large range of linear velocities. In HPLC, smaller-sized stationary phase particles will yield a van Deemter curve that is essentially flat (refer to Figure 11.6). Thus, a range of flow rates will produce optimum analyte separation.

11.3.8 Detectors

There are a variety of detectors available for use with HPLC. These range from near-universal detectors that will respond to any organic chemical structure to specialty detectors that only respond to a few analytes. Both have their advantages. Universal detectors provide economic use for a large diversity of chemicals but specific analyte detection can be hampered if two or more

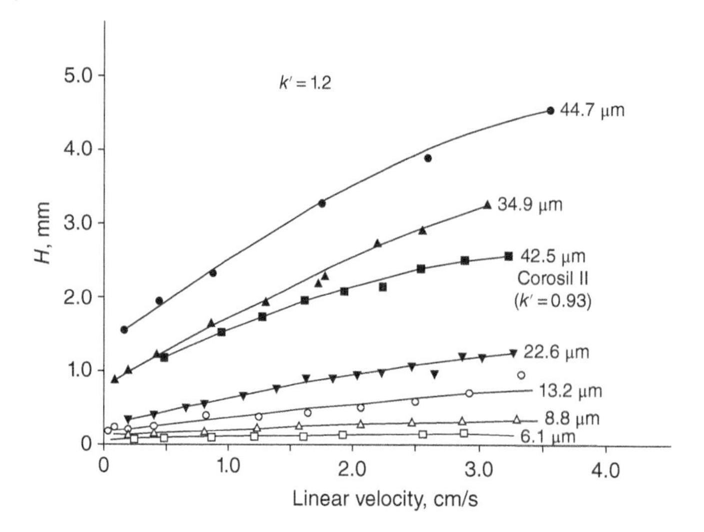

Figure 11.6 The effect of stationary phase particle size on the optimum flow rate in HPLC (Source: Ronald et al. (1973)/with permission of Oxford University Press).

Table 11.1 Common commercially available HPLC detectors, applications, and detection limits.

Detector	Application(s)	Detection Limits
UV–visible absorbance	For compounds that absorb in the UV or visible range	pg quantities
Fluorescence	For compounds capable of fluorescence (especially polyaromatic hydrocarbons)	fg quantities
Refractive index (RI)	For alcohol, sugar, saccharide, fatty acid, and polymer analysis with refractive indices different from the solvent	ng quantities
Electrochemical	For analyzing a wide range of compounds	High pg quantities
Conductivity for IC	Mainly for inorganic ions	~ng quantities
Evaporative light scattering (ELS)	For a wide variety of compounds that lack UV–vis chromophores including triglycerides, sugars, and natural products	μg quantities
Fourier transfer infrared	For compounds with vibrational functional groups	μg quantities
Mass spectrometry	A universal detector	μg to pg quantities depending on the type of mass spectrometer

analytes elute from the analytical column at the same retention time. Specialty detectors provide near-interference-free detection of only a few analytes. Common detectors, along with their use and approximate detection limits, are given in Table 11.1.

The most common detector is the UV–vis detector that comes standard on basic HPLC systems. This is a near-universal detector since most organic compounds have a chromophore capable of adsorbing UV or visible wavelengths. The focus of this textbook, mass spectrometer detectors, will be discussed in the next chapter.

11.4 Advanced and Specialty LC Systems

11.4.1 Ultrahigh-pressure Liquid Chromatography (U-HPLC)

HPLC technology made a significant advance with the development of ultrahigh-pressure systems. These systems, currently operating at 15 000–20 000 psi, offer significant time and cost savings per analysis. The increased pressure allows samples to pass through the column faster and results in significant increases in column efficiency (and decreases in theoretical plate height) by minimizing longitudinal diffusion processes. These systems offer higher column efficiencies (resolution per column length) and allow the systems to operate at a much wider range of linear velocities, flow rates, and backpressures. Nanobore-sized columns offer up to 95% decreases in solvent use and interface well with mass spectrometers. Higher-pressure systems are under development that will possibly operate at up to 100 000 psi and yield even greater column efficiencies.

11.4.2 Ion Chromatography (IC)

Ion chromatography (IC) is a form of HPLC where the system is modified to analyze mainly inorganic ions such as nutrient ions (i.e. nitrate, sulfate, phosphate, etc.) as well as many organic anions and metal cations. The systems require a special ion exchange analytical column that is specific to the analytes of interest, a relatively simple conductivity detector, and a unique ion suppressor column for removing ions other than the analytes that would generate electrical conductivity in the detector. Columns and tubing are usually composed of PEEK to avoid reactions on metal surfaces. Analytical columns are mainly divided into columns for cations and anions and are similar in diameter to standard HPLC columns but slightly longer in length. Cationic exchange columns have active strong acid sites such as sulfonic acid ($-SO_3^-H^+$) or weak acid sites such as carboxylic acid groups ($-COO^-H^+$) that preferentially exchange with analyte cations. Common anionic exchange sites include the strongly basic tertiary amines ($-N(CH_3)_3^+OH^-$) and the weakly basic primary amine group ($-NH_3^+OH^-$). Both exchange groups are usually placed on porous microparticles of silica. As the injected sample passes through the analytical column, shown below for the sulfonic acid exchange surface, the following reaction occurs:

$$-R{-}SO_3^-H^+{}_{solid} + \text{Metal Cation}^+{}_{solution} \leftrightarrow (-R{-}SO_3^-)_n\text{Metal Cation}^+{}_{solid} + nH^+{}_{solution}$$

Similarly, for anion analysis, the reaction would be:

$$-NH_3^+OH^-{}_{solid} + \text{Anion}^-{}_{solution} \leftrightarrow (-NH_3^+)_n\text{Anion}^-{}_{solid} + nOH^-{}_{solution}$$

At appropriate flow rates, an equilibrium migration front is set up for each analyte passing through the column. Since each metal cation will have a different affinity for the stationary ion exchange resin, each analyte will elute from the column at a different time. As in all types of chromatographs (in the absence of mass spectrometry detection), identification is largely based on retention time.

The detector in IC is a simple conductivity detector where a potential is placed across two electrodes. In the absence of ions in the solution passing through the detector, minimal current is transmitted and no signal is generated by the electronics. As each packet of analytes, cations or anions, pass through the detector, a current is generated that is proportional to the concentration of analytes. In order to detect low concentrations of analytes, the background signal (current) of the mobile phase must be maintained as low as possible. Since all samples have counterions (an equal concentration of cations or anions), half of the ionic components must be removed prior to entering the detector. For example, detecting the presence of cations requires the removal of anions before the column effluent reaches the detector. The removal of these ions is accomplished with an ion suppressor device shown in Figure 11.7. This device is positioned within the IC system as shown in Figure 11.8.

As presented in Figure 11.8, the sample first passes through the analytical column where the cations or anions are separated relative to their affinity for the solid-phase exchange resin. The effluent from the analytical column then enters the ion suppressor device, along with acid or base to

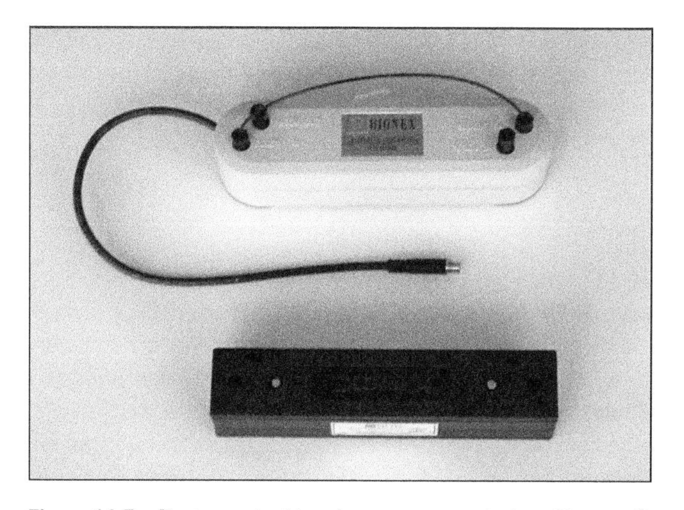

Figure 11.7 Photograph of two ion suppressor devices (Source: Dunnivant and Ginsbach (Book Authors)).

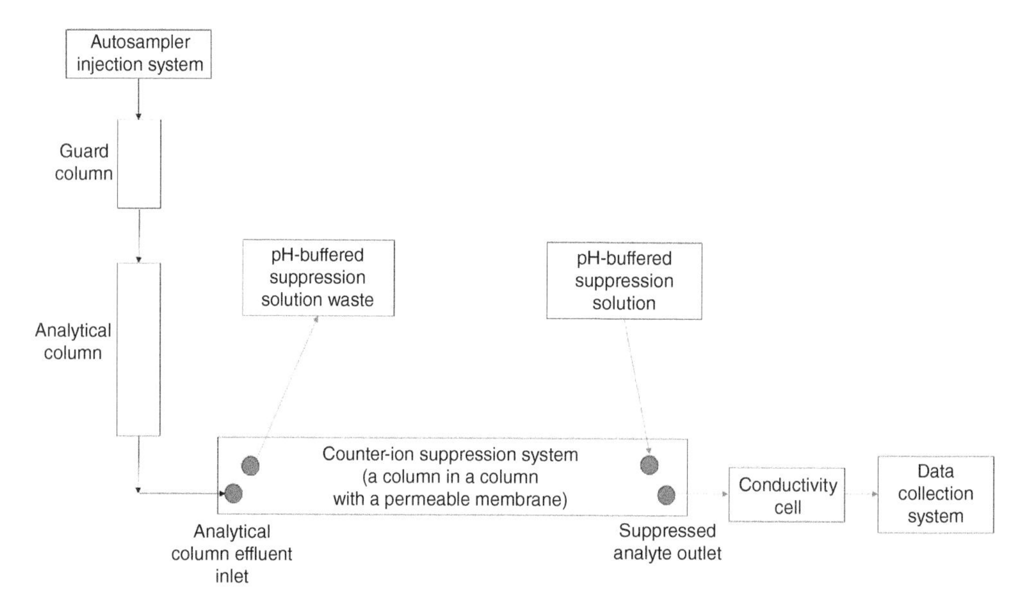

Figure 11.8 Overview of the components used to separate analytes in an ion chromatography system (Source: Dunnivant and Ginsbach).

keep the ion suppressor column active. For anionic analysis, strong acid (H^+) is used in the suppressor device; for cationic analysis, strong base (OH^- or COO^-) is used in the device. In both cases, the analytes of interest are unchanged and unretained. As a consequence, the following suppression reactions occur depending on the mode of operation (cation or anion analysis).

In anion analysis, cations in the sample are exchanged for H^+ in the suppressor column and neutralized in the center flow-through portion of the suppressor column by the following reaction:

$$H^+{}_{solution} + Cl^-{}_{solution} + resin^+OH^-{}_{solid} \rightarrow resin^+Cl^-{}_{solid} + H_2O$$

Thus, only anions, such as Cl^-, contribute to the conductivity in the detector.

In cation analysis, anions in the sample are exchanged with HCO_3^- (or OH^-) in the suppressor column and neutralized in the center flow-through portion of the suppressor column by the following reaction:

$$cation^+{}_{solution} + HCO_3^-{}_{solution} + resin^-H^+{}_{solid} \rightarrow resin^-cation^+{}_{solid} + H_2CO_{3solution}$$

Thus, only cations, such as Ca^{2+}, contribute to the conductivity in the detector.

Again, note that these reactions convert all ionic species, except for the analytes, to nonconductive chemicals. These reactions in relation to the flow through the ion suppressor are illustrated in Figure 11.9. In both cases, nonionic species are produced and the interfering ions are removed (suppressed) from the solution. This significantly decreases the background ions in solution and allows for low detection limits (parts per trillion, ppt to parts per billion, ppb in the injected sample). IC systems also operate well with MS detectors.

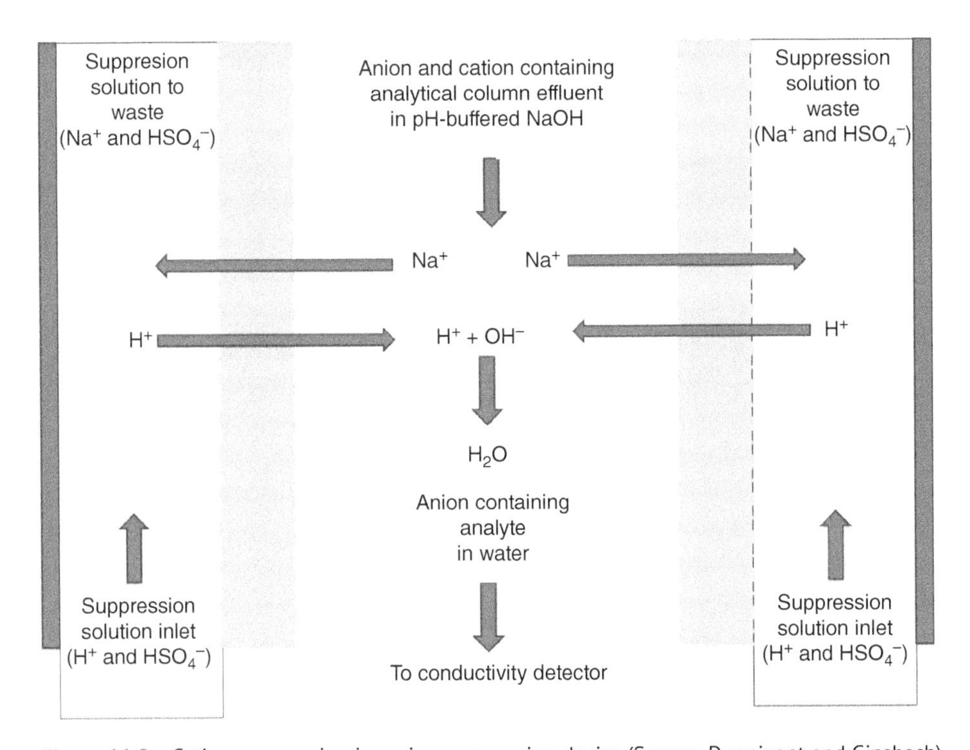

Figure 11.9 Cation suppression in an ion suppression device (Source: Dunnivant and Ginsbach).

11.4.3 Supercritical Fluid Chromatography

A liquid turns into an SCF when its temperature rises above the critical temperature—the temperature where it can no longer exist as a liquid no matter how much pressure is applied. SCFs exist in a state between a liquid and a gas and have the penetration abilities of a gas and the dissolving power of a liquid. As a result, SCF is a mixture of GC and HPLC and, in some cases, is superior to GC or HPLC. Supercritical CO_2, the same matrix used to selectively extract caffeine from coffee beads and nicotine from tobacco products, can be used as a mobile phase in chromatography. SCF chromatography is a special form of HPLC where a near-identical system is used but the mobile phase, as noted above, is supercritical CO_2. The system is therefore pressured and temperature-controlled to maintain the SCF. SCF is a form of normal-phase chromatography that is used for the analysis of thermally labile molecules. The same types of packed and capillary columns that are used in HPLC are utilized in SCF. Due to the nature of the SCF, packed columns can actually contain more

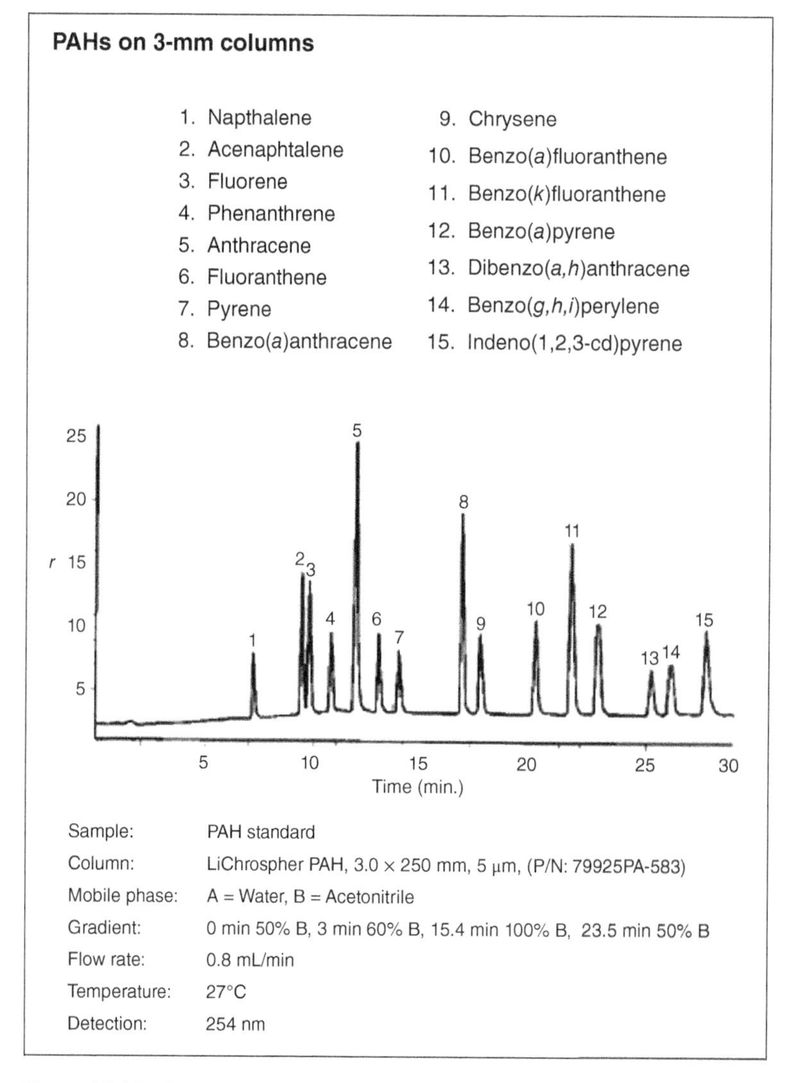

Figure 11.10 Separation of PAHs by HPLC (Source: With permission of Agilent Technologies, Inc., 2006).

theoretical plates than capillary columns. Also, the shape of the van Deemter curve is different from those observed in GC and HPLC in that a minimum plate height exists over a very broad range of linear velocities. SCF can be used for a variety of separations, but it is most commonly used in the separation of chiral compounds in the pharmaceutical industry. In place of a solvent gradient, the chromatographer uses pressure programming and the affinity of the stationary phases to separate complex mixtures of analytes. Pressure programming in SCF is analogous to temperature programming in GC and gradient programming in HPLC. Methods of detection include UV–vis, mass spectrometry, FID (as in GC but unlike in HPLC), and evaporative light scattering.

11.5 Applications/Case Studies

HPLC extends the capabilities of chromatographic separations past the abilities of capillary GC by allowing the analysis of thermally unstable analytes but at the expense of poorer resolving power

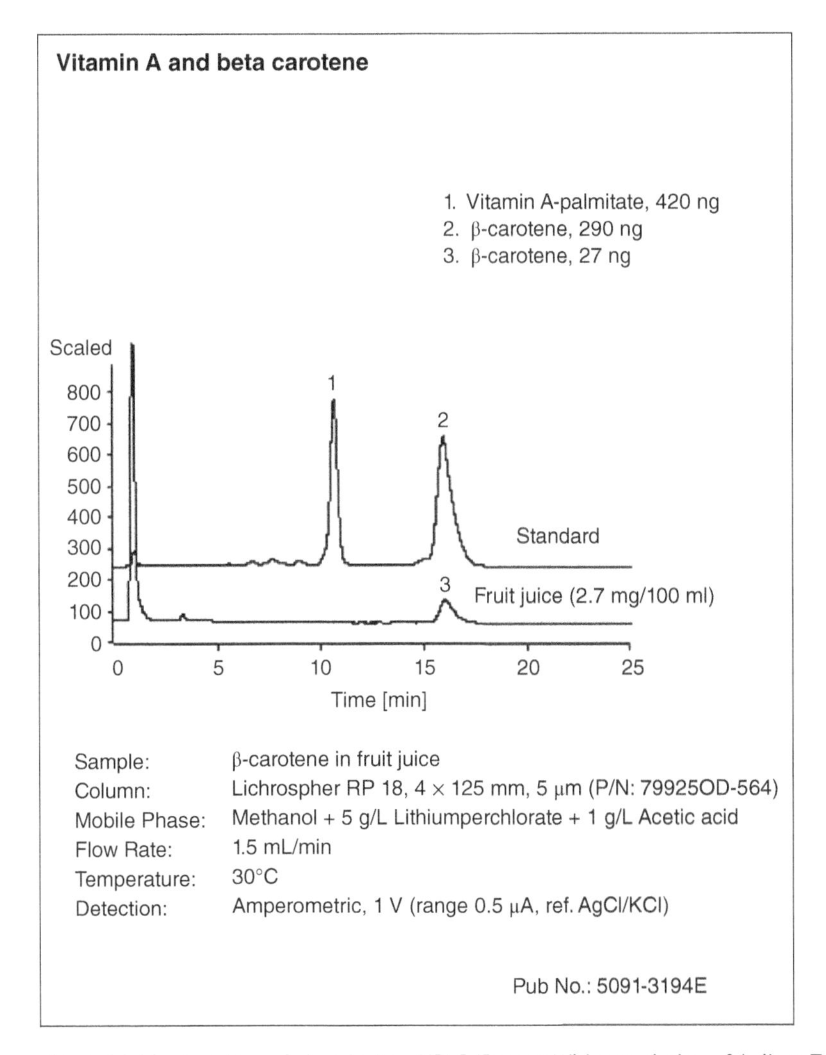

Figure 11.11 Isolation of vitamin A by HPLC (Source: With permission of Agilent Technologies, Inc., 2006).

due to the packed column nature of the system. These limitations are being slowly overcome with the use of capillary columns and ultrahigh-pressure systems. This section highlights a few HPLC applications. As with GC, major column manufacturers and distributors provide very useful websites that contain example chromatograms for common analytes. These sources can be extremely helpful in selecting the appropriate column for your needs. In addition, technical help is available from professional chromatographers at these companies with more complex separations. The chromatograms below were selected from numerous ones that are available from Agilent technologies at http://www.chem.agilent.com/en-us/Search/Library/Pages/ChromatogramSearch.aspx

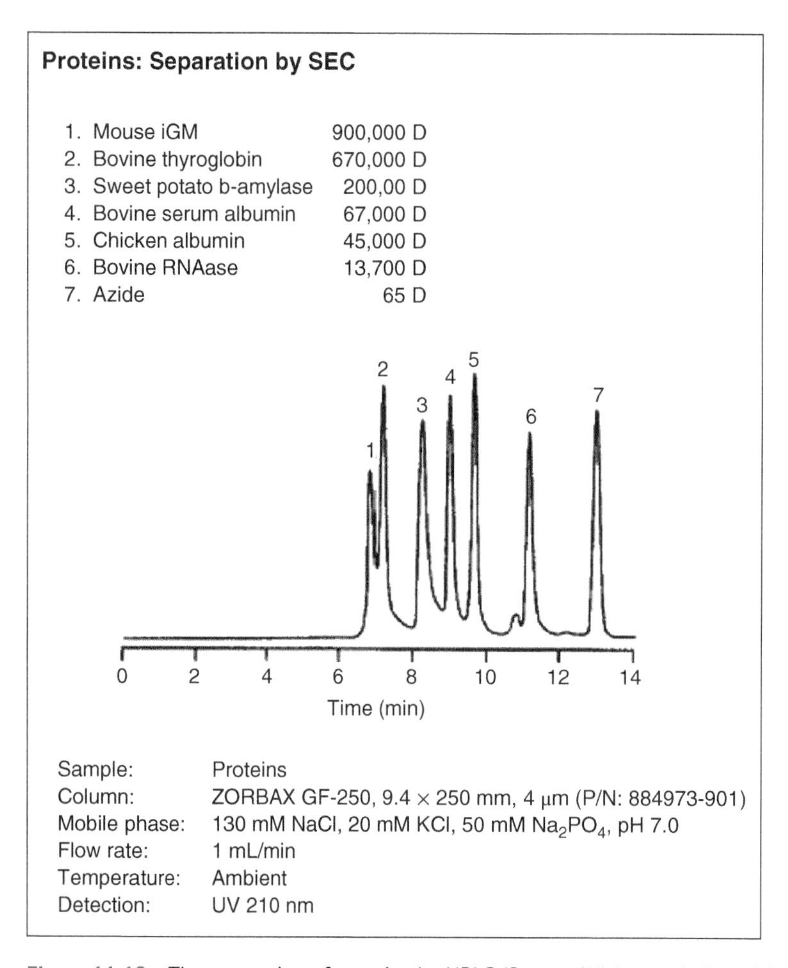

Figure 11.12 The separation of proteins by HPLC (Source: With permission of Agilent Technologies, Inc., 2006).

and Thermo Scientific Dionex Corporation at https://www.thermofisher.com/us/en/home/industrial/chromatography/ion-chromatography-ic.html.

The first example we will illustrate is for polyaromatic hydrocarbons that can be analyzed very efficiently by GC. While HPLC also offers adequate separation (as shown in Figure 11.10) as compared to GC, HPLC allows improved detection limits with the use of fluorescence detection.

Applications that cannot be performed by GC include the isolation of vitamin A (Figure 11.11), the separation of proteins (Figure 11.12), and the separation of food coloring agents (Figure 11.13).

A separation of anions in drinking water by IC is shown in Figure 11.14.

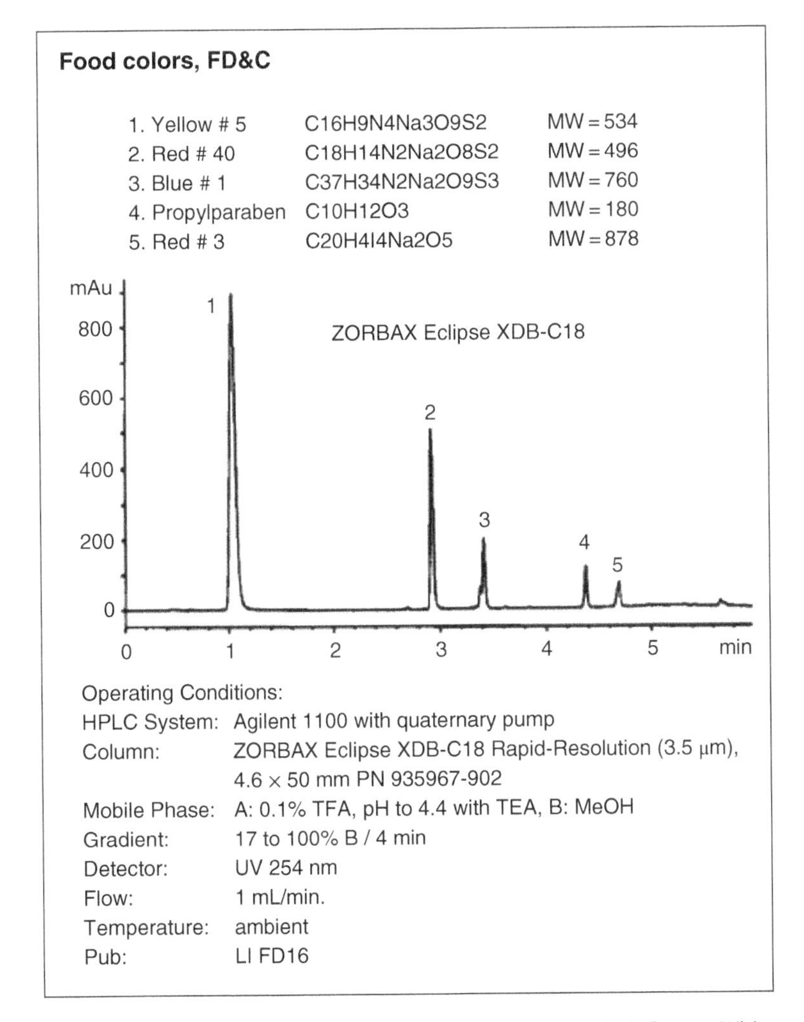

Figure 11.13 The separation of food coloring agents by HPLC (Source: With permission of Agilent Technologies, Inc., 2006).

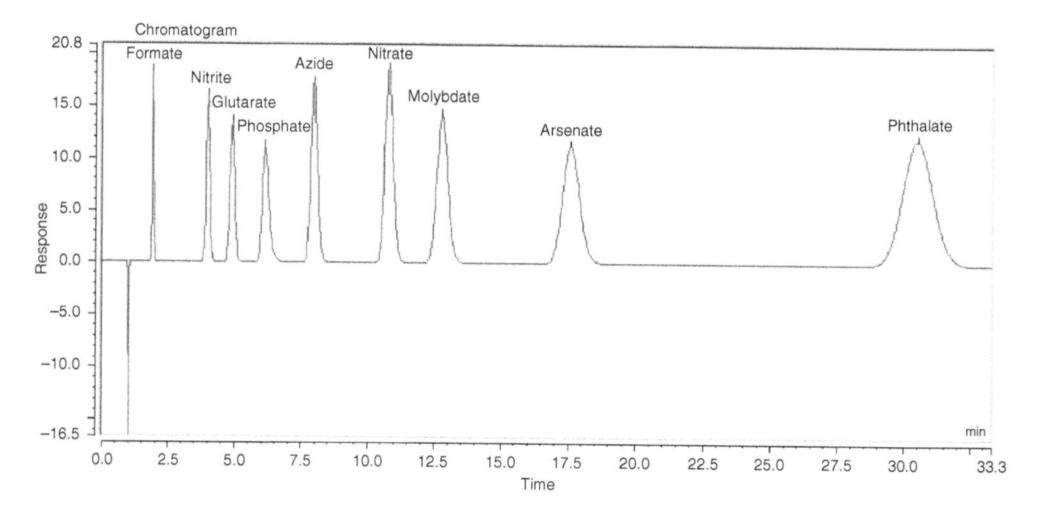

Figure 11.14 Separation of anions by IC (Source: Daviddamoore/Wikimedia commons/Public Domain).

11.6 Summary

This chapter described the basic applications and operation of an HPLC system. Modern HPLC systems significantly expand the capabilities of chromatography to include the relatively delicate biomolecules that are so abundant in nature. Current research in ultrahigh-pressure systems may result in HPLC performances that match those of capillary column GC in the near future. Basic nonconfirmatory detector systems were described in this chapter and confirmatory identification by mass spectrometer techniques that are applicable to all forms of chromatography will be covered in Chapter 13.

11.7 Questions

1. Contrast the advantages and disadvantages of thin-layer chromatography (TLC) versus modern HPLC.
2. What does HPLC stand for?
3. What are the advantages of dual reciprocating pumps over syringe pumps?
4. How much does a basic HPLC system cost?
5. What are the subcategories of liquid chromatography?
6. What is the difference between normal-phase HPLC and reverse-phase HPLC? Which is most commonly used today?
7. What chemical factors determine if a chemical will be analyzed in a GC or LC?
8. Can moderately volatile, thermally stable chemicals be analyzed on an LC?

9 Why do we filter analyte solutions before injection into an HPLC?

10 Draw a basic HPLC system and label all of the components.

11 Why are pressurized gases used in HPLC?

12 What two preparatory steps must be taken before a solvent can be used as an HPLC mobile phase?
In general, what is the maximum pressure limit of standard HPLC systems?

13 What is the purpose of the proportioning valve? How does this reduce the cost of an HPLC?

14 What is the difference between isocratic and gradient programming? Why is gradient programming sometimes necessary?

15 Why are dual-piston pumps preferred over single-piston pumps?

16 What is the purpose of a pulse damper?

17 Why are six-port valves used for injecting samples in HPLC?

18 Draw and explain how a six-port valve works.

19 Why are in-line filters used in HPLC systems?

20 What is the composition of the stationary phase and purpose of the guard column?

21 What are common stationary phases used in reverse-phase HPLC?

22 Why do chromatographers purchase their analytical columns instead of self-packing their own?

23 How will a poorly packed column affect performance?

24 What is the relationship between performance (resolution) and stationary phase particle size?

25 Compile a list of HPLC detectors and provide a list of chemicals each can be used to analyze.

26 Name three advanced types of LC.

27 Why is U-HPLC superior to standard HPLC?

28 How does IC differ from standard HPLC?

29 What is the purpose of the suppressor column in IC?

30 Draw a suppressor column for cation analysis in IC. Explain how it works. Write all requisite chemical reactions.

31 Although not shown in this textbook, attempt to draw a suppressor column for anion analysis in IC. Explain how it works. Write all requisite chemical reactions.

32 What is supercritical fluid?

33 What types of compounds are usually separated by SCF chromatography?

34 What type of gradient is used in SCF chromatography?

35 Review each of the chromatograms in Section 11.5 and determine what intermolecular force is involved in these attractions.

Supporting Information

Additional supporting information may be found online in the Supporting Information section in the HTML rendition of this article.

References

Agilent Technologies Internet site at http://www.home.agilent.com

Thermo Scientific Dionex Corporations Internet site at https://www.thermofisher.com/us/en/home/industrial/chromatography/ion-chromatography-ic.html

Ronald Majors. Effect of Particle Size on Column Efficiency in Liquid-Solid Chromatography Feb., Vol. 11, 1973 Figure 6, page 93.

Supelco-Aldrich Internet site at http://www.sigmaaldrich.com

Waters Corporation Internet site at http://www.waters.com

12

Capillary Electrophoresis by Nicole James

12.1 Introduction

The separation of compounds based on their movement when exposed to an electric field was first observed in 1807 by Ferdinand Friedrich Reuß, who noticed the movement of clay particles in water when a constant electric field was applied. The theory of electrophoresis was refined in the early 1900s by Marian Smoluschowski, and further refined in 1937 by Arne Tiselius, who won the 1948 Nobel Prize in Chemistry for his work.

Electrophoresis was initially conducted in polyacrylamide or agarose gels on a slab—a technique called *slab gel electrophoresis*—where charged molecules are applied to one end of the slab and an electric field is applied over the length of the slab. The molecules migrate down the slab at different rates according to the charge-to-size (m/z) ratio. Slab gel electrophoresis is still widely used in the fields of biology and biochemistry on large molecules such as nucleic acids and proteins to show relative concentrations of molecules, the purity of a sample, or, when used in conjunction with a standard, to identify compounds. Slab gel electrophoresis generally has long analysis times (often 20–40 minutes per sample), low efficiency, difficulties in analysis, and is unable to definitively identify compounds in a sample as migration time is not necessarily unique to each compound. Additionally, slab-gel electrophoresis is difficult to automate, making it very time-intensive to run multiple samples.

Capillary electrophoresis (CE) is becoming widely used as an alternative to slab gel electrophoresis. Gel media are not necessary in CE as capillary tubes are themselves anticonvective. Stellan Hjérten performed the first work in CE in 1967, using millimeter–diameter capillary tubes. By the early 1980s, the diameter of capillary tubes had been reduced to 75 μm. CE generally runs faster than slab electrophoresis, provides better precision and accuracy, uses fewer reagents, and is more easily automated. CE can also analyze smaller molecules than slab electrophoresis, thus expanding the range of possible analytes. By pairing CE with mass spectroscopy, it is possible to obtain confirmatory results. This makes CE an intensely useful and powerful instrument for many scientific disciplines. An overview of a CE system is shown in Figure 12.1.

12.2 Electrophoresis and Capillary Electrophoresis

The separation of compounds in CE depends on the velocity of individual compounds. The velocity is given by the electrophoretic mobility (μ_e) of the compound and the applied electric field (E)

$$\text{velocity} = \mu_e E \tag{12.1}$$

Essential Methods of Instrumental Analysis, First Edition. Frank M. Dunnivant and Jake W. Ginsbach.

Companion Website: www.wiley.com/go/essentialmethodsofinstrumentalanalysis1e

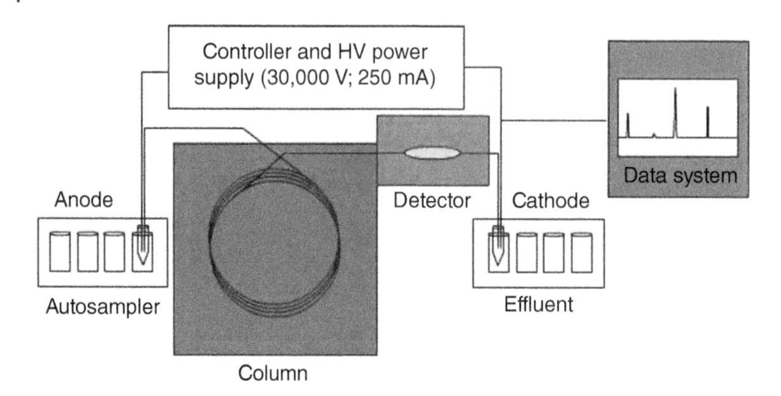

Figure 12.1 Overview of a modern CE system (Source: Nicole James (Author)).

Electrophoretic mobility is a measure of the particle's tendency to move through the medium given the applied electrical field and thus changes with the medium and the particle; tabulated values of electrophoretic mobility often differ from the experimentally observed electrophoretic mobility. Consequently, experimentally determined mobilities are called "effective mobilities," and can change radically with different solvents and with different solution pH. The electrophoretic mobility depends on the frictional drag (F_f) the medium exerts on the particle and the electrical force (E_f) exerted to move the particle such that

$$\mu_e = \frac{E_f}{F_f} \tag{12.2}$$

where the electrical force is dependent on the charge of the ion (q) and the strength of the electrical field (E)

$$E_f = qE \tag{12.3}$$

The frictional force can be described as

$$F_f = -6\pi\eta r v \tag{12.4}$$

where r is the ion radius, v is the ion velocity, and η is the solution viscosity.

In slab electrophoresis, the sample is placed on the gel, and then the electrical field is applied. The force of the electrical field causes the sample to progress down the slab, accelerating until the drag—or friction—experienced by intermolecular interactions with the medium equals the force caused by the electrical field, and then the sample proceeds at a constant rate. Once the sample has reached a steady velocity, it can be shown that

$$\mu_e = \frac{q}{6\pi\eta r} \tag{12.5}$$

This equation shows that highly charged, small particles move quickly—especially with a low viscosity fluid—whereas less-charged, larger particles move more slowly, especially through a more viscous medium.

12.2.1 Electro-osmotic Flow

Also called electroendosmotic flow, electro-osmotic flow (EOF) is the movement of liquid in a porous material (such as a capillary tube) caused by a difference in potential across the material.

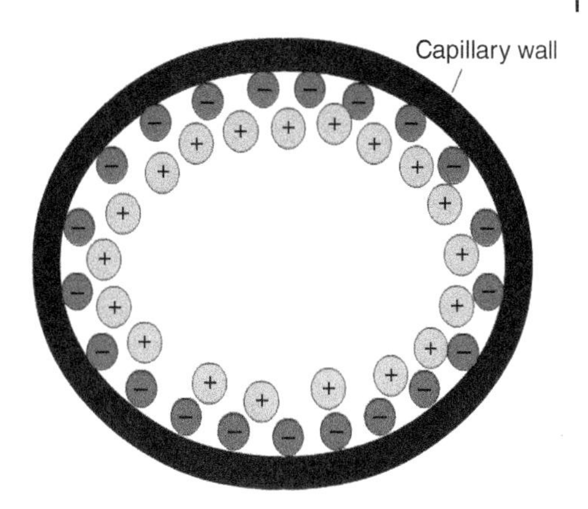

Figure 12.2 Capillary double-layer (Source: Nicole James (Author)).

When a charged particle is put in contact with a liquid in a capillary tube, a double-layer—or electrical double layer (EDL)—forms at the wall of the capillary (see Figure 12.2); this occurs at the interface of the glass capillary wall and the bulk solution. The first layer is surface charge and can be positive or negative depending on the material. As capillaries are generally borosilicate glass, the numerous silanol (SiOH) groups cause the charge of the first layer to be negative. This layer is also sometimes called the *Stern layer* or *Helmholtz layer*. The second layer is made up of ionic particles in solution that are electronically attracted to the charge of the capillary surface. As the particles in this layer are not fixed, but move as a result of electrical and thermal energy, it is called the diffuse layer.

The diffuse layer has a net charge, whereas the bulk, uncoordinated solution generally does not. When an electrical potential is placed across the capillary tube, the diffuse layer is pulled to one side. As the diffuse layer progresses to one side of the capillary tube, it drags the bulk solution along with it, creating a flow (specifically, the EOF) of the solution through the cathode. Because the species in the solution are charged, they separate in solution as discussed above; for most applications,[1] positively charged analytes move toward the anode given normal (negatively charged capillary) conditions.

The EOF can be described in terms of velocity or mobility. The mobility can be calculated from Equation 12.6, where ε (unitless) is the dielectric constant, a measure of a material's ability to respond to a dielectric's ability to store charge; ζ (V) is the zeta potential, discussed later; and η (Pas $= \frac{\text{kg}}{sm}$) is the solution viscosity.

$$\mu_{\text{EOF}} = \frac{\varepsilon\zeta}{\eta} \tag{12.6}$$

Since the mobility is the ability to respond to electric potential, the velocity is the mobility multiplied by the electric potential

$$V_{\text{EOF}} = E\frac{\varepsilon\zeta}{\eta} \tag{12.7}$$

The zeta potential (ζ) is the potential due to the double layer on the capillary wall and is directly correlated with the charge on the capillary. Because the charge on the capillary varies as a function of pH, the zeta potential also varies with pH, meaning the mobility and velocity of the EOF are highly pH-dependent. At high pH, the SiOH groups are deprotonated, meaning the net charge of

1 For small monoatomic ions, the solute mobility can sometimes be greater than EOF; for these conditions, anions and cations will move in opposite directions.

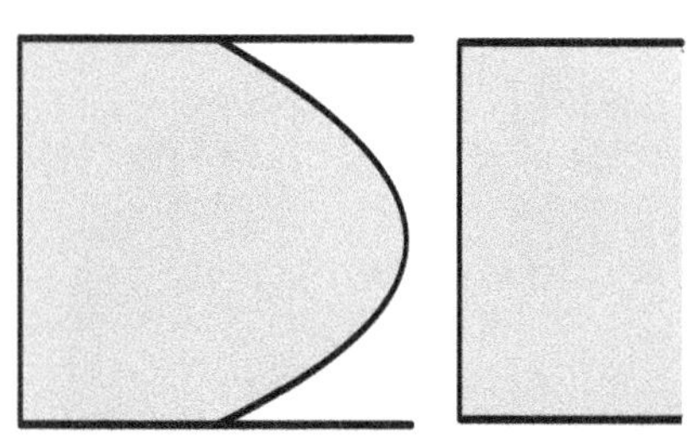

Figure 12.3 Pump-induced flow versus electro-osmotic-induced flow (Source: Nicole James (Author)).

Table 12.1 Variables that control electro-osmotic flow.

Variable	Effect on EOF	Possible Problems
pH	Directly proportional	Can alter structure and charge of analyte
Ionic strength	Inversely proportional	Sample heating and adsorption
Temperature	Inversely proportional	Changes in viscosity
Electric field	Directly proportional	Can cause heating
Organic modifiers	Usually inversely proportional	Can alter selectivity of instrument
Hydrophilic polymer	Usually inversely proportional	Dependent on polymer
Covalent coating	Dependent on coating	Coating often unstable

the capillary wall is greater. Thus, the EOF is significantly greater at high pH than at low pH; in fact, the EOF can sometimes vary more than an order of magnitude between a pH of 2 and a pH of 12.

A marked difference between EOF and flow induced by a pump is the cross-section of the flow. Because a pump applies a force across the entire cross-sectional area of the liquid in a capillary or tube, the friction between the wall and the liquid will cause the liquid at the center of the cross-section to move faster than the liquid closer to the wall. Thus, the liquid velocity gradient flows parabolically down through the capillary. However, if a liquid is moved by EOF, the bulk flow is caused primarily by the acceleration of the cations near the capillary wall, and the force on the liquid is uniformly distributed. This results in a flat flow profile. As seen in (Figure 12.3), the flat flow profile drops off sharply at the capillary wall; this is caused by friction against the direction of flow, but to a much lower extent than observed in laminar flow caused by pumps. An even (or flat) flow results in compact peaks whereas a parabolic laminar flow results in a broad peak because the analyte is spread over a larger area. "Tighter" peaks are easier to integrate and facilitate the isolation of peaks that elute at similar times. A summary of factors affecting EOF are summarized in Table 12.1.

12.2.2 Peak Resolution and Solute Zones

Peak resolution in CE is dependent on many variables, including longitudinal diffusion, injection length, sample adsorption to capillary walls, electrodispersion, Joule heating, and detector cell size. The loss of peak resolution is often referred to as "zone broadening" in CE. As in other forms of chromatography, this results in reduced separation of compounds and increased difficulty in accurate peak integration.

12.2.2.1 Longitudinal Diffusion

Under normal conditions, diffusion along the capillary tube (longitudinal diffusion) is one of the more substantial contributors to zone broadening. A solute's propensity to diffuse throughout the solvent depends on the solute and the solvent. A longitudinal diffusion coefficient can be provided for a given system, with a lower coefficient meaning the solute will form narrower zones.

The equation for the number of theoretical plates in a capillary tube is given in Equation 12.8, where N is the number of theoretical plates, μ_a is the solute mobility, E is the electric field, I is the effective capillary length, and D is the diffusion coefficient of the solute.

$$N = \frac{\mu_a EI}{2D} \tag{12.8}$$

This equation only accounts for longitudinal diffusion, although there are often other variables that affect zone broadening.

12.2.2.2 Injection Length

When introducing a sample to a CE system, the volume of liquid that enters the capillary is called the sample plug (see Figure 12.4). If the sample plug is very spread out, analytes will not separate into tight peaks and tailing or broadening can occur. A practical injection should be between 1% and 2% of capillary length.

12.2.2.3 Sample Adsorption

Interactions of the analyte with the capillary wall negatively affect zone broadening. In some cases, complete adsorption of the analyte to the capillary wall can occur. As most capillary tubing is made of borosilicate glass, this is most substantial with positively charged analytes that can interact strongly with the hydroxide groups on the surface of the glass. Prolonged sample–surface interactions can result in peak tailing and zone broadening.

12.2.2.4 Electrodispersion

The capillary in a CE system contains an ambient level of ions, called the background electrolyte (BGE). For ideal separation, the BGE should be uniform in and between all solute zones. Therefore, the background solvent is generally always a buffer.

12.2.2.5 Joule Heating

Joule heat is generated by passing electric current through a conductor and is directly proportional to the amount of power used. Joule heating can result in nonuniform temperature gradients

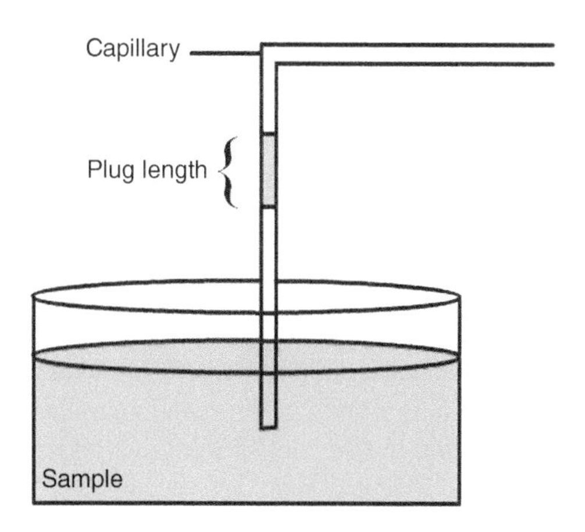

Figure 12.4 Diagram of sample introduction and plug length (Source: Nicole James (Author)).

throughout the capillary tube. Temperature increases are often around 10 °C, although increases of up to 70 °C or higher can occur. While uniform increases in temperature generally do not affect zone broadening, uneven increases in temperature leading to temperature gradients can cause substantial peak broadening and/or tailing. Heat can relatively easily dissipate through capillary walls, which can result in a large difference in temperature between the center of the capillary and the internal edges of the capillary. These gradients can result in differences in solvent viscosity, and thus air or liquid cooling of the capillaries is often necessary. Using thicker capillary tubes with a large outer radius and a smaller inner radius can also help combat this problem. Smaller sample volumes and higher surface-to-volume ratios limit the Joule heating, even when hundreds of volts per centimeter are applied.

12.2.2.6 Detector Cell Size

Detectors on CE systems average a signal over a finite volume. Naturally, if the volume averaged is too large, it can combine signals from different analytes. In this way, zone broadening due to the size of the detector cell is similar to zone broadening due to a long injection plug. Generally, any variance due to detector cell size can be ignored if the length of the detection cell is roughly one-third the sample zone length.

12.3 Samples

12.3.1 Sample Introduction

For loading a sample, a stable DC power supply of about 30 kV (200–300 mA) is required. Instruments are made with the ability to switch polarity to facilitate a variety of methods of operation. Usually, current is monitored and maintained to compensate for some temperature-induced changes in buffer viscosity.

Sample volumes are generally incredibly small. Volumes as low as 5 μL can be sufficient for several injections. This can be disadvantageous if the sample is very dilute because being limited in the loading volume of a dilute sample can reduce sensitivity. Loading too much of a sample can result in bad peak shapes due to field inhomogeneities if the conductivity of the sample is different from that of the buffer. Additionally, if the injection length is greater than the diffusion-controlled zone, peak broadening will occur.

12.3.1.1 Types of Sample Injection

There are several methods for introducing samples into CE systems. It is important to note that some sample is always injected when inserting a capillary into the sample reservoir, due to capillary action. This volume is called the zero injection, and the effect of this initial injection on sample migration and analysis is called the zero-injection effect. In most cases, this phenomenon is insignificant, but it can be important in systems being employed for quantitative analysis.

12.3.1.1.1 Hydrodynamic Injection Hydrodynamic injection introduces a sample by applying pressure on the sample reservoir (Figure 12.5). This physically forces solution through the capillary. The volume of sample loaded is independent of the sample matrix; instead, the volume depends on the details of the system, such as capillary dimensions, the viscosity of the buffer, the pressure applied, and the amount of time that pressure was applied. Pressures and times generally range from 25 to 100 mbar and 0.5–5 seconds.

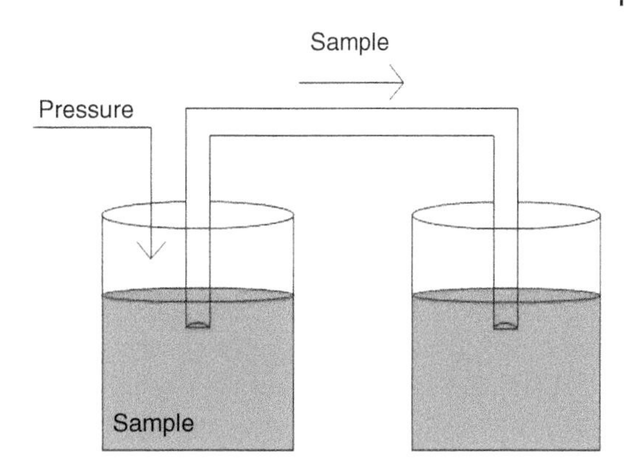

Figure 12.5 Hydrodynamic injection setup (Source: Nicole James (Author)).

Similarly, hydrodynamic injection can be performed by siphoning, or raising the input reservoir 5–10 cm above the output reservoir while keeping the levels of each equal. This technique is generally employed when the equipment necessary for pressure injection is unavailable.

12.3.1.1.2 Electrokinetic Injection In electrokinetic injection, a voltage 3–5 times lower than that required for separation is applied between the input and output reservoirs for approximately 10–30 seconds (Figure 12.6). Sample is loaded according to the difference in voltage across the capillary and the amount of each individual analyte that is loaded is dependent on the analyte's electrophoretic mobility. As such, electrokinetic injection cannot be used quantitatively. This technique is useful in situations where the media in capillary is very viscous making hydrodynamic injection difficult. Additionally, it requires no extra instrumentation.

12.3.1.1.3 Field-amplified Sample Stacking and Field-amplified Sample Injection Sample zones are sharpened when the conductivity of the sample matrix is substantially lower than that of the buffer. Essentially, this is the reverse of electrodispersion, described previously. Thus, samples with low salt content can be actively sharpened and/or large quantities of them can be loaded onto a capillary before zone-broadening effects begin to occur. When this phenomenon is utilized to allow larger sample lengths to be loaded hydrodynamically, the concentration is called field-amplified sample stacking (FASS). When the injection is performed electrokinetically, it is called field-amplified sample injection (FASI).

In the case of FASI, if the conductivity of the sample varies or is unstable, results can be unreliable. In such a case, FASS can be used instead and, while peak resolution may be affected by variability in sample conductivity, the peak areas are still reliable.

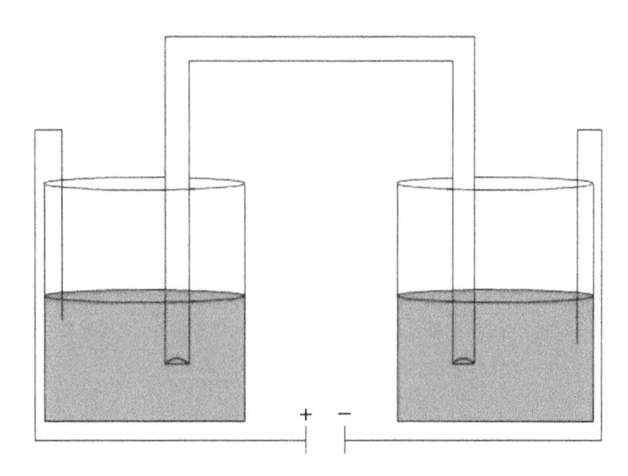

Figure 12.6 Electrokinetic injection setup (Source: Nicole James (Author)).

Often 100 times more samples can be loaded by FASS before separation is greatly affected. The actual amount of sample that can be loaded by FASS is regulated by the sample's conductivity and factors such as Joule heating. As the conductivity of the sample must necessarily be much lower than that of the running buffer, the applied electric field can induce a large degree of heating within the sample plug, which can change the viscosity of the sample and result in degassing and/or boiling.

Additionally, due to the difference in ionic strength between the sample plug and the running buffer, the electro-osmotic mobility will vary between the sample plug and the buffer. This can result in a parabolic flow profile between the plug and the running buffer that may cause zone broadening. Similarly, a substantially different pH between the sample plug and the buffer can also result in differences in analyte mobility. So, a buffer with a similar pH to the sample solution should be chosen.

12.3.2 Sample Purification

CE is attractive as a preparatory purification source because of its separatory strength. However, CE can only be used to purify small amounts of analyte, as the sample-loading limit is low. However, small-volume fractions can be collected for sequencing DNA fragments, establishing biological activity of substances, and ascertaining the purity of a collected peak by reinjection.

Autosamplers can be programmed to run a sequence of samples so that, in the event that multiple runs must be made and fractions collected and combined, the instrument does not need to be attended for the entire period.

12.4 Methods of Operation

12.4.1 Capillary Zone Electrophoresis

Capillary zone electrophoresis (CZE) is the standard method of CE. Separations are performed by a difference in migration time as previously described.

12.4.2 Micellar Electrokinetic Chromatography

Micellar electrokinetic chromatography is a combination of electrophoresis and chromatography. If one dissolves surfactants (for example, 8–9 mM SDS) in the running buffer, micelles—often anionic—form and travel against the EOF. The EOF, however, is still faster and the net motion of the micelles is in the same direction as the EOF. Solutes partition between the micelles and the buffer through hydrophobic and electrostatic interactions. In this way, the micelles act as a stationary phase which is not necessarily stationary, but slower with respect to the EOF. The more hydrophobic compounds migrate slowly through the capillary due to their interactions with the micelles. All neutral compounds will elute between the EOF, which elutes first, and the micelles, which elute last (see Figures 12.7 and 12.8).

The surfactants can be anionic, cationic, nonionic, zwitterionic, mixtures of these, salts, or microemulsions. Deciding which surfactant to use is based on the specific interactions necessary to separate target compounds.

Surfactants may affect the EOF and solute-wall interactions. Generally, using a high pH buffer will maintain the EOF and diminish changes in solute–wall interactions. Additionally, a consistent

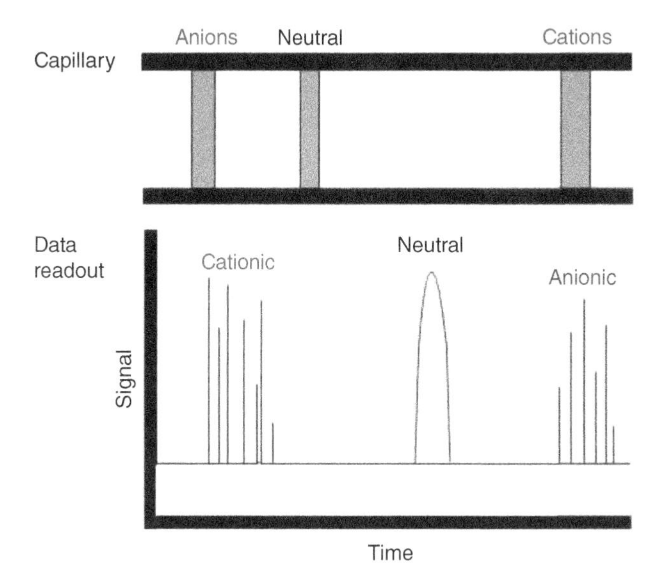

Figure 12.7 Diagram of a CZE system (Source: Nicole James (Author)).

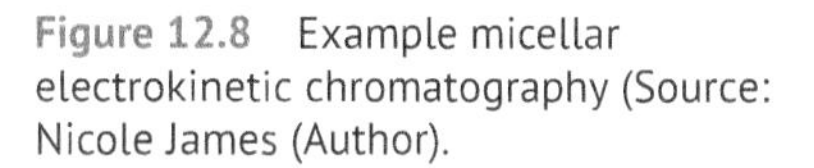
Figure 12.8 Example micellar electrokinetic chromatography (Source: Nicole James (Author).

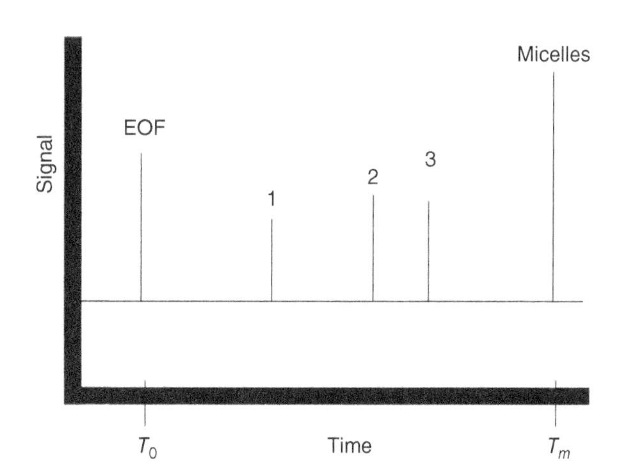

temperature is especially important as changes in temperature will affect the concentrations of surfactant required to form micelles as well as the partition coefficient(s) of the system.

12.4.3 Capillary Electrochromatography

Capillary electrochromatography (CEC) is miniaturized liquid chromatography. An applied electric field pumps liquid through a packed capillary column. The silica capillaries are packed with silica-based reversed-phase particles of 1–5 μm diameter. The migration times of different solutes vary as a function of their partitioning between the mobile and stationary phases. If the packed phase is charged, there is also an electrophoretic component to the partitioning. On average, there are 100 000 theoretical plates per peak in a CEC system.

As the only difference in a CEC setup is the capillary tubing used, CZE systems can be easily set up for CEC analysis.

12.4.4 Capillary Gel Electrophoresis

Capillary gel electrophoresis (CGE) is essentially gel electrophoresis conducted in a capillary tube (Figure 12.9). The "gel" is not necessarily stiff and solid as it is in slab gel electrophoresis. To avoid confusion, it is usually referred to as a polymer matrix. Many polymer matrices are available and offer different separating properties.

The polymer is dissolved in buffer and loaded into the capillary tube. Generally, the polymer concentration necessary for adequate separation is inversely proportional to the size of the analyte. As low-viscosity polymer solutions do not experience the same capillary action as most regular buffer solutions, the capillary gel is often pumped with pressure and the capillary wall is coated to eliminate EOF.

There are generally 10 million theoretical plates per meter of capillary tubing. CGE is most commonly used to separate single-stranded oligonucleotides, proteins, and DNA fragments.

As in CZE, it is possible to add agents such as chiral selectors and ion-pairing agents by covalently bonding them to the gel or the running buffer. This can greatly increase the separation possibilities of a single system.

12.4.5 Capillary Isoelectric Focusing

Capillary isoelectric focusing (CIEF) separates peptides and proteins based on their isoelectric point (pI)—the point at which they carry no net charge. A pH gradient is formed within the capillary using ampholytes—zwitterionic molecules with both an acidic and basic moiety.

The capillary is filled with a mixture of solutes and ampholytes, and a pH gradient is formed with a basic solution at the cathode and an acidic solution at the anode. When an electric field is applied, ampholytes and solutes migrate through the pH gradient until they reach the pH that corresponds to their isoelectric point, where they become uncharged. The solute zones will maintain a narrow breadth as any solute which migrates into the zone of a different pH becomes charged and migrates back into the pH of its isoelectric point. After this equilibrium occurs, a steady state is reached and hydraulic pressure is used to pass the zones through to the detector.

EOF must be minimized to prevent flushing ampholytes through the capillary.

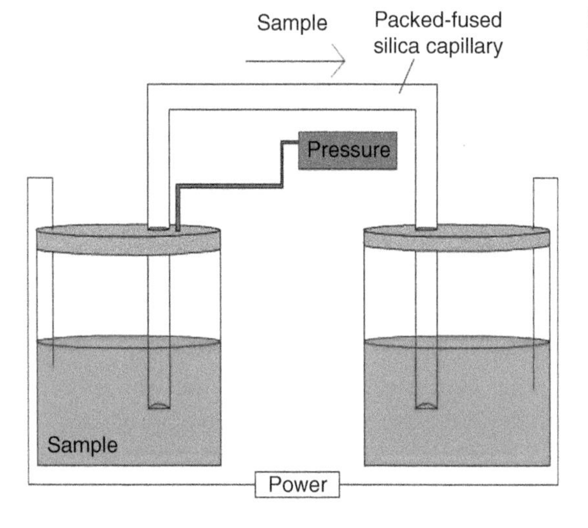

Figure 12.9 Capillary electrochromatography (CEC) (Source: Nicole James (Author)).

12.4.6 Capillary Isotachophoresis

Capillary isotachophoresis (CITP) is a moving boundary technique using two buffer systems. The buffer systems trap the analyte zones and all zones lay in between "leading" and "terminating" electrolytes, where the leading electrolytes are those in the buffer with a higher EOF and the terminating electrolytes are those in the buffer with a lower EOF.

This bracketing affects the electric field experienced by each zone, creating different electric fields for each analyte. The electric field varies such that each analyte moves at the same velocity as that established by the leading electrolyte. Because velocity is defined as the product of the mobility and the electric field (velocity = mobility•field), this varying field acts to complement the mobility such that all analytes move at the same velocity. Zones are formed and focused much in the same way as they are in CIEF; as soon as an analyte drifts away from its zone, it experiences a new electric field such that it either accelerates or decelerates until it migrates back to its zone and the steady state is re-established.

According to the Kohlrausch Regulating Function, the concentration of analyte in a zone is determined by the concentration of the leading electrolyte. Analytes with a higher concentration than the leading electrolyte experience broadened until the concentrations match, whereas analytes with a lower concentration than the leading electrolyte sharpen. In this way, CITP can be used as a preconcentration step prior to other methods of analysis. Generally, up to 30–50% of the capillary can be filled with samples while still maintaining good separation and concentration.

Also due to the Kohlrausch Regulating function, increasing the concentration of the leading and/or trailing electrolytes can result in better separation and can improve detection of analytes that can be difficult to separate. However, when applying CITP, it can be difficult to identify a buffer system with leading and trailing electrolytes that have the appropriate mobilities while still maintaining the appropriate pH for the sample analysis. Additionally, only cations *or* anions can be analyzed by a given buffer system.

12.5 Detectors

Detection in CE can be a significant challenge as the small injection volumes require the samples to be relatively concentrated for reasonable detection. Many of these techniques are similar to those employed in liquid chromatography.

12.5.1 UV–Vis Spectrophotometry

UV–vis absorption is the most common detection technique for CE. Detection is nearly universal, and with fused silica as the sample holder, wavelengths from 200 nm through the visible spectrum can be used without interference. Generally, the capillary itself is used as the sample holder and passes through a UV–vis detector.

Because the capillary tube is so thin compared to conventional path lengths, care needs to be taken in the design of the detector. The beam of light must be of a small radius and focused directly on the capillary for accurate results. A disadvantage of this is that, correspondingly, few photons reach the detector and a generally high baseline is obtained. The optical beam is generally passed through a slit to restrict its width and center it on the capillary tube.

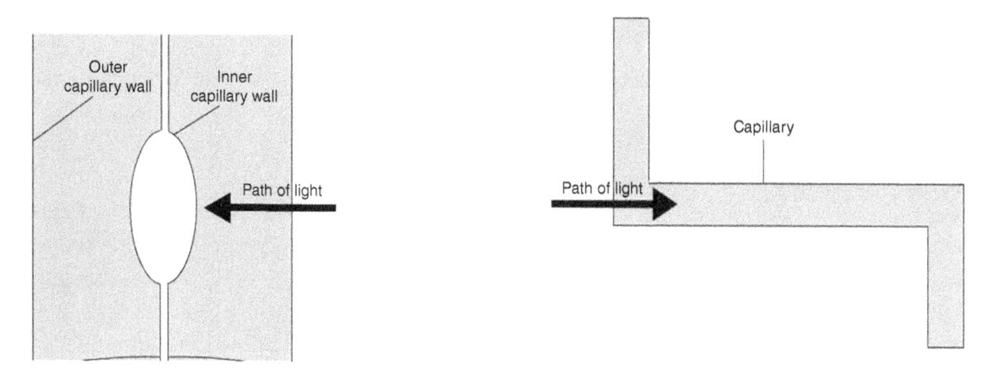

Figure 12.10 (a, left) Capillary bubble cell for UV–vis detection and (b, right) capillary cell (Source: Nicole James (Author)).

One approach to decrease the detection limit is by increasing the path length of the flow cell. This can be done by increasing the diameter of the capillary, but by doing this, there is also an increase in the current and resulting Joule heating. Two ways of increasing path length without increasing capillary diameter are depicted in Figure 12.10. The method shown on left side of Figure 12.10, called the "bubble cell," does not require an increase in current, as the detection occurs after separation. When a zone enters the bubble, the velocity decreases as it expands to fill the capillary, it contracts along the capillary, and overall concentration remains the same, even with increasing path length. There is a third method, which is a combination of Figure 12.10 (left-hand figure) and Figure 12.10 (right-hand figure); the capillary bends to increase the path length, but the capillary is also bubbled along the path.

When trying to detect a given analyte, care must be taken to ensure the running buffer is not optically active in the range of the analyte. Most peptides and carbohydrates absorb around 200 nm, and, consequently, many biological buffers, such as HEPES, CAPs, and tris, should not be used in systems analyzing peptides and carbohydrates.

A normal signal to noise ratio for a 75 mm ID capillary is 62.5, and up to 650 on an increased path length cell.

12.5.2 Indirect Photometry

Indirect photometry utilizes a monitoring ion that is UV-active to trace the presence of a non-UV-active compound. An ion of the running buffer is chosen, generally to have a high UV absorbance intensity, and its absorbance is monitored. When the sample zone passes through the detector, it will be accompanied by a corresponding drop in the UV absorption due to the monitoring ion.

A very simplified explanation for this phenomenon is that the monitoring ion is displaced by an analyte with the same charge based on electroneutrality. However, the displacement ratio can be greater than one and analytes with charges opposite to the monitoring ion can be detected. This can be better explained by a closer analysis of the Kohlrausch Regulating Function.

12.5.3 Laser-induced Fluorescence

Laser-induced fluorescence (LIF) adapts fluorescence spectrometry to CE in the same manner as UV–vis spectrometry. A laser is used to excite the analyte, and fluorescence is measured via a PMT. Often the laser is aligned noncollinearly, and a ball lens is used to focus the light on the center of the capillary. Generally, a reflective ellipsoid is glued to the capillary to direct fluorescence toward the PMT.

A drawback of LIF is that most compounds of interest in CE are not fluorescent. However, a number of derivatizing agents can be used to act as markers for the compound. These are generally organic aromatics and require excitation between 250 and 500 nm.

12.5.4 Contactless Conductivity

Contactless conductivity detectors (CCDs) place an actuator electrode and receiver electrode along the capillary tube, as depicted in Figure 12.11. An oscillating frequency is applied at the actuator electrode and detected at the receiver electrode with the liquid inside acting as a resistor. As a sample zone passes, the conductivity changes. Since analytes are charged, the conductivity generally increases—except for when samples have lower ionic strength than the running buffer.

CCDs are universal and work well for inorganic analytes that are difficult to measure by other methods. CCDs function well with low sample volume, which is advantageous for use in CE. Additionally, it generally has a detection limit in the ppm–ppb range for small organic ions.

12.5.5 Mass Spectrometry

Mass spectrometry is a particularly attractive detecting strategy as it is confirmatory and can give valuable structural information about analytes. However, it is difficult to transition from a liquid-phase separatory technique to a gaseous, vacuum-based detection technique. To an extent, this problem is alleviated by the small amount of solvent and the low flow rate associated with CE.

12.5.5.1 Electrospray Ionization

Electrospray ionization (ESI) is an ideal technique for interfacing CE and MS, as it accepts small volumes at a low flow rate and results in gaseous, ionized particles. Additionally, ESI is well-suited to larger, biological molecules, which are frequently analyzed by CE. However, CE and ESI are both dependent on electric fields. In order to couple them in the simplest of cases, the end electrode of the CE instrumentation is shared with the ESI electrode. However, in some situations, the difference in the magnetic field can be several orders of magnitude and the electrodes may be of the opposite sign.

Some situations can be remedied by grounding the end electrolyte of the CE system, and providing voltage to the electrospray system from the MS. If the potential applied to the ESI is negative, then positive ion will enter the MS and this is called positive ion mode. If the CE voltage is much greater than the ESI voltage, a ground path for the current is provided by a resistor sink. In general, when the ESI voltage is much different from the CE voltage, the variability in the electric field can affect the sample separation in the CE system. The coupling of CE and MS will be discussed in Chapter 13.

12.5.5.1.1 Hydraulic Interfacing A hydraulic flow rate is necessary to form the electrospray. The amount of hydraulic flow necessary is dependent on many variables, including the EOF and

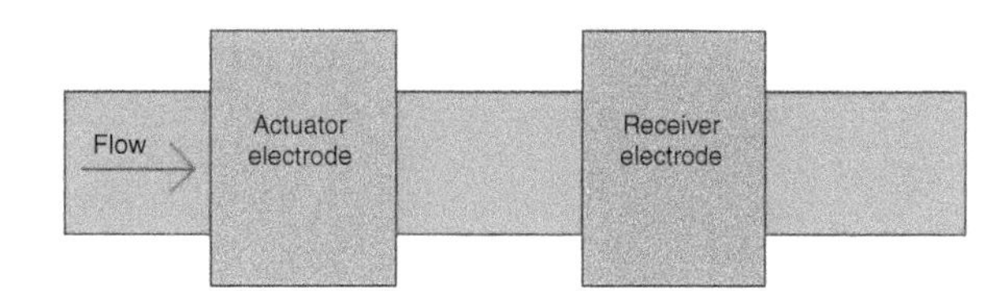

Figure 12.11 CCD setup (Source: Nicole James (Author)).

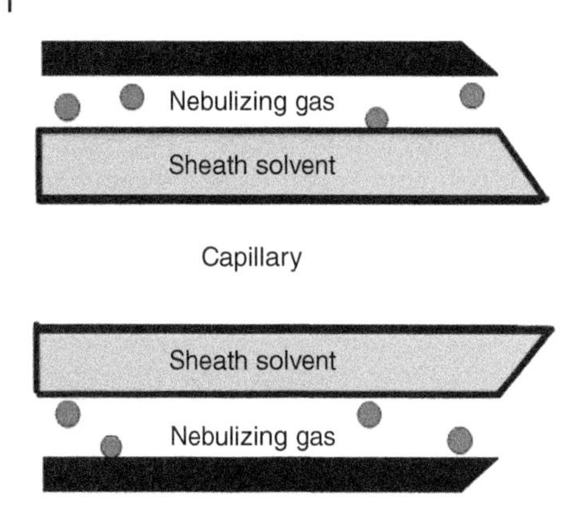

Figure 12.12 Sheath flow interface (Source: Nicole James (Author)).

capillary dimensions. In general, there are two methods for introducing a hydraulic flow: sheath flow and sheathless flow.

Sheath Flow Interface Smith, et al. invented the triaxial capillary sprayer, similar to the one shown in Figure 12.12.

Solvent flows between the capillary and the sprayer tip and combines with the capillary solution immediately before the nebulizing gas. In this way, a hydraulic flow independent of the EOF is provided by the sheath solvent.

The major disadvantage of the sheath flow interface is the fact that the sample is necessarily diluted by the sheath solvent. Despite this, the sheath flow interface is one of the most reliable methods of interfacing CE and MS, although several alternative methods exist but are not "as robust," according to Agilent literature. (It should be noted that Agilent possesses the distribution rights on the triaxial sheath.)

Sheathless Flow Interface Several methods of uniting the CE capillary with the ESI system exist. However, in all of these systems, electrochemical processes such as oxygen and hydrogen formation can occur. The formation of gas bubbles that are not frequently vented can cause increased resistance. In the triaxial system, evolved gases are frequently flushed due to sheath flow and/or the nebulizer gas, but in sheathless systems, increased resistance can form a substantial problem.

In 2007, Moini published a technique that does not require sheath flow and does not create electrochemical processes at the electrodes. A thin, porous segment of glass acts as a liquid and electrical contact outside the capillary. The tube is either grounded or placed at the ESI voltage. This system has recently become available commercially through Beckman-Coulter.

12.5.5.2 Atmospheric Pressure Chemical Ionization and Atmospheric Pressure Photoionization

ESI is highly efficient for many systems. However, for systems where the running buffer is not especially volatile and/or the molecules are not strongly charged, alternative ionization techniques may be more efficient. Atmospheric pressure chemical ionization (APCI) and atmospheric pressure photoionization (APPI) can both be used for systems like these.

APCI relies on the evaporation of an analyte-containing aerosol, where a charged reagent gas collides with analyte molecules and confers the charge. In APPI, an analyte aerosol is again evaporated to a vapor, and then photons ionize the analyte. In situations where the analyte is not easily photoionized, a buffer gas may be photoionized and then transfer the charge to the analyte.

APCI and APPI both require substantial heating of the sample and therefore are not appropriate for analytes that easily degrade at higher temperatures. Typically, APCI and APPI systems are heated to between 200 and 300 °C. Additionally, low molecular weight (>1 kDa) analytes are better suited to APPI and APCI as they are more volatile.

12.6 Application/Case Studies

Figures 12.13–12.16 show applications of CE for a variety of analytes.

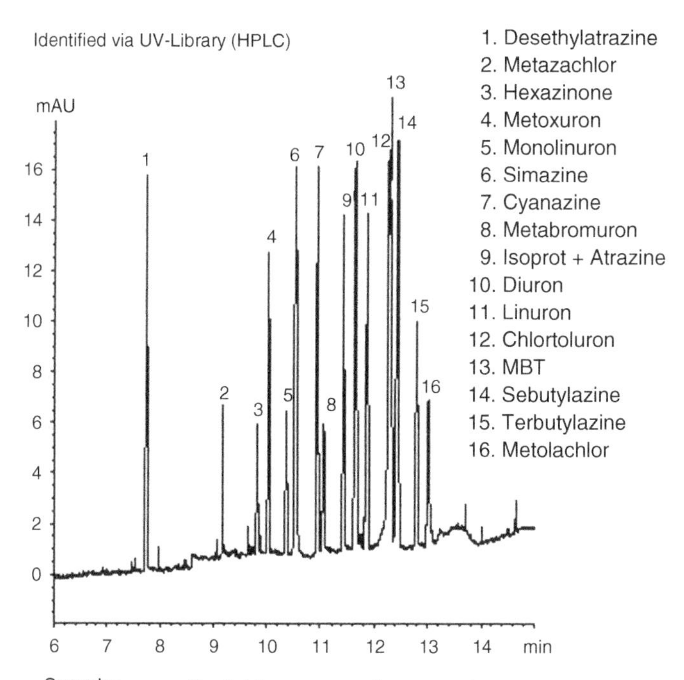

Figure 12.13 Analysis of pesticides by CE (Source: With permission of Agilent Technologies, Inc., 2006).

Separation of Cu(II), Ni(II), Cr(III) and Fe(III) complexes with EDTA

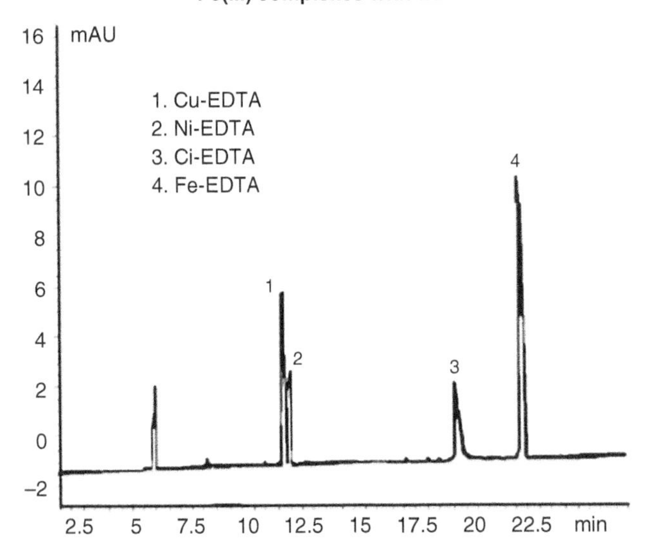

Sample:	10 mg/l each
Preconditioning:	4 min at run buffer
Capillary:	CEP coated, l/L 72/80.5 cm, 75 μm id, (PN: G1600–62318)
Buffer:	50 mM Sodium acetate, pH 5.5
Injection:	150 mbar•sec
Voltage:	–25 kV, (reversed polarity)
Temperature:	20 °C
Detection:	UV 230/8 nm, ref. = off

Figure 12.14 Analysis of metals and chelates by CE (Source: With permission of Agilent Technologies, Inc., 2006).

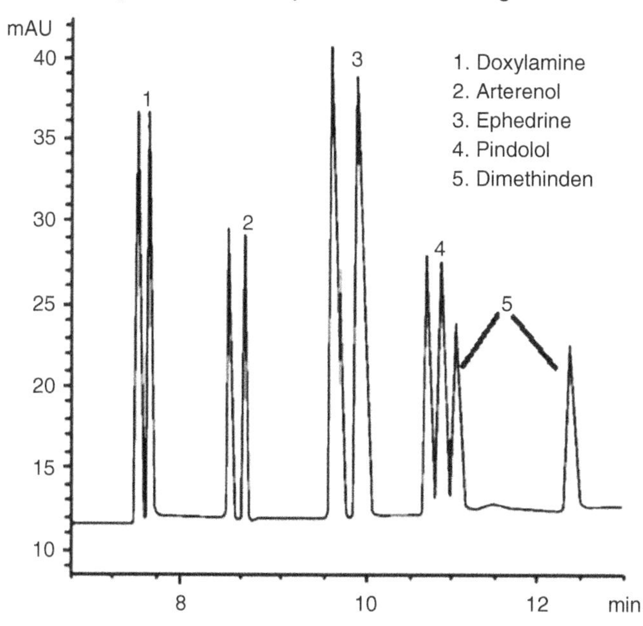

Capillary: Bare fused silica, I/L 56/64.5 cm, 75 μm id, (PN: G1600–61311)
Buffer: 20 mM citrate, pH 2.5, 2% Carboxymethyl-β-CD
Injection: 200 mbar•sec
Electric field: 300 V/cm
Temperature: 20 °C
Detection: UV 214/20 nm, ref. 450/80 nm

Figure 12.15 Analysis of chiral drugs by CE (Source: With permission of Agilent Technologies, Inc., 2006).

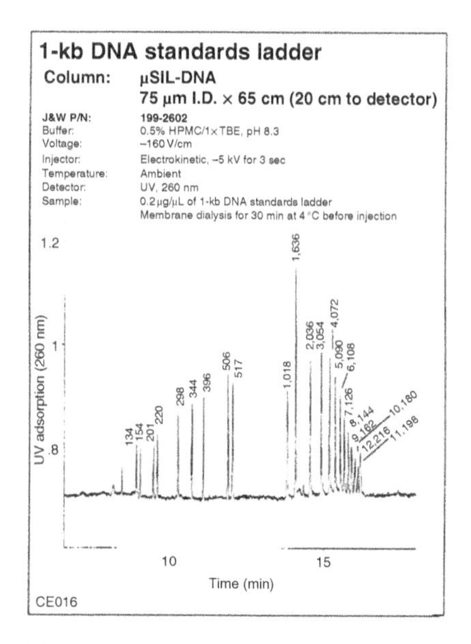

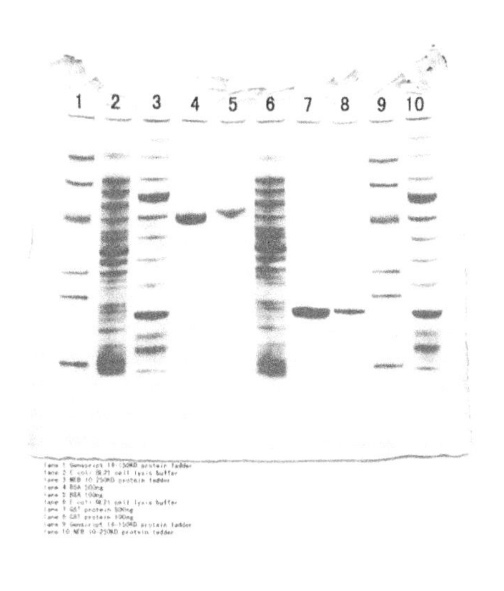

Figure 12.16 Analysis of DNA by CE and slab gel electrophoresis. Left CE (Source: With permission of Agilent Technologies, Inc., 2006).

12.7 Summary

CE is an analytical technical under constant development and improvement. While CE's excellent analyte resolution with UV–vis detection offers many advantages, its coupling with various forms of mass spectrometry, discussed in Chapter 13, offers numerous advantages in chemical, environmental, and biological applications. These applications are far superior to slab gel electrophoresis. One of the notable achievements of CE is the unique confirmatory identification of a single individual's DNA in forensic science.

References

Dolnik, V. (1999). DNA sequencing by capillary electrophoresis (review). *Journal of Biochemistry and Biophysics* 41: 103–119.

Lauer, H.H. and Rozing, C.P. (2010). *High Performance Capillary Electrophoresis*. Germany: Agilent Technologies.

Petersen, J., Okorodudu, A., Mohammad, A., and Payne, D. (2003). Capillary electrophoresis and its application in the clinical laboratory. *Clinica Chimica Acta* 330: 1–30.

Tagliaro, F., Turrina, S., and Smith, F. (1996). Capillary electrophoresis: principles and applications in illicit drug analysis. *Forensic Science International* 77: 211–229.

Tagliaro, F., Manetto, G., Crivellente, F., and Smith, F. (1998). A brief introduction to capillary electrophoresis. *Forensic Science International* 92: 75–88.

13

Mass Spectrometry

13.1 Introduction and History

The earliest forms of mass spectrometry go back to the observation of canal rays by Goldstein in 1886 and again by Wien in 1899. Thompson's later discovery of the electron also used one of the simplest mass spectrometers (MSs) to bend the path of the cathode rays (electrons) and determine their charge to mass ratio. Later, in 1928, the first isotopic measurements were made by Aston. These basic experiments and instruments were presented to most readers in first-year general chemistry. More modern aspects of mass spectrometry are attributed to Arthur Jeffrey Dempster and F.W. Aston in 1918 and 1919. Since this time, there has been a flurry of activity [not only concerning minor advances in components of MSs such as different types of instrument interfaces (direct injection; gas chromatography, GC; and high-performance liquid chromatography, HPLC)] to different ionization sources (electron and chemical ionization) but also new types of ion separators. For example, double-focusing magnetic sector mass filters were developed by Mattauch and Herzog in 1934 (and recently revised into a new type of mass filter), time-of-flight (TOF) MS by Stephens in 1946, ion cyclotron resonance MS by Hipple and Thomas in 1949, quadrupole MS by Steinwedel in 1953, and ion trap MS by Paul and Dehmelt in the 1960s.

Mass spectrometry was first coupled with GC as a means of sample introduction by Golhke et al. (1956) and with HPLC via electrospray ionization (ESI) in the mid-1980s (Blakely and Vestal, 1983; Yamashita and Fenn, 1984). New methods of mass spectrometry are constantly under development and even as recently as 1985, Hillenkamp and Michael Karas developed the MALDI technique (a laser-based sample introduction device) that radically advanced the analysis of protein structures and more types of mass analyzers will certainly be developed. This chapter will deal only with basic MS instruments used in the analysis of organic chemicals exiting GC, HPLC, and CE systems, and is also applicable to effluents from ion chromatographic systems. ICP–MS interfaces and quadrupole MS were covered in Chapter 7. One of the most comprehensive Internet summaries of the history of mass spectrometry can be found at https://masspec.scripps.edu/learn/ms/.

Essential Methods of Instrumental Analysis, First Edition. Frank M. Dunnivant and Jake W. Ginsbach.

Companion Website: www.wiley.com/go/essentialmethodsofinstrumentalanalysis1e

13.2 Sample Introduction from GC and Analyte Ionization

The purpose of coupling GC with MS is to provide confirmatory identification with minimal effort. Prior to the common availability of MSs, confirmatory identification was possible but required twice the effort. GC analysis alone can provide confirmatory analysis, but it is usually necessary to analyze a sample using two different columns. With capillary systems, it is possible to perform two independent analyses by installing two different capillary columns into one injector system and monitoring each column effluent with a separate detector. If the same retention time and concentration are obtained from external standards, the identity of a compound is determined and the results are considered confirmatory.

Capillary column systems are more easily interfaced with an MS than packed columns. The high flow rate of packed columns (30–60 mL/min) created problems in maintaining the necessary low pressure of an MS. In contrast, capillary columns typically have a flow rate between 1 and 5 mL/min, which has a minimal effect on the low-pressure MS requirements. The GC and MS are interfaced by inserting the effluent end of the capillary column into the MS with a standard nut and ferrule system near the ionization source. Since GC analytes are volatile, the interface and MS must be maintained at temperatures and pressures that keep the analyte (or ionized form) in a volatile form.

As implied in the previous paragraph, MS systems require a low operating pressure, typically 10^{-5} to 10^{-6} Torr throughout the system (ionization source, mass analyzer, and detector). This is necessary to avoid collisions between ionized molecules. If collisions are prominent, the mass resolving capabilities will be effected, which decreases the detection limit and the resolution. Collisions also affect the interpretative value of the mass spectrum preventing identification.

The MS works by (1) ionizing each analyte as it exits the GC column, (2) accelerating and focusing the ionized compound and its fragments into the mass analyzer, (3) separating the fragments in the mass analyzer based on mass to charge (m/z) ratios, and (4) detecting the fragments as they exit the mass analyzer. There are a variety of ionization systems and mass analyzers that achieve these results. The following sections are dedicated to a description of most common ones.

13.2.1 Analyte Ionization

Analytes can be introduced into the ionization zone of an MS in two states, a solid or a vapor. Solids can be introduced by depositing milligram quantities of pure analyte onto a metal probe or in a matrix that is inserted into the ionization chamber. These more direct forms of ionization do not require the interfacing of a separatory instrument such as GC or LC since relatively pure analytes are directly placed into the MS. More commonly, analytes enter the MS system in a pure form (a peak) after separation by a capillary column GC. The MALDI technique, an increasingly popular tool described below, does not neatly fit into either of these categories but is included below due to its powerful applications for biological systems. Irrespective of the samples state, analytes must be ionized into positively or negatively charged ions, and are in some cases broken into fragments before they can be detected. Almost every compound has a unique fragmentation pattern that can subsequently be used for conclusive identification purposes. This pattern is dependent on the type of ionization source used and the stability of the energized analyte molecule. Below we will divide the ionization techniques into those for solid, nonvolatile analytes and volatile analytes entering the MS from a GC.

13.2.1.1 Ionization Techniques for Solid Nonvolatile Analytes

13.2.1.1.1 Field Desorption Field desorption (FD) techniques are relatively simple and do not require analyte separation in a GC since only one compound is introduced into the MS at a time. As noted in the heading above, compounds analyzed by this technique tend to be nonvolatile, have high molecular weights, and degrade at higher temperatures. Analytes are introduced to the system on a probe made of carbon fiber that has been lightly coated with pure analyte. A high current is applied between the probe and an adjacent electrode. The current moves the ionized analyte toward the end of the carbon fibers by charge attraction, where the molecules are ionized into the vapor (plasma) phase. Then they enter the mass analyzer and then the detector. The breaking of bonds within the analyte (fragmentation) is rare in FD techniques, thus the spectrum only contains the molecular ion. Many older inexpensive benchtop systems used to come with a direct probe built into EI systems. However, this feature has been removed due to the high number of service calls to clean out the MS units when too much analyte was placed on the probe. Service technicians refer to these analyte-rich probes as having "peanut butter" placed on them.

13.2.1.2 Ionization Techniques for Volatile Analytes Entering the MS from a GC

13.2.1.2.1 Electron Ionization or Electron Impact Electron ionization (EI) of analytes is referred to as a hard ionization technique since it causes bonds to be broken within a sample molecule (fragmentation). Neutral, radical, and positively or negatively charged species are produced from fragmentation. Neutral and radical species are not affected by the accelerator plates or mass analyzer and are removed by the vacuum. Positive ions are accelerated toward the mass analyzer and some either (1) collide with a surface in the source (typically the accelerator plate) or (2) enter into the mass analyzer through the slit in the electronic lens. The ions that collide with any surface are neutralized and removed by the vacuum. The ions that enter into the mass analyzer are separated by mass to charge ratios. The high degree of fragmentation can be an advantage in compound identification. When more ion fragments are created, the more unique the fragmentation pattern and the more confirmatory analyte identification will be. On the other hand, the detection of the molecular ion in EI can be difficult, which is often a goal of MS analysis in organic chemistry synthesis.

EI works by forcing the stream of pure analytes exiting the GC through a beam of high-energy electrons in the MS. Electrons are created by heating a metal filament, usually tungsten, to a temperature high enough to expel electrons. Electrons are drawn toward an anode, passing through the stream of analyte molecules. It is important to note that electrons do not actually impact analyte molecules as implied by the name "electron impact." The high energy of the electron (70 eV) is actually transferred to an analyte when the electronic transition of the analyte matches the frequency of the electron. The exact electron energy was selected through experimentation. It was found that a 70 eV electron energy source resulted in the most reproducible spectra and a high degree of fragmentation. This 70 eV condition is now the standard and all computer libraries of fragmentation are based on this energy level.

Video S.13.1 shows a beam of electrons that is generated by a heated filament at the bottom of the figure that is accelerated toward the anode at the top of the figure. When different analytes (in this case butane) exit the GC column (the brown column on the left) and cross through the electron beam, an electron from the sample molecules is removed. Once the molecular ion is formed, they are forced to the right by repulsion from a positively charged accelerator plate on the left (not shown) and drawn toward the negatively charged accelerator plate to the right. Some butane molecules also fragment into smaller ions. The prevalence of this process is underestimated by the animation due to space restraints. The molecular ion and fragments would next enter the mass analyzer (shown later).

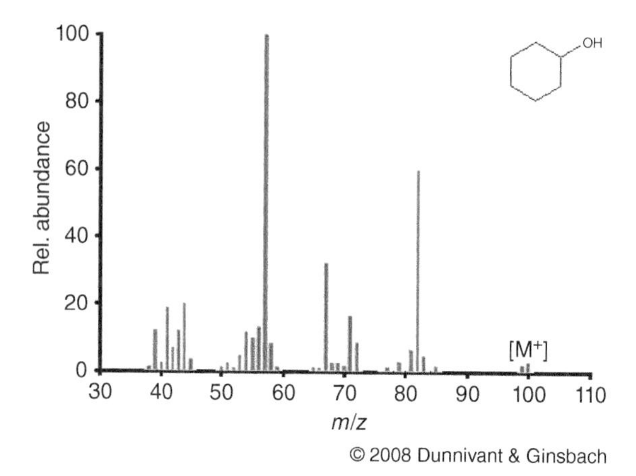

Figure 13.1 Fragmentation of cyclohexanol by EI (Source: Dunnivant and Ginsbach (Authors)).

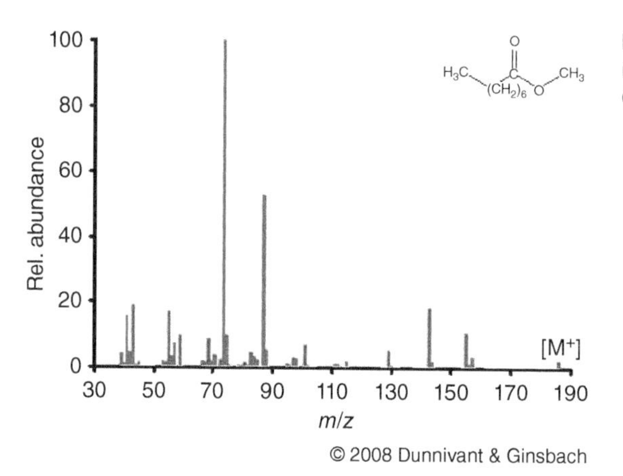

Figure 13.2 Fragmentation of decanoic acid methyl ester by EI (Source: Dunnivant and Ginsbach (Authors)).

After the energy transfer between the electron beam and the analyte, the energy causes the molecule to become unstable and frequently cleave bonds. The fragmentation patterns are predictable because the types of bond cleavages a molecule undergoes are related to its structure (Chapter 14). The ionization rate is predicted to be between one in a thousand and one in a million of the molecules entering the ionization chamber, typically one in ten thousand. This level of successful ionization should be noted since MS detection limits are approximately one part per million and below (injected analyte concentration). In early systems, the instrument only ionized and detected approximately one-millionth of the number of molecules that were injected; today this has been improved to about one in a thousand or more. Two examples of EI spectra are shown in Figures 13.1 and 13.2; note the extensive fragmentation of each analyte.

13.2.1.2.2 Chemical Ionization Today, most MSs can perform both EO and chemical ionization (CI), with different interchangeable ionization units. The CI unit is less open to diffusion of the reagent gas in order to contain the reagent gas longer and promote chemical ionization. Several reagent gases are used including methane, propane, isobutane, and ammonia, with the most common being methane. CI is referred to as a soft ionization technique since less energy is transferred to the original analyte molecule, and hence, less fragmentation occurs. In fact, one of the main purposes of using CI is to observe the molecular ion, represented by $M^{\cdot+}$ or M^{-}, or a close adduct of it, such as MH^{+}, MH^{+2}, or M plus the chemical ion (i.e. $M{+}CH_3$ with methane as the reagent gas or $M + NH_3$ with ammonia as the reagent gas). Notice again that neutral, negative, and positive fragments are produced but only the positive fragments are of use in positive CI detection, while negative ion fragments are detected in negative CI mode.

This section will limit its discussion to CI and methane, the most common reagent gas. Methane enters the ionization chamber at about 1000 times the concentration of the analyte molecules. While the electron beam in EI is usually set at 70 eV, in CI lower energy levels are used near the range of 20–40 eV. This energy level produces electrons that react with methane to form CH_4^+, CH_3^+, and CH_2^+. These ions rapidly react with unionized methane in the following manner (Reactions 13.1 and 13.2):

$$CH_4^{+\cdot} + CH_4 \rightarrow CH_5^+ + CH_3 \tag{13.1}$$

then

$$CH_3^{+\cdot} + CH_4 \rightarrow C_2H_5^+ + H_2 \tag{13.2}$$

The CH_5^+ and $C_2H_5^+$ ions collide with the analytes (represented by M) and form MH^+ and $(M-1)^+$ by proton and hydride transfer (Reactions 13.3–13.6).

$$CH_5^+ + M \rightarrow MH^+ + CH_4 \text{ Proton transfer;} \tag{13.3}$$

$$C_2H_5^+ + M \rightarrow MH^+ + C_2H_4 \text{ Proton transfer;} \tag{13.4}$$

$$C_2H_5^+ + M \rightarrow (M-1)^+ + C_2H_6 \text{ Hydride transfer.} \tag{13.5}$$

Note that several types of ions can occur, $(M+1)^+$ or MH^+ from proton transfer, $(M-1)^+$ from hydride transfer, and $M + CH_3^+$ and even $M+C_2H_5^+$ from additions. By inspecting the mass spectrum for this pattern, the molecular mass of the analyte can be deduced. Similarly, if other reagent gases are used, such as propane, isobutene, and ammonia, similar proton and hydride transfer and adduct formations can occur. The usual goal of CI is to obtain a molecule weight for the molecular ion that would usually not be present in an EI spectrum.

A relatively simple illustration of a CI chamber and its reactions is shown in Figure 13.3. This figure is similar to the EI animation, but the continuous addition of a reagent gas (from

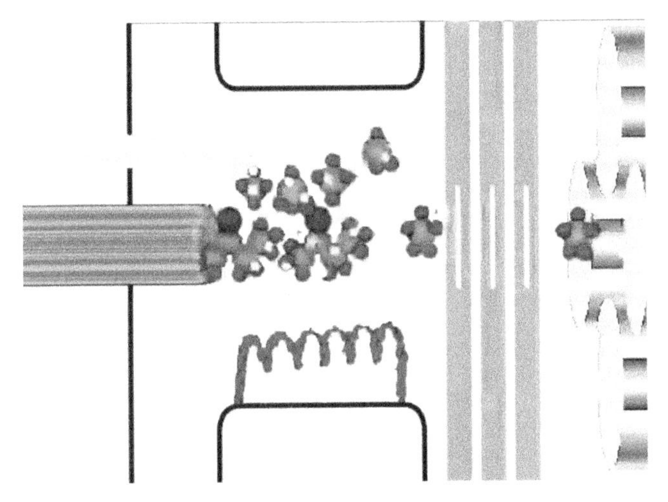

Figure 13.3 Illustration of a CI chamber and reagent gas-analyte reactions (Source: Dunnivant and Ginsbach (Authors)).

the top tube), methane, causes the gas to be ionized by the beam of electrons. Subsequently, the ionized methane reacts with analytes exiting the GC column (brown tube on the right). Methane is preferentially ionized by the beam of electrons due to its significantly higher concentration as compared to analytes from the GC. Positively charged fragments are drawn into the focusing lens and mass analyzer by a positively charged repeller plate (not shown) and the negatively charged accelerator plate, then enter the quadrupole MS.

Chemical ionization is most commonly used to create positive ions, but some analytes, such as those containing acidic groups or electronegative elements (i.e. chlorinated hydrocarbons) will also produce negative ions that can be detected by reversing the polarity on the accelerator and detector systems. Some of these analytes produce superior detection limits with CI as opposed to EI, while others only give increased sensitivity (slope of the response to concentration line). Negative ions are produced by the capture of thermal electrons (relatively slower electrons with less energy than those common in the electron beam) by the analyte molecule. Thermal electrons are present from the low-energy end of the distribution of electrons produced by the lower-energy CI source (~20 eV as opposed to 70 eV in EI). These low-energy electrons arise mostly from the chemical ionization process but also from analyte/electron collisions. Analyte molecules react with thermal electrons in the following manner, where *R–R′* is the unreacted analyte molecule and *R* represents an organic group.

$$R - R' + e^- \rightarrow R - R'^- \text{ by associative resonance capture} \tag{13.6}$$

$$R - R' + e^- \rightarrow R\cdot + R'^- \text{ by dissociative resonance capture} \tag{13.7}$$

$$R - R' + e^- \rightarrow R^+ + R'^- + e^- \text{ by ion pair production} \tag{13.8}$$

The identification of negative ion fragmentation patterns of analytes can be used in the same manner as in EI or positive ion CI. But note that extensive fragmentation libraries exist only for 70 eV EI. Many analysts create their own reference libraries with the analysis of reference materials that will later be used for the identification of unknown analytes extracted from samples.

Figures 13.4 and 13.5 contain CI spectra for the same compounds analyzed by EI in Figures 13.1 and 13.2, respectively. Note the obvious lack of fragmentation with the CI source and the presence of molecular ions (M^+ or M+1) in the CI spectra.

To summarize, for GC–MS systems, individual analytes exit the GC column, are ionized, and fragmented using electron or chemical impact (ionization). Since the detector in an MS is universal (responds to any positively charged ion), it is necessary to separate the molecular ion and its fragments by their mass or mass to charge ratio. This process is completed in a mass analyzer, which is explained in the section below. But first, some mass analyzers require the beam of ion fragments to be focused and all require the ion fragments to be accelerated in a linear direction.

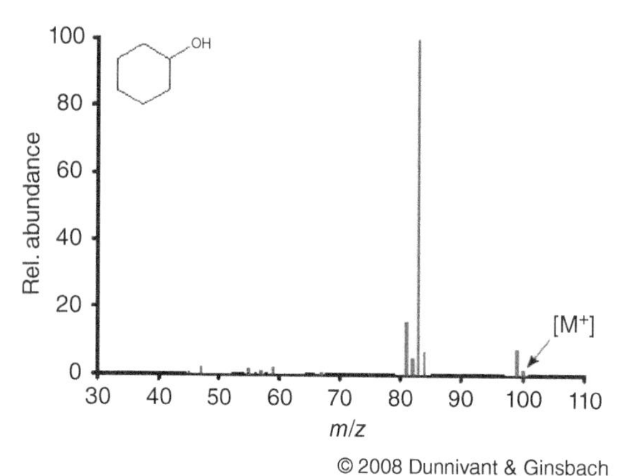

Figure 13.4 Fragmentation of cyclohexanol by CI (Source: Dunnivant and Ginsbach (Authors)).

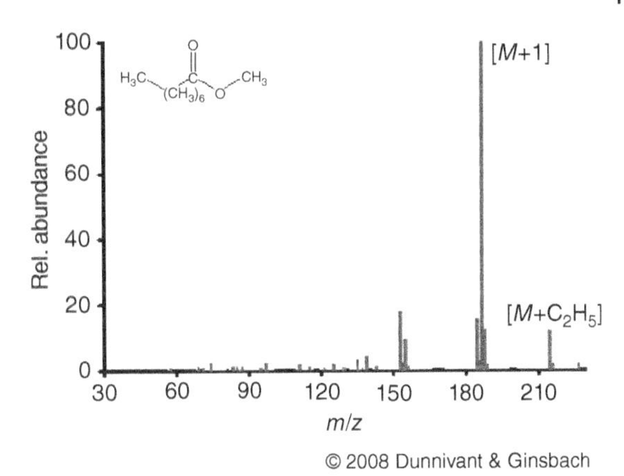

Figure 13.5 Fragmentation of decanoic acid methyl ester by CI (Source: Dunnivant and Ginsbach (Authors)).

13.2.1.3 Repulsion and Accelerator Plates, Slits, and Electronic Focusing Lens

Ions, regardless of the way they are generated, need to be accelerated into the mass filter/analyzer in order to separate ions of different masses. Since the majority of the ionization sources produce positively charged species, the most common way of accelerating ions is to place a positively charged plate on the "upstream" side of the system. This plate repels the cations toward the mass filter/analyzer. Most systems require ions to have a minimum velocity. So, negatively charged plates are placed on the "downstream" side of the instrument, just prior to the mass filter, to accelerate the ion in that direction (shown earlier in the EI and CI animations). The accelerator plates also act as slits since a relatively small hole is present in the middle of the plates that allow some of the ions to pass through the plate/slit and into the mass filter.

Accelerator plates/slits can also act as "gates" to the mass filter. This is accomplished by placing a positive charge on the slit that will repel the entry of an ion fragment or packet of ions into the system. Gates are used to hold up the entry of new ions to the mass filter until all of the ions have passed through to the detector. After this, the polarity on the gate is returned to negative and a new set of ion fragments is allowed to enter the mass filter. This type of gating system is important in the TOF mass filters discussed in Section 13.5.4.

Some systems, especially the quadrupole mass filter, require the stream of ions to be focused into a narrow point in order to allow successful mass to charge separation. One such electrical lens is the Einzel lens, which is analogous to a focusing lens in an optical spectrophotometer. Figure 13.6 illustrates, in two dimensions, how an Einzel lens works. Six plates are in parallel, three on each side, and are exposed to the potentials shown below. These potentials set up a set of electrical field lines that act to bend the ions near the outside of the plates toward the center. Ions are focused on a small point for entry into the mass filter. The series of lenses stretch the length of a given beam of ions since ions on the outside (near the plates) have to travel a longer distance to reach the focal point.

The Einzel lens above is shown and explained as six horizontal plates. In practice, Einzel lenses are vertical plates with a hole in each plate. Thus, the applied electrical potential creates

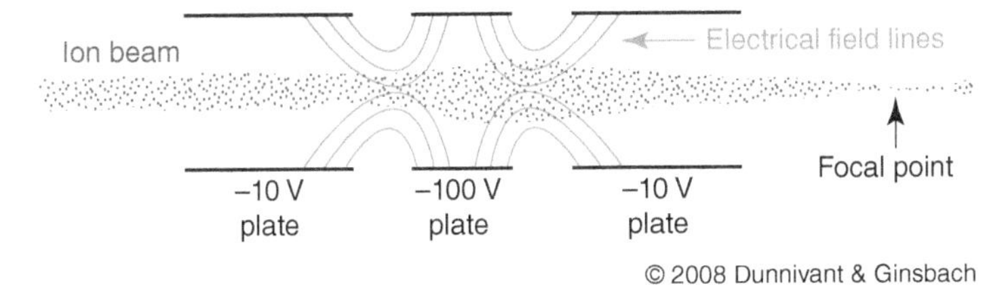

Figure 13.6 An Einzel lens (electronically focusing lens). *Note*: The figure shows the 2D representation of a 3D ring lens (Source: Dunnivant and Ginsbach (Authors)).

three-dimensional field lines that focus the ion beam to a point where the entrance slit/hole to the next component is located.

Electrostatic, magnetic, and TOF instruments have only repulsion and accelerator plates. In addition to these plates, quadrupole instruments have a focusing lens to help introduce the ions toward the center of the mass filter/analyzer.

13.3 Introduction of Samples from HPLC

At this point, it is noteworthy to recall the differences between GC and LC. Chapter 10 defined GC as a technique applicable to relatively volatile, thermally stable compounds. These restrictions greatly limited the types and number of compounds that could be analyzed by GC, and GC–MS. LC, discussed in Chapter 11, uses a liquid mobile phase in the analysis of many of the compounds analyzed by GC, and also can be used to analyze the plethora of biomolecules that are nonvolatile and thermally unstable at even slightly elevated temperatures. While the conditions used in LC greatly extend the applications of chromatography, it has historically suffered difficulties with mass spectrometry interfaces. Most of the various forms of LC, especially HPLC types discussed in Chapter 11, can be interfaced with MS today.

The largest difficulties in interfacing LC with MS are the removal of the mobile phase solvent prior to introduction to the MS mass analyzer and the transfer and ionization of nonvolatile analyte molecules into the gas phase. The first attempt at an LC–MS interface was to place the effluent droplets from the LC onto a supposed chemical-resistant conveyor belt that transported the liquid into the MS ionization chamber. The conveyor belt was then cleaned and returned to the HPLC effluent for more sample. However, these early attempts resulted in inefficient removal of the analytes from the conveyor belt and analyte residue being left on and released from the belt during subsequent MS runs. This problem was significantly compounded with 4.5 mm diameter HPLC columns with *liquid* flow rates in the range of ~1 mL/min. The later use of 300–75 mm long capillary columns improved flow rate problems. The invention of ESI solved all of the major problems associated with sample introduction to MS. ESI was first conceived in the 1960s by Malcolm Dole at Northwestern University, but it was not put into practice until the early 1980s by John B. Fenn of Yale University (and resulted in his Noble prize in 2002). Its common use today has been one of the most important advances in HPLC and today allows routine identification of biological macromolecules.

13.3.1 Electrospray Ionization Sample Introduction

Today, the most common form of LS–MS interface is the ESI sample introduction system. An overview of this system is shown in Figure 13.7. Samples can be introduced via a syringe or an HPLC system (convention or capillary column type). A restriction in the syringe needle or HPLC column causes the solvent containing the analytes to form droplets. An electrical potential, discussed in the next paragraph, is placed between the sample inlet and the first cone. This cone separates the sample introduction from the vacuum chamber in the MS. For high-flow HPLC applications, N_2 gas is used to evaporate the solvent or mobile phase and desolvate the analyte molecules. This is usually unnecessary for capillary columns or nanoapplications. After desolvation and charge formation occur, as discussed below, the charged molecules enter a slightly heated transfer capillary tube and pass through two more cones that are used to control the vacuum. Finally, the positively charged ions enter a mass analyzer such as the quadrupole shown in Figure 13.7.

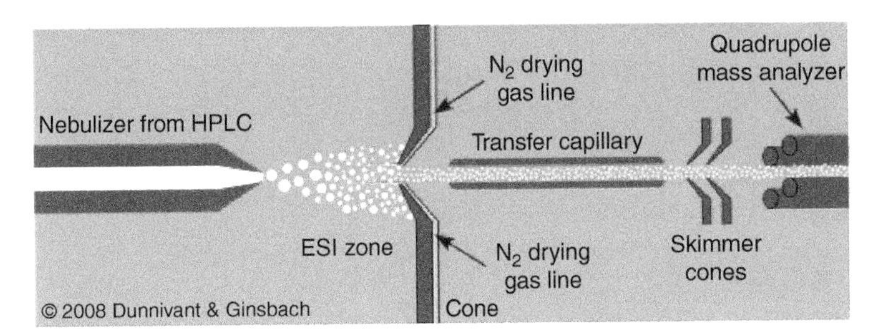

Figure 13.7 Overview of an ESI (LC–MS) interface. One of the most important advancements in LC–MS interfaces in the last 10 years has been the replacement of the transfer capillary with an ion funnel. The ion funnel allows more carrier gas to be removed and ions to be focused to enter the MS. The new ion funnel results in a 10-fold improvement in detection limits but is present only in high-end MS systems (Source: Dunnivant and Ginsbach (Authors)).

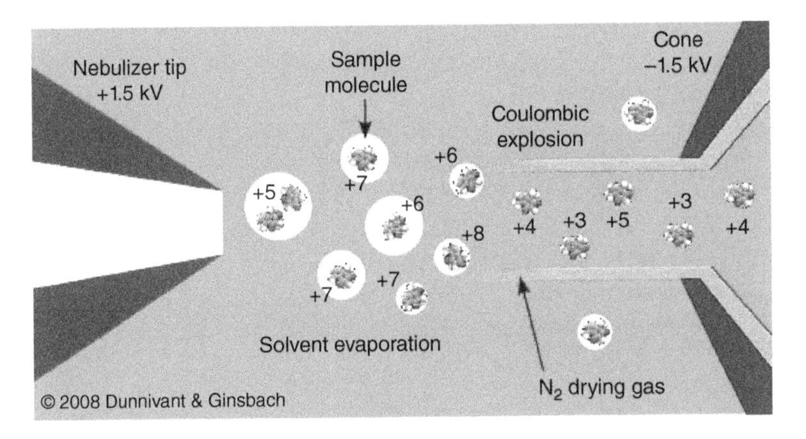

Figure 13.8 Charge formation in ESI (Source: Dunnivant and Ginsbach (Authors)).

The heart of ESI is the desolvation and charge formation shown in Figure 13.8. “Ionization” in ESI is referred to as soft ionization and is really not ionization but charge formation since no real ionization source is present. Charge formation occurs by evaporating the solvent by passing a dry gas counter current to the movement of droplets while at the same time, the droplets are passed along a charged field (from 2.5 to 4 kV) between the tip of the sample introduction point and the first cone. Charge formation occurs by one of two proposed mechanisms: (1) ion evaporation model where the droplet reaches a certain radius such that the field strength at the surface of the droplet becomes large enough to assist the field desorption of solvated ions and (2) charged residue model where electrospray droplets undergo evaporation and fission cycles, resulting in gas-phase ions that form after the remaining solvent molecules evaporate.

The charged residue model is the most accepted theory and is explained in the following. As the droplets pass from left to right, desolvation occurs in the presence of the dry N_2 gas. At the same time, the charged field results in the collection of a positive charge on the droplet. As this process continues, from left to right, the droplet shrinks until it reaches a point where the surface tension can no longer sustain the charge accumulation, this point is referred to as the Rayleigh limit. Above the Rayleigh limit, Rayleigh fission (also known as Coulombic explosion) occurs and the droplet is ripped apart forming smaller charged droplets containing the analyte molecules. This process continues until desolvation is complete and the charge is transferred to the ionized and now gaseous analyte molecule. The resulting charged molecules can be singly or multiply charged (refer to Figure 13.8). The positively charged ions enter the mass analyzer. Simple molecules result in a single mass to charge ion, while complex molecules result in a Gaussian distribution of mass to charge ions yielding a single-molecule molecular mass for identification purposes. As noted above, the ionization process is considered to be soft ionization. Thus, if structural identification is required, the parent ion is usually analyzed by tandem MS where it is fragmented into smaller fragments for identification. Nanospray versions of this process have recently become available.

13.3.2 Matrix-assisted Laser Desorption/Ionization

The matrix-assisted laser desorption/ionization (MALDI) technique has revolutionized the analysis of large-molecular-weight nonvolatile compounds, especially synthetic polymers and biopolymers with molecular weights up to 300 000 Da. Unlike the field desorption technique that desorbs and ionizes pure analyte from a probe, MALDI volatilizes a mixture of a matrix and analyte in order to “transport” the nonvolatile analyte into a vapor phase.

The MALDI technique is completed in two steps. First, a solution of solvent, analyte, and matrix compound is thoroughly mixed and placed on a disk to dry. As the solvent evaporates crystals of matrix containing evenly dispersed analyte molecules are formed. For the second step, the coated disk is placed in the vacuum chamber of the MS. Then the disk is repeatedly pulsed with a laser in the UV or visible spectrum depending on the matrix (Table 13.1). During each laser pulse, the matrix molecules are rapidly volatilized (sublimated/ablated) and carry the individual analyte molecules into a low-pressure plasma. The wavelength of the laser is selected to heat and volatize the matrix and to avoid significant heat or degrade the analyte molecules. Analyte molecules are mostly ionized in the vapor phase by photoionization, excited-state proton transfer, ion-molecule

Table 13.1 Frequently used matrix compounds.

Matrix Compound	Active Wavelength (nm)
Nicotinic acid	220–290
Benzoic acid derivatives such as Vanillic acid	266
Pyrazine-carboxylic acid	266
3-Aminopyrazine-2-carboxlic acid	337
Cinnamic acid derivatives such as Caffeic acid	266–355
3-Nitrobenzylalcohol	266

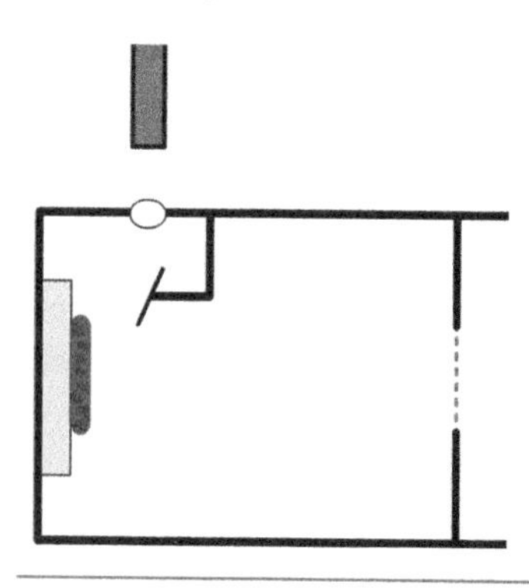

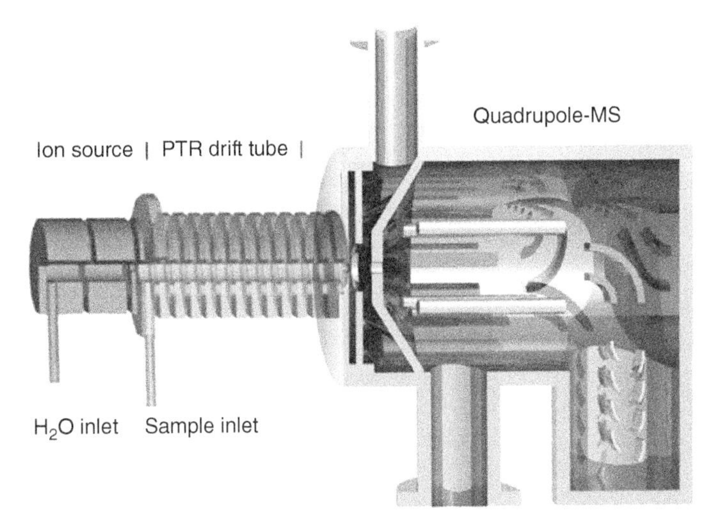

Figure 13.9 Illustration of a proton transfer reaction–MS system (Source: Reproduced with permission from Ionicon Analytik Gesellschaft, Innsbruck, Austria).

reactions, desorption of preformed ions and most commonly by gas-phase proton transfer in the expanding plume by photoionized matrix molecules.

After the analyte molecules are ionized (to cations), they are drawn toward the negative accelerator plate and into the mass filter. A TOF mass filter is always used because of its rapid scanning abilities and large mass range. The introduction of ions into the flight tube is controlled so that all ions reach the detector before the next group enters into the TOF tube. This requires carefully spacing the laser pulses and electric gates. The spectrum of the analysis is considerably "clean" since only pure analyte is introduced into the MS and essentially no fragmentation occurs (matrix molecules/ions can be ignored by the mass filter due to their relatively low mass). Ionized analytes can acquire +1, +2, and +3 charges and multiple molecules can form dimer and trimer peaks (combined fragments of two or three molecular ions), so the confirmational molecular weights can easily be determined. A relatively simple illustration of a MALDI–TOF MS (the most common combination) is shown in Video S.13.2.

13.3.3 Proton Transfer Reaction Ionization

Proton transfer reaction (PTR) is a relatively recent addition to mass spectrometry (1995) that was originally developed for GC and LC. There is no reason that it cannot be used for CE. It was developed at the Institut für Ionenphysik at the Leopold-Franzens University in Innsbruck, Austria, by Hansel et al. (1995). As shown in Figure 13.9, the PTR consists of a reaction chamber where water vapor is ionized to gas phase ions by hollow cathode discharge via the following reactions

$$e^- + H_2O \rightarrow H_2O^+ + 2e^- \tag{13.9}$$

$$e^- + H_2O \rightarrow H_2^+ + O + 2e^- \tag{13.10}$$

$$e^- + H_2O \rightarrow H^+ + OH + 2e^- \tag{13.11}$$

$$e^- + H_2O \rightarrow O^+ + H_2 + 2e^- \tag{13.12}$$

These products undergo ion-water vapor reactions in a short drift tube to form

$$H_2^+ + H_2O \rightarrow H_2O^+ + H_2 \tag{13.13}$$

$$H^+ + H_2O \rightarrow H_2O^+ + H \quad (13.14)$$

$$O^+ + H_2O \rightarrow H_2O^+ + O \quad (13.15)$$

$$H_2O^+ + H_2O \rightarrow H_3O^+ + OH \quad (13.16)$$

The hydronium ion (H_3O^+) is end product and the primary reacting ion that ionizes organic analytes in the reaction drift tube via the reaction

$$H_3O^+ + R \rightarrow RH^+ + H_2O \quad (13.17)$$

Unlike in TOF or ion mobility MS, reaction ions are not subjected to an electrical potential in the drift tube but are moved through the system by placing a low-pressure vacuum pump at the interface of the PRT drift tube and the inlet to the mass filter (refer to Figure 13.9). Analyte cations created in the drift tube enter a mass filter where they are separated by the operating parameters of each mass filter and are detected with an electron multiplier (EM).

A PTR–MS is illustrated via the link to Illustration of a Proton Transfer Reaction–Mass Spectrometer http://www.uibk.ac.at/ionen-angewandte-physik/umwelt/research/pics/animation.gif.

Advantages of the PTR–MS include (1) low fragmentation with allows improved detection limits due to the formation of more molecular ions, (2) direct sampling of atmospheric gases (no sample preparation), (3) real-time measurements, and (4) high mobility due to the lack of gas cylinders, relative ease of operation only requiring electrical power and distilled water, and part-per-billion detection limits.

13.3.4 Fast Ion Bombardment

Another technique for ionizing large biomolecules (up to and greater than 10 000 Da) is to bombard them with ions of argon or xenon; this is also referred to as a liquid secondary ion source. First, analytes are embedded in a matrix such as glycerol, thioglycerol, *m*-nitrobenzyl alcohol, crown ethers, sulfolane, 20-nitrophenyloctyl ether, diethanolamine, or triethanolamine. An electronic impact source similar to that described in the GC ionization section is used to ionize Ar or Xe gas at a pressure of 10^{-5} Torr. Ar and Xe ions are accelerated toward the matrix containing the analytes and their impact sputters off positive and negative analyte ions (mostly molecular ions) that enter an MS for mass determination.

13.4 Introduction of Samples from a Capillary Electrophoresis System

Years ago, if you wanted to own a CE–MS system, you had to purchase the CE and MS separately and hire the MS manufacturer or vender to interface the two instruments. Later (~2008), you are now able to purchase off-the-shelf interfaced instruments from chromatography vendors. CE–MS interfaces are designed and operate in much the same way as the HPLC–MS interface, with two exceptions. While HPLC columns can be composed of metal that readily conduct the electrical potential to ionize the analytes, the CE columns are only composed of fused silica. As a result the effluent of the CE column must be coated with a conducting metal sheath. Also, as you will recall from Chapter 12 on CE, minimal solvent flow results in CE, only from the dragging of solvent by the electrophoretic mobility of the buffer ions. Thus, CE is almost ideal for MS interfaces and is far superior to HPLC interfacing since very little solvent must be removed prior to entry into the MS vacuum system. Other than these two differences, CE–MS operates like HPLC–MS. Solvent droplets, containing analytes, are created at the end of the fused silica column, and are charged by the electrical potential placed between the metal sheath and the metal cone at the entry to the MS system (Figure 13.10a,b). Solvent is evaporated with a drying gas that flows counter current to the movement of the solvent droplets. Charge transfer occurs through Coulombic explosion and the de-solvated and ionized anions or cations (depending on the potential) are accelerated through the MS interface cone. CE–MS has finally reached a level of maturity and dependability that promises significant advances in many areas of analytical separation and quantification, especially protein studies.

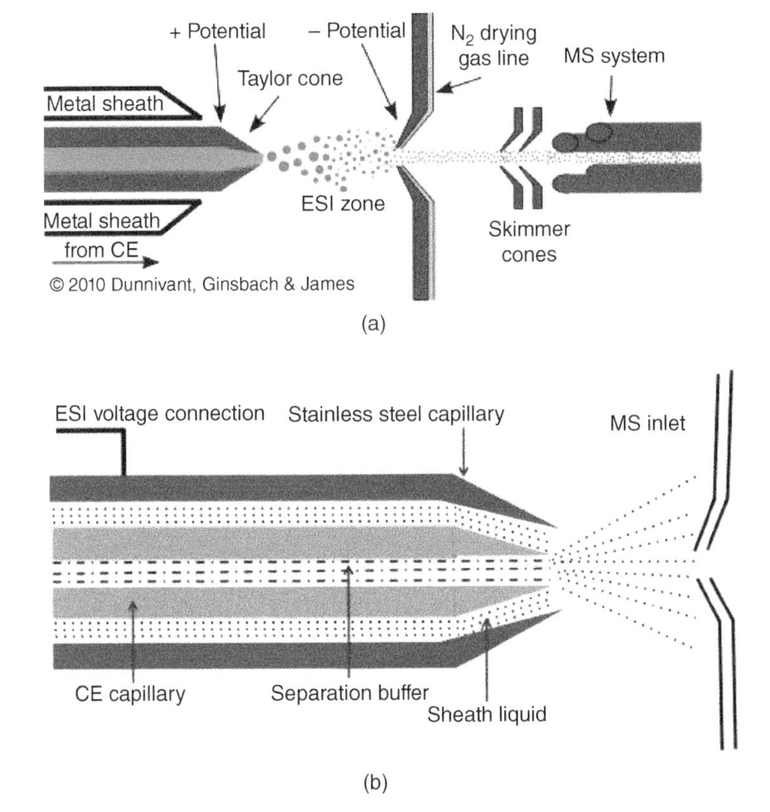

Figure 13.10 (a) An older CE–MS interface (Source: Dunnivant and Ginsbach (Authors)) and (b) a newer CE–MS interface (Source: Banstola.bijay/WikiCommons/https://en.wikipedia.org/wiki/Capillary_electrophoresis%E2%80%93mass_spectrometry#/media/File:Sheath_Flow_Interface.jpg (accessed February 06, 2024)).

13.5 Common Mass Filters (Mass Analyzers)

Mass analyzers separate the molecular ion and its fragments by ion velocity, mass, or mass to charge ratio. A number of mass filters/analyzers are available for GC, LC, and CE interfaces, but not all are commercially available. These can be used individually or coupled in a series of mass analyzers to improve mass resolution and provide more conclusive analyte identification. This text will discuss the most common MSs.

The measure of "power" of a mass analyzer is resolution, the ratio of the average mass (m) of the two adjacent peaks being separated to the mass difference (Δm) of the adjacent peaks, represented by

$$R_s = m/\Delta m \tag{13.1}$$

Resolution (R_S) is achieved when the midpoint between two adjacent peaks is within 10% of the baseline just before and after the peaks of interest (the valley between the two peaks). Resolution requirements can range from high-resolution instruments that may require discrimination of a few ten thousands (1/10 000) of a gram molecular weight (0.0001 amu) to low-resolution instruments that only require unit resolution (28 versus 29 Da). Resolution values for commonly available instruments can range from 500 to 500 000.

Before introducing the various types of mass analyzers, remember the current location of the mass analyzer in the overall MS system. In the previous sections, the analyte was being ionized, underwent fragmentation, was accelerated, and in some cases, focused on a focal point with a velocity toward the mass analyzer. Now the packet of ion fragments needs to be separated based on their momentum, kinetic energy (KE), or mass to charge ratio (m/z). Often the terms mass filter and mass analyzer are used interchangeably, as is done in this text. But, first a controversy in the literature needed to be addressed with respect to how a mass filter actually separates ion fragments.

Some resources state that all mass analyzers separate ions with respect to their mass to charge ratio while others are more specific and contend that only quadrupoles separate ions by mass to charge ratios. The disagreement in textbooks lies in what components of the MS are being discussed. If one is discussing the effect of the accelerator plates and the mass filter, then all mass filters separate based on mass to charge ratios. This occurs because the charge of an ion will be a factor that determines the velocity a particle of a given mass has after interacting with the accelerator plate in the electronic, magnetic sector, and TOF mass analyzers. But after the ion has been accelerated, a magnetic section mass filter actually separates different ion-based momentums and kinetic energies while the TOF instrument separates different ions based on ion velocities (arrival times at the detector after traveling a fixed length). In the other case, no matter what the momentum or velocity of an ion, the quadrupole mass analyzer separates different ions based solely on mass to charge ratios (or the ability of the ion to establish a stable oscillation in an oscillating electrical field). These differences may seem semantic but some MS users insist on their clarification. For the discussions below, in most cases, mass to charge will be used for all mass analyzers.

13.5.1 Magnetic Sector Mass Filter

It has been known for some time that the trajectory of a point charge, in our case, a positively charged molecular ion or fragment can be altered by an electrical or magnetic field. Thus, the first MS systems employed permanent magnets or electromagnets to bend the packets of ions in a semi-circular path and separate ions based on their momentum and KE. Common angles of deflection are 60°, 90°, and 180°. The change in trajectory of the ions is caused by the external force of the magnetic field. The magnitude of the centripetal force, which is directly related to the ion velocity, resists the magnetic field's force. Since each mass to charge ratio has a distinct KE, a given magnetic field strength will separate individual mass to charge ratios through space. A slit is placed in front of the detector to aid in the selection of a single mass to charge ratio at a time.

A relatively simple mathematical description will allow for a better understanding of the magnetic field and the ion's centripetal force. First, it is necessary to compute the KE of an ion with mass m possessing a charge z as it moves through the accelerator plates. This relationship can be described by

$$\mathrm{KE} = \frac{1}{2}mv^2 = z\mathrm{eV} \tag{13.2}$$

where e is the charge of an electron (1.60×10^{-19} C), v is the ion velocity, and V is the voltage between the two accelerator plates (shown in Video S.13.3). Fortunately, in EI and CI in GC systems, most ions have a charge of +1. As a result, an ion's KE will be inversely proportional to its mass. The two forces that determine the ion's path, the magnetic force (F_M) and the centripetal force (F_C), are described by

$$F_\mathrm{M} = Bz\mathrm{eV} \tag{13.3}$$

and

$$F_\mathrm{C} = \frac{mv^2}{r} \tag{13.4}$$

where B is the magnetic field strength and r is the radius of curvature of the magnetic path. In order for an ion of a particular mass and charge to make it to the detector, the forces F_M and F_C must be equal. This obtains

$$Bz\mathrm{eV} = \frac{mv^2}{r} \tag{13.5}$$

which upon rearrangement yields

$$v = \frac{Vzer}{m} \tag{13.6}$$

Substituting this last equation into our first KE equation yields

$$\frac{m}{z} = \frac{B^2r^2\mathrm{e}}{2V} \tag{13.7}$$

Since e (the charge of an electron) is constant and r (the radius of curvature) is not altered during the run, altering the magnetic field (B) or the voltage between the accelerator plates (V) will vary the mass to charge ratio that can pass through the slit and reach the detector. By holding one constant and varying the other throughout the range of m/z values, the various mass to charge ratios can be separated. One option is to vary the magnetic field strength while keeping the voltage on the accelerator plates constant.

However, it is difficult to quickly vary the magnetic field strength. The resulting slow scan rate is especially problematic with capillary column GCs since the peak width is narrow. Using a magnetic sector instrument could complicate identification of a compound if two or more peaks emerge from the GC during a single scan, especially in the relatively fast elution of peaks from a capillary column GC. Generally, several complete mass to charge scans are desired for accurate analyte identification. This can be overcome in modern magnetic sector instruments by rapidly sweeping the voltage between the accelerator plates, in order to impart different momentums on the ion fragments, as opposed to sweeping the field strength. Due to the operational advantages of this technique, most electromagnets hold the magnetic field strength (B) and vary the voltage (V) on the accelerator plates.

The magnetic sector mass filter is illustrated in Video S.13.4. Although B and r are normally held constant, this modern design is difficult to illustrate. So, we will illustrate a magnetic sector MS where B, the magnetic field, is varied to select for different ions. As a particular peak (compound) enters the MS from a GC, it is ionized/fragmented by an EI in the animation. The ions are then uniformly accelerated by the constant voltage between the two accelerator plates/slits on the left side of the figure. As the different ions travel through the electromagnet, the magnetic field is varied to select for different m/z ratios. Ions with the same momentum or KE (and therefore mass) pass through the exit slit together and are measured by the detector, followed by the next ion, and so on.

While magnetic sector mass filters were once the only tool to create a mass spectrum, they are becoming less common today. This is due to the size of the instrument and its weight. As a result, many magnetic sector instruments have been replaced by quadrupole systems that are much smaller, lighter, and able to perform extremely fast scans. Magnetic sector instruments are still used in cases where extremely high resolution is required such as double-focusing instruments (Section 13.5.6).

13.5.2 Quadrupole Mass Filter

Quadrupole mass filters have become the most common type of MS used today due to their relatively small size, light weight, low cost, and rapid scan times (<100 ms). This section is a repeat of Section 7.2.6.1 for an ICP–MS but is repeated here due to its common use in GC, LC, and CE applications. This type of mass filter is most commonly used in GC applications and to some extent in LC systems because they are able to operate at a relatively high pressure (5×10^{-5} Torr). The quadrupole has also gained widespread use in tandem MS applications (a series of MS analyzers).

Despite the fact that quadrupoles produce the majority of mass spectra today as mentioned earlier, they are not true MSs. Actual MSs produce a distribution of ions either through time (TOF MS) or space (magnetic sector MS). The quadrupole's mass resolving properties are instead a result of the ion's stability/trajectory within the oscillating electrical field.

A quadrupole system consists of four rods that are arranged at an equal distance from each other in a parallel manner. Paul and Steinwegen theorized in 1953 that hyperbolic cross-sections were necessary. In practice, it has been found that circular cross sections are both effective and easier to manufacture. Each rod is less than a cm in diameter and is usually less than 15 cm long. Ions are accelerated by a negative voltage plate before they enter the quadrupole and travel down the center of the rods (in the z direction). The ions' trajectory in the z direction is not altered by the quadrupole's electric field.

The various ions are separated by applying a time-independent dc potential as well as a time dependent AC potential. The four rods are divided up into pairs where the diagonal rods have an identical potential. The positive DC potential is applied to the rods in the X–Z plane and the negative DC potential is applied to the rods in the Y–Z plane. The subsequent AC potential is applied to both pairs of rods, but the potential on one pair is the opposite sign of the other and is commonly referred to as being 180° out of phase (Figure 13.11).

Mathematically the potentials that ions are subjected to are described by Equations 13.8 and 13.9:

$$\Phi_{X-Z} = +(U + V\cos\omega t) \quad (13.8)$$

and

$$\Phi_{Y-Z} = -(U + V\cos\omega t) \quad (13.9)$$

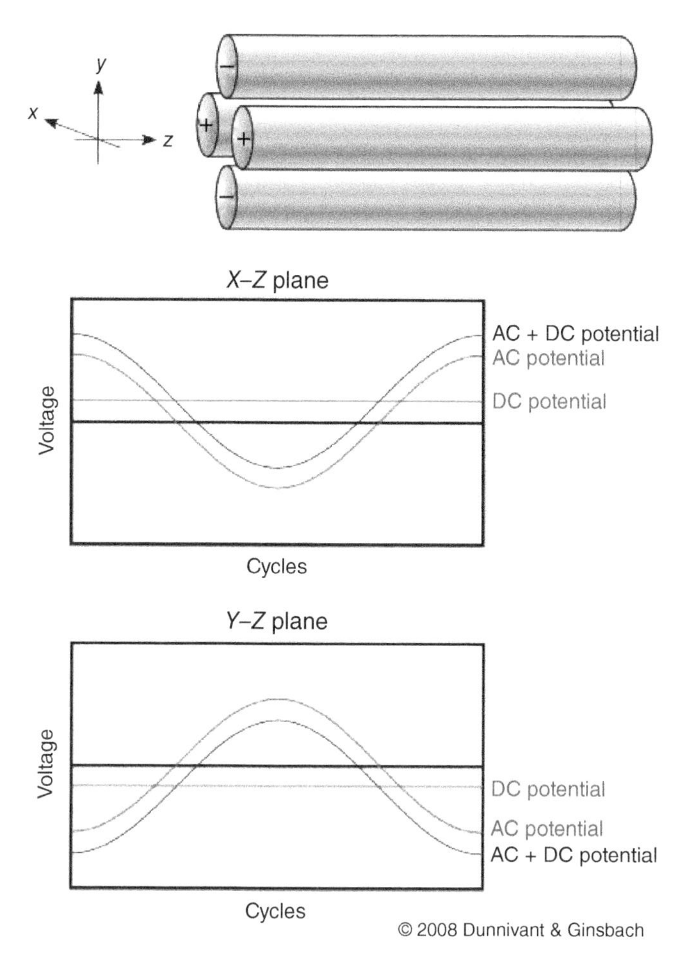

Figure 13.11 AC and DC potentials in the quadrupole MS (Source: Dunnivant and Ginsbach (Authors)).

where Φ is the potential applied to the X–Z and Y–Z rods, respectively, ω is the angular frequency (in rad/s) and is equal to $2\pi\nu$ where ν is the radiofrequency of the field, U is the DC potential, and V is the zero-to-peak amplitude of the radiofrequency voltage (AC potential). The positive and negative signs in the two equations reflect the change in polarity of the opposing rods (electrodes). The values of U range from 500 to 2000 V and V in the above equation ranges from 0 to 3000 V.

To understand the function of each pair, consider the rods in the X–Z plane in isolation. For now, imagine that only an AC potential is applied to the rods. Half the time when the potential was positive, ions (cations) would be repelled by the rod's charge and would consequently move toward the center of the rods. Likewise, when the potential was negative, ions would accelerate toward the rods in response to an attractive force. If during the negative AC potential, an ion comes into contact with the rod, it is neutralized and is removed by the vacuum. The factors that influence whether or not a particle strikes the rod during the negative cycle include the magnitude of the potential (its amplitude), the duration of time the ions are accelerated toward the rod (the frequency of the AC potential), the mass of the particular ion, the charge of the ion, and its position within the quadrupole.

Now imagine that a positive DC potential (at a fraction of the magnitude of the AC potential) is applied to the rod in the X–Z plane. This positive DC potential alone would focus all of the ions toward the center of the pair of rods. When the AC and DC potentials are applied at the same time to the pair of rods in the X–Z plane, ions of different masses respond differently to the resulting potential. Heavy ions are largely unaffected by the AC and as a result, respond to the average potential of the rods. This results in heavy ions being focused toward the center of the rods. Light ions, on the other hand, will respond more readily to the AC. Ions that are sufficiently light will have an unstable trajectory in the X–Z plane and will not reach the detector. Only ions heavier than a selected mass will not be filtered out by the X–Z electrodes. As a result, the X–Z plane electrodes only filter light ions and form a high pass mass filter (Figure 13.12).

Now look at the other pair of rods and the converse of the AC/DC potentials. The rods in the Y–Z plane have a negative DC voltage and the AC potential is the same magnitude but the oppose sign as the potential applied to the X–Z plane. Heavy ions are still mostly affected by the DC potential, but since it is negative, they strike the electrode and are unable to reach the detector. The lighter ions respond to the AC potential and are focused toward the center of the quadrupole. The AC potential can be thought of as correcting the trajectories of the lighter ions, preventing them from striking the electrodes in the Y–Z plane. These electrical parameters result in the construction of a low pass mass filter.

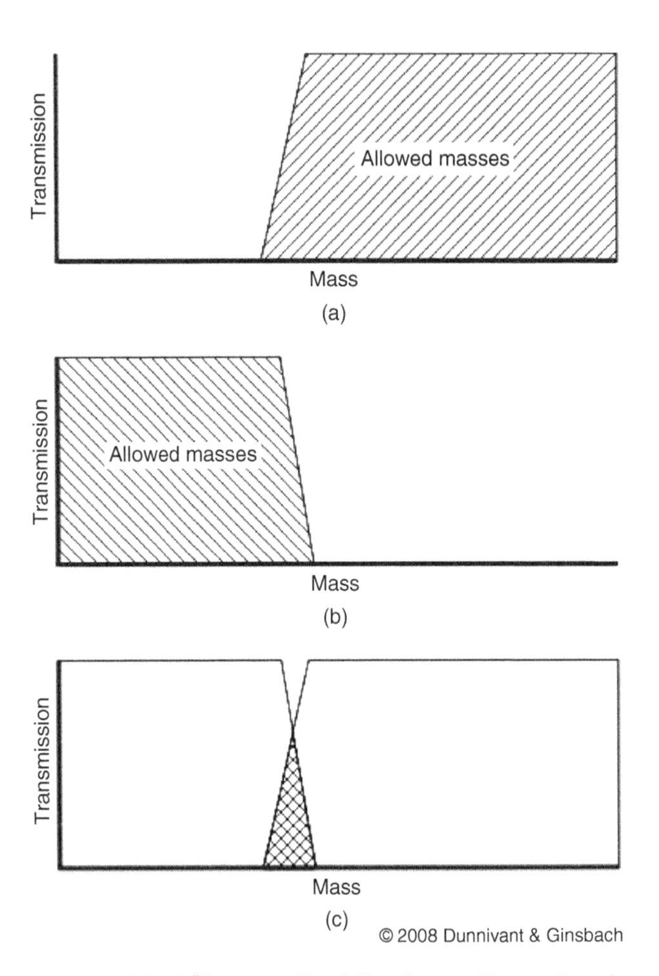

Figure 13.12 A "Conceptual" stability diagram (Source: Dunnivant and Ginsbach (Authors)).

When both the electrodes are combined into the same system, they are able to selectively allow a single mass to charge ratio to have a stable trajectory through the quadrupole. *Altering the magnitude of the AC and DC potential changes the mass to charge ratio that has a stable trajectory resulting in the construction of mass spectra.* Different ions possess a stable trajectory at different magnitudes and reach the detector at different times during a sweep of the AC/DC magnitude range. The graph of the combined effect, shown in Figure 13.12c, is actually a simplification of the actual stability diagram.

One way to generate an actual stability diagram is to perform a series of experiments where a single mass ion is introduced into the quadrupole. The DC and AC voltages are allowed to vary and the stability of the ion is mapped. After performing a great number of experiments, the resulting plot would look like Figure 13.13. The shaded area under the curve represents values of AC and DC voltages where the ion has a stable trajectory through the potential and would reach the detector. The white space outside the stability diagram indicates AC and DC voltages where the ion would not reach the detector.

While any AC and DC voltages that fall inside the stability diagram could be utilized, in practice, quadrupoles keep the ratio of the DC to AC potential constant, while the scan is performed by changing the magnitude of the AC and DC potential. The result of this is illustrated as the mass scan line intersecting the stability diagram in Figure 13.13. The graphs below the stability diagram correspond to specific points along the scan and help to illustrate the ions' trajectories in the X–Z and Y–Z plane (Figure 13.13). While the mass to charge ratio of the ion remains constant in each pair of horizontal figures, the magnitude of the applied voltages changes while their ratio stays constant. As a result, examining points along the mass scan line in Figure 13.12 is equivalent to

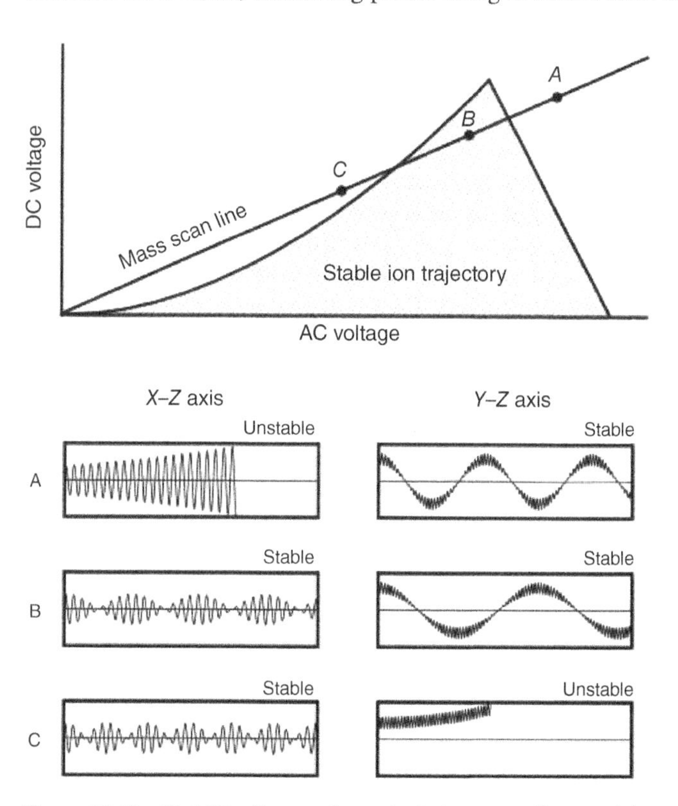

Figure 13.13 Stability diagram for a single ion mass (Source: Adapted from Steel and Henchman (1998)).

shifting the position of the high and low pass mass filters with respect to the x-axis illustrated in Figure 13.13. Even though the mass is not changing for the stability diagram discussed here, the mass that has a stable trajectory is altered.

In the above figure, the graph corresponding to point A indicates that the ion is too light to pass through the X–Z plane because of the high magnitude of the AC and DC potentials. As a result, its oscillation is unstable, and it eventually impacts the electrode/rod. The motion of the particle in the Y-axis is stable because the combination of the AC potential as well as the negative DC potential yields a stable trajectory. This is the graphical representation of the AC potential correcting the trajectory of the light ions in the Y–Z plane. At point B the magnitude of voltages has been altered so the trajectories of the ion in both the X–Z and Y–Z plane are stable and the ion successfully reaches the detector. At point C, the ion has been eliminated by the low-mass pass filter. In this case, the AC potential is too low to allow the ion to pass through the detector and it strikes the rod. This is caused by the ion's increased response to the negative DC potential in the Y–Z plane. The trajectory in the X–Z axis is stable since the DC potential focusing the ion toward the center of the poles overwhelms the AC potential.

Until this point, the stability diagram shown above is only applicable to a single mass. If a similar experiment were to be performed using a different mass, the positions of the AC and DC potential on the x and y axes would be altered but the overall shape of the curve would remain the same. Fortunately, there is a less time-consuming way to generate the general stability diagram for a quadrupole mass filter using a force balance approach. This derivation requires a complex understanding of differential equations and is beyond the scope of an introductory text, but it can be explained graphically (Figure 13.14). The parameters in the axes are explained below the figure.

While this derivation is particularly complex, the physical interpretation of the result helps explain how a quadrupole is able to perform a scan. The final solution is dependent on six variables, but the simplified two-variable problem is shown above. Utilizing the reduced parameters, a and q, the problem becomes a more manageable two-dimensional problem. While the complete derivation allows researchers to perform scans in multiple ways, we will focus only on the basic mode that makes up the majority of MSs. For the majority of commercially available MSs, *the magnitude of the AC potential (V) and the DC potential (U) are the only parameters that are altered during run time* and allow a sweep of mass to charge ranges. The rest of the parameters that

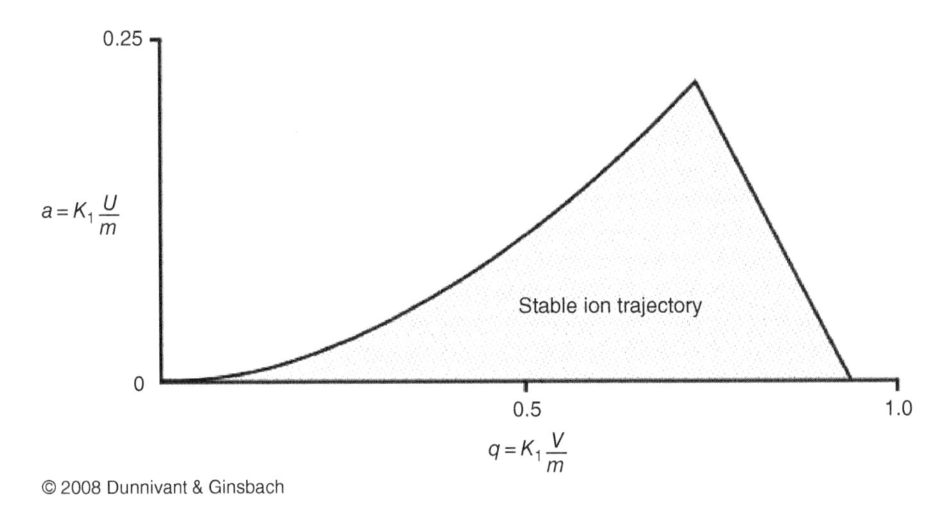

Figure 13.14 The general stability diagram (Source: Adapted from Steel and Henchman (1998)).

describe K_1 and K_2 are held constant. The values for K_1 and K_2 in the general stability diagram can be attributed to Equations 13.10 and 13.11

$$K_1 = \frac{2e}{r^2\omega^2} \tag{13.10}$$

$$K_2 = \frac{4e}{r^2\omega^2} \tag{13.11}$$

The parameters that make up K_1 and K_2 are exactly what we predicted when listing the variables earlier that would affect the point charge. Both K terms depend upon the charge of the ion e, its position within the quadrupole r, and the frequency of the AC oscillation ω. These parameters can be altered, but for the majority of applications remain constant. The charge of the ion (e) can be assumed to be equivalent to a positive one for almost all cases. The distance from the center of the quadrupole (r) is carefully controlled by the manufacturing process and an Einzel lens that focuses the ions into the center of the quadrupole and is also a constant. Also the angular frequency (ω) of the applied AC waveform can be assumed to be a constant for the purposes of most spectrometers and for this discussion.

The first important note for the general stability diagram is the relationship between potential and mass. The general stability diagram (in Figure 13.14) is now illustrated where there is an inverse relationship between the two. Figure 13.15 shows the lighter ions ($m - 1$) are higher on the mass scan line and the heavy ions ($m + 1$) are lower on the line. This is why, in Figure 13.15 at point A, the molecule was too light for the selected frequencies, and it was too heavy at point C.

From the general stability diagram, it is also possible to explain how an instrument's resolution can be altered. The resolution is improved when the mass scan line intersects the smallest area at the top of the stability diagram (Figure 13.15). The resolution can be improved when the slope of the mass line is increased. The resolution will subsequently increase until the line no longer intersects the stability diagram. While it would be best for the line to intersect at the apex of the stability diagram, this is impractical due to fluctuations in the AC (V) and DC (U) voltage. As a result, the line intersects a little below this point allowing the quadrupole to obtain unit resolution.

Once the resolution has been determined, the ratio of the AC to DC potential is left unchanged throughout the scan process. Again, to perform a scan, the magnitude of the AC and DC voltages is altered while their ratio stays constant. This places a different mass to charge inside the stability diagram. For example, if the AC and DC voltages are doubled, the mass to charge ratio of the selected ion would also be doubled as illustrated in the second part of Figure 13.15. By scanning throughout a voltage range, the quadrupole is able to create the majority of mass spectra produced in today's chemical laboratories. But it should be noted that quadrupole mass filters have an upper limit of approximately 650 amus.

Now that we have given a detailed description of the factors influencing the movement of a charged particle through the quadrupole, it is advantageous to summarize the entire process as a physicist would do in the form of a force balance. This is the origin of the governing equation where the French scientist E. Mathieu balanced the equations for the motion of ionized particles to the potential forces (electrical potentials) encountered in a quadrupole mass analyzer. This six-parameter differential equation, known as the Mathieu equation, is represented by

$$\frac{d^2u}{d\xi^2} + [a_u + 2qu\cos 2\xi]u = 0 \tag{13.12}$$

where

$$a = \frac{4eU}{\omega^2 r_0^2 m} \tag{13.13}$$

and

$$q = \frac{2eV}{\omega^2 r_0^2 m} \tag{13.14}$$

and where u is either the x or y directional coordinate, ξ is the redefining of time ($t/2$), e is the charge of the ion, U is the magnitude of the DC potential, ω is the angular frequency (2pf) of the applied AC waveform, r_o is the distance from the center axis (the z-axis) to the surface of any electrode (rod), m is the mass of the ion, and V is the magnitude of the applied AC or radiofrequency waveform. By using the reduced terms, a and q, the six-parameter equation (e, w, r_0, m, U, and V) can be

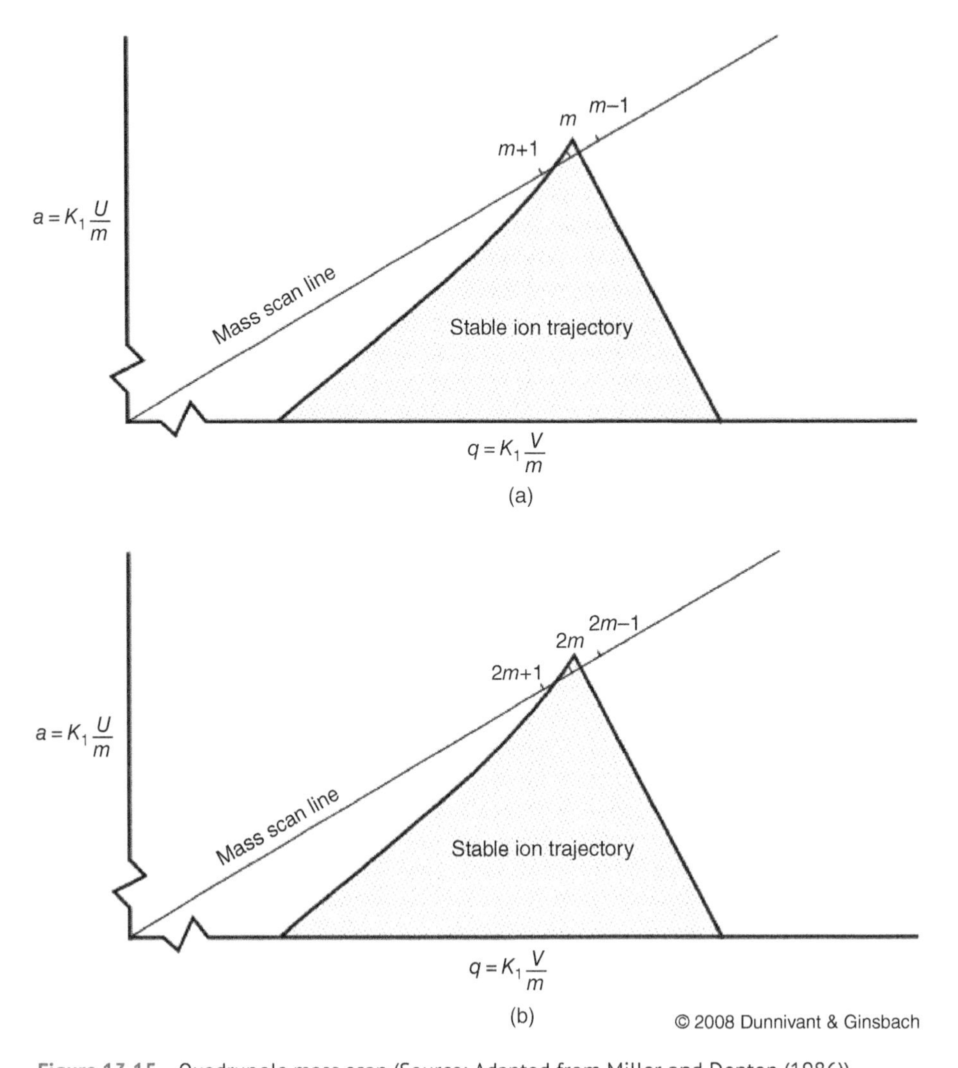

Figure 13.15 Quadrupole mass scan (Source: Adapted from Miller and Denton (1986)).

simplified to a two-parameter equation (involving a and q). Thus, when the two opposing forces are balanced, the movement of a charged particle in an electrical field, the particle will pass through the quadrupole and strike the detector.

13.5.3 Quadrupole Ion Trap Mass Filter

While the operation of the ion trap was characterized shortly after the linear quadrupole in 1960 by Paul and Steinwedel, its application in the chemical laboratory was severely limited. This was due to difficulties associated with manufacturing a circular electrode and performance problems. These performance problems were overcome when a group at Finnigan MAT led by Stafford discovered two breakthroughs that led to the production of a commercially available ion trap mass filter. The first ion trap developed used a mode of operation where a single mass could be stored in the trap when previously all of the ions had to be stored. Their next important discovery was the ability of 1 mTorr of helium gas to improve the instrument resolution. The helium molecules' collisions with the ions reduced their KE and subsequently focused them toward the center of the trap.

After these initial hurdles were cleared, many new techniques were developed for a diverse set of applications especially in biochemistry. This is a result of its comparative advantage over the quadrupole when analyzing high-molecular-mass compounds (up to several thousand m/z units) to unit resolution in commonly encountered instruments. The ion trap is also an extremely sensitive instrument which allows a molecular weight to be determined with a small number of molecules. The ion trap is also the only mass filter that can contain ions that need to be analyzed for any significant duration of time. This allows the instrument to be particularly useful in monitoring the kinetics of a given reaction. The most powerful application of the ion trap is its ability to be used in tandem mass spectrometry (Section 13.5.7).

The ion trap is made up of a single ring electrode that is placed in the X–Y plane between two endcap electrodes (Figure 13.16). Both an AC and DC voltage can be applied to the ring electrode while only an AC voltage can be applied to the endcap electrodes. The two endcap electrodes and the ring electrode ideally have a hyperbolic shape to establish an ideal field however in practice,

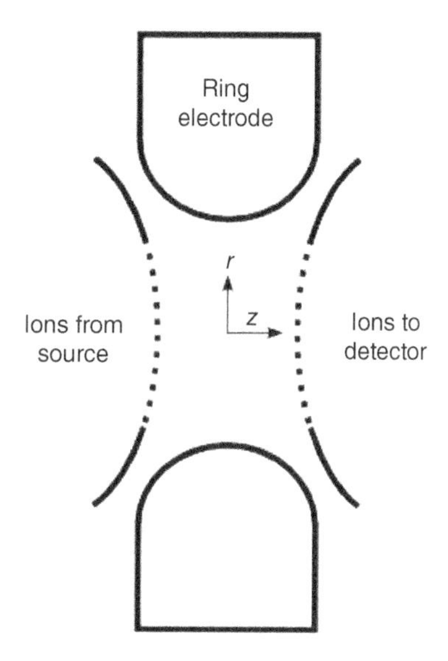

Figure 13.16 A cross-section of the ion trap (Source: Dunnivant and Ginsbach (Authors)).

nonideal fields can operate effectively. While the ion trap is applying force to the charged ions in three directions, the problem can be simplified into a two-dimensional problem. Since the ring is symmetrical, the force in any direction is always the same. As a result of this symmetry, movement of the molecules can be expressed in terms of r and z where $r = \sqrt{x^2 + y^2}$ where x and y are coordinates. For commercially available instruments, r_0 (the distance from the center of the trap to the ring electrode is either 1.00 or 0.707 cm.

After the sample molecules have been ionized by the source, they enter into the ion trap through an electric gate located on a single-endcap electrode. This gate functions in the same fashion as the one that is utilized in TOF mass spectrometry (Section 13.5.4). The gate's purpose is to prevent a large number of molecules from entering into the trap. If too many sample molecules enter into the trap, the interaction with other molecules becomes significant resulting in space-charge effects, a distortion of the electrical field that minimizes the ion trap's performance. Once the ions enter the trap, their collisions with the helium gas focus the ions toward the center of the trap. An AC frequency is also applied to the ring electrode to assist in focusing the ions toward the center of the trap.

In the ion trap, the ring electrode oscillates with a very high frequency (typically 1.1 MHz) while both the endcap electrodes are kept at a ground potential ($U = 0$ V). This frequency causes the ion to oscillate in both the r and z direction (Figure 13.17). The oscillation in the r direction is an expected response to the force generated by the ring electrode. The oscillation in the z direction, on the other hand, may seem counterintuitive. This is a response to both the grounded endcap electrodes and the shape of the ring electrode. When the AC potential increases, the trajectory of the ion becomes unstable in the z direction. The theoretical basis for this motion will be discussed

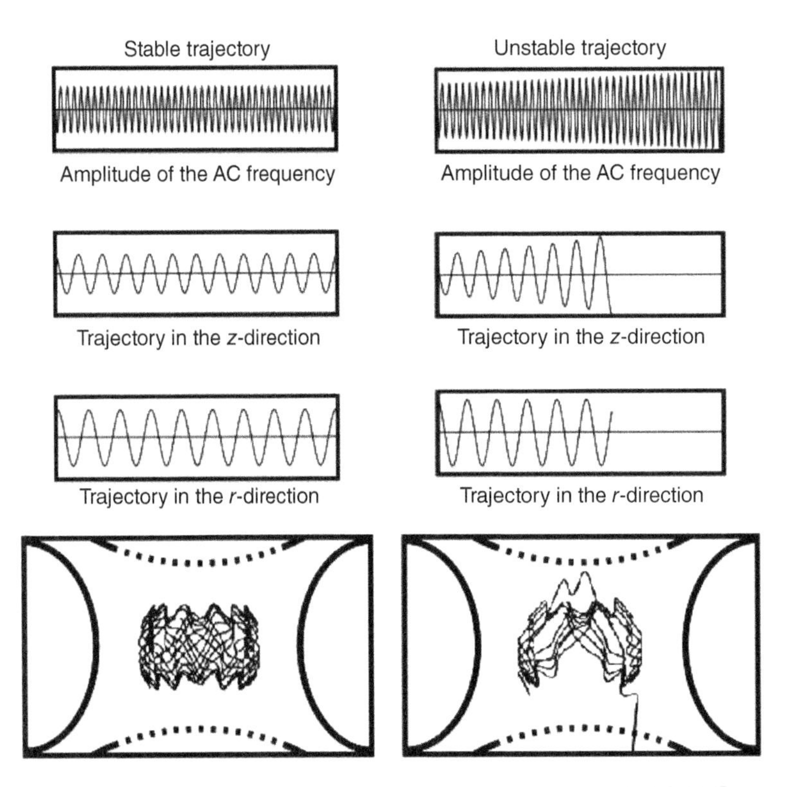

Figure 13.17 The trajectories of a single mass within the electrical field (Source: Adapted from Wong and Cooks, 1997).

later. While it would be convenient to describe the ion trap's function as a point charge responding to an electrical field, the complexity of the generated field makes this impractical.

The simplest way to understand how the ion trap creates mass spectra is to study how ions respond to the electrical field. It is necessary to begin by constructing a stability diagram for a single ion. Imagine a single mass to charge ratio being introduced into the ion trap. Then, the AC and DC voltages of the ring electrode are altered and the ion stability in both the z and r directions are determined simultaneously. If this experiment was performed multiple times, the stability diagram for that single mass would look similar to Figure 13.18.

The yellow area indicates the values of the AC and DC voltages where the given mass has a stable trajectory in the z direction but the ion's trajectory in the r direction is unstable. As a result, the ion strikes the ring electrode, is neutralized, and removed by the vacuum. The blue area is voltages where the ion has a stable trajectory in the r direction, but not in the z direction. At these voltages, the ion exits the trap through the slits in the endcap electrode toward a detector. The detector is on if the analyst is attempting to generate a mass spectrum, and can be left off if the goal is to isolate a particular mass to charge ratio of interest. The gray–purple area is where the stability in both the r and z direction overlap. For these voltages, the ion has a stable trajectory and remains inside the trap.

Similar to the quadrupole mass filter, differential equations are able to expand the single mass stability diagram to a general stability diagram. The derivation of this result requires an indepth understanding of differential equations, so only the graphical result will be presented here (Figure 13.19). As with the linear quadrupole mass filter, the solution here is simplified from a six-variable problem to a simpler two-variable problem.

From the general stability diagram, it becomes visible how scans can be performed by just altering the AC voltage on the ring electrode. But before we discuss the ion trap's operation, it is necessary to understand the parameters that affect ions stability within the field. The terms K_1 and K_2 are characterized by the following equations:

$$K_1 = \frac{4e}{r_0^2\omega^2} \tag{13.14}$$

$$K_2 = \frac{-8e}{r_0^2\omega^2} \tag{13.15}$$

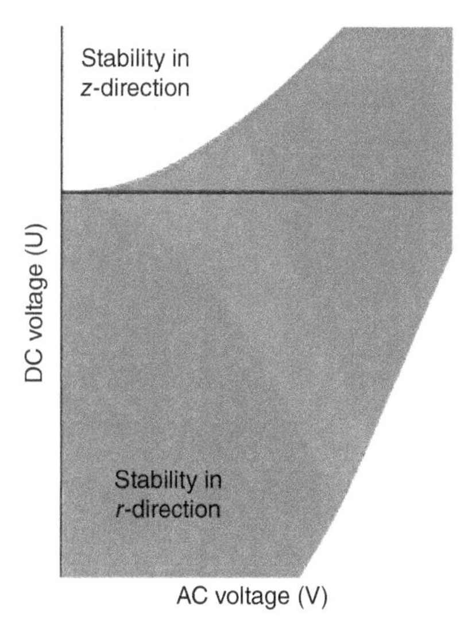

Figure 13.18 A single mass stability diagram for an ion trap (Source: Adapted from Wong and Cooks, 1997).

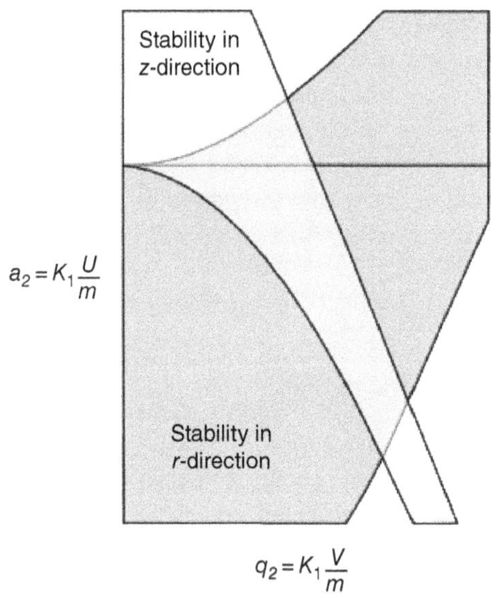

Figure 13.19 A general stability diagram (Source: Adapted from Wong and Cooks, 1997).

As expected, these parameters are very similar to the ones that resulted from the general stability diagram for the quadrupole mass filter. These parameters, like in the quadrupole, are also kept constant during a scan. The charge of the particle (e), the distance from the center of the trap to the ring electrode (r_0), and the radial frequency of the AC voltage (ω) are all kept constant during the run. While it would be possible to alter both the AC and DC voltages, in practice, it is only necessary to alter the AC voltage (V) of the ring electrode. The DC voltage (U) on the other hand, is kept at zero. If the DC voltage is kept at a ground potential, increasing the AC voltage will eventually result in an unstable trajectory in the z direction. When AC voltage creates a q_z value that is greater than 0.908, the particle will be ejected from the trap toward a detector through the endcap electrode. As illustrated below, the q_z value is dependent on the mass to charge ratio of the particle, each different mass has a unique AC voltage that causes them to exit the trap.

For example, let us place four different ion masses into the ion trap where each has a single positive charge. The general stability diagram in Figure 13.19 is identical to Figure 13.20 except that it is focused around a DC voltage (U) of zero and the scale is enlarged; thus, a_x is equal to zero through a scan. A mass scan is performed by starting the ring electrode out at a low AC voltage. Each distinct mass has a unique q_z value, which is visually illustrated by placing these particles on the stability diagram. As the AC frequency begins to increase, the q_z values for these masses also increase. Once the q_z value becomes greater than 0.908, the ions still have a stable trajectory in the r direction but now have an unstable trajectory in the z direction. As a result, they are ejected out of the trap through the endcap electrode toward the detector.

The stability diagram above at A, B, and C was the result of taking a snapshot of the AC voltage during the scan and placing each mass at its corresponding q_z values for that particular voltage. In this mode of operation, the lightest masses (m_1) are always ejected from the trap (Figure 13.20B) before the heaver masses (m_2). The heaviest masses (m_3 and m_4) still remain in the trap at point C. To eject these ions, a very large AC voltage is necessary. This voltage is so high that it becomes extremely difficulty to eject ions over an m/z value of 650. Since it is impractical to apply such high voltages to the electrode and its circuits, a new method of operation is needed to be discovered so the ion trap can analyze more massive molecules.

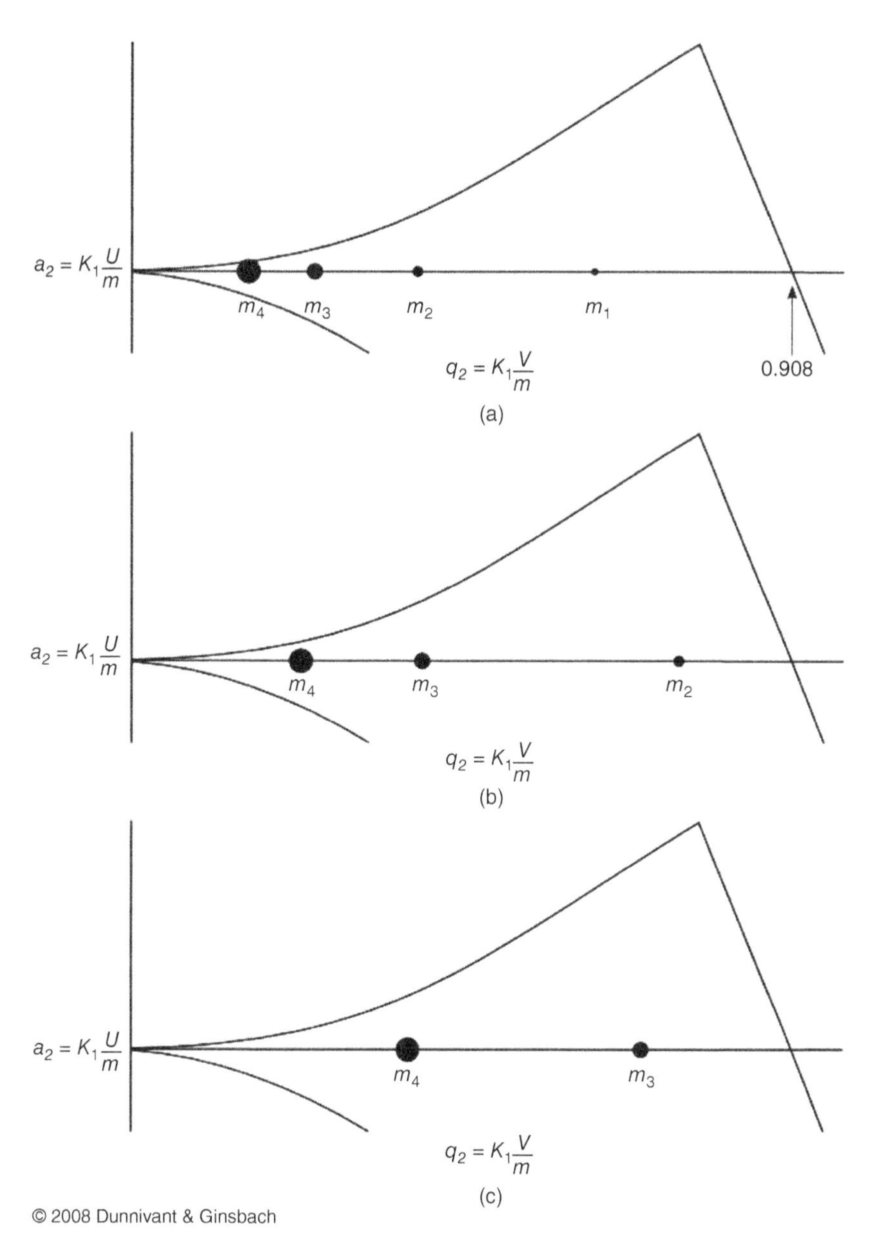

Figure 13.20 A stability diagram during a sample scan (Source: Dunnivant and Ginsbach (Authors)).

As a result, resonance ejection was developed to extend the mass range of the ion trap to an m/z value of several thousand. Under normal scanning conditions, ions oscillate at a given frequency depending on their q_z value which is a function of its mass, charge, and the amplitude of the AC voltage. This frequency is referred to as the ion's secular frequency. It was discovered that an AC voltage applied to the endcap electrodes would only affect one ion's secular frequency. The effected ion's oscillation in the z direction would increase linearly until it was ejected from the trap. Resonance ejection can be conceptualized as a "hole" inside the stability diagram at any chosen q_z value. Then the AC voltage of the ring electrode can be altered so any mass can have the same q_z value as

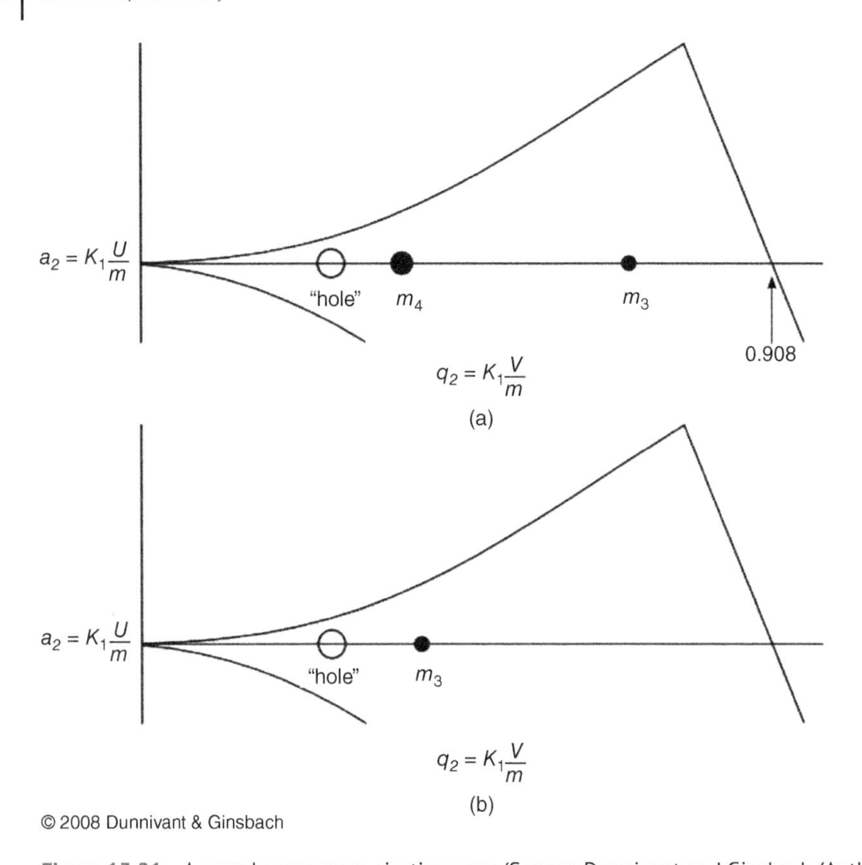

Figure 13.21 A sample resonance ejection scan (Source: Dunnivant and Ginsbach (Authors)).

the "hole" and exit the trap in the *z* direction (Figure 13.21). This mode of operation not only extended the mass analyzer's mass range, but it also made it possible to eject ions from the trap in any order. Before this mode of operation existed, it was only possible to eject the ions in order from lightest to heaviest. Figure 13.21 illustrates how it is possible to eject the heaviest ion (m_4) before the lighter ion (m_3).

The resonance ejection mode of operation is one reason why the ion trap is such a valuable tool. It not only greatly extends the mass range of the mass analyzer, but it also increases its applications in tandem spectroscopy (Section 13.5.8). The ability to isolate any given mass under several thousand amu is an extremely powerful tool. Through the use of both modes of operation, the ion trap has become a valuable tool in performing many specialized mass separations.

View Video S.13.5 at this time for an illustration of how an ion trap mass filter contains and ejects ions of given mass to charge ratios.

13.5.4 Time-of-flight Mass Filter

While TOF mass filters were one of the first MS systems to be developed, they had limited use due to their need for very fast electronics to process the data. Developments in fast electronics and the need for mass filters capable of resolving high mass ranges (such as in MALDI systems) have renewed interest in TOF systems. TOF is used exclusively with MALDI systems and also has other applications, as in HPLC where high molecular weight compounds are encountered.

Entry into the TOF mass filter is considerably different than with other mass filters. The entry has to be pulsed or intermittent in order to allow for all of the ions entering the TOF to reach the detector before more ions are created. With sources that operate in a pulsing fashion such as MALDI or field desorption, the TOF functions easily as a mass analyzer. In sources that continually produce ions such as a GC system or an EI source, the use of a TOF is more difficult. In order to use a TOF system with these continuous sources, an electronic gate must be used to create the necessary pulse of ions. The gate changes the potential on an accelerator plate to only allow ions to enter the TOF mass filter in pulses. When the slit has a positive charge, ions will not approach the entryway to the mass analyzer and are retained in the ionization chamber. When all of the previously admitted ions have reached the detector, the polarity on the accelerator(s) is again changed to negative and ions are accelerated toward the slit(s) and into the TOF mass analyzer. This process is repeated until several scans of each chromatographic peak have been measured. (This type of ionization and slit pulsing will be shown in the animation below.) The other way to interface EI with TOFs is to operate the EI source in a pulsing mode. This is achieved by maintaining a constant negative polarity on the accelerator plate/slit, and pulsing the EI source. This method can also periodically introduce packets of ions into the TOF mass filter.

Whichever type of ionization and entry into the TOF mass filter is used, the remainder of the process is the same. Prior to developing the mathematics behind TOF separations, a simple summary is useful. Mass to charge ratios in the TOF instrument are determined by measuring the time it takes for ions to pass through the "field-free" drift tube to the detector. The term "field-free" is used since there is no electronic or magnetic field affecting the ions. The only force applied to the ions occurs at the repulsion plate and the acceleration plate(s) where ions obtain a similar KE. All of the ions of the same mass to charge ratio entering the TOF mass analyzers have the same KE and velocity since they have been exposed to the same voltage on the plates. Ions with different mass to charge ratios will have velocities that will vary inversely to their masses. Lighter ions will have higher velocities and will arrive at the detector earlier than heavier ones. This is due to the relationship between mass and KE.

$$\mathrm{KE} = \frac{mv^2}{2} \tag{13.16}$$

The KE of an ion with a mass m and a total charge of $q = ze$ is described by

$$\frac{mv^2}{2} = qV_\mathrm{S} = z\mathrm{e}V_\mathrm{S} \tag{13.17}$$

where V_S is potential difference between the accelerator plates, z is the charge on the ion, and e is the charge of an electron (1.60×10^{-19} C). The length (d) of the drift tube is known and fixed. Thus, the time (t) required to travel this distance is

$$t = \frac{d}{v} \tag{13.18}$$

By solving Equation 13.17 for v and substituting it into Equation 13.18, we obtain

$$t^2 = \frac{m}{z}\left(\frac{d^2}{2V_\mathrm{S}\mathrm{e}}\right) \tag{13.19}$$

In a TOF mass analyzer, the terms in parentheses are constant, so the mass to charge of an ion is directly related to the time of travel. Typical times to traverse the field-free drift tube are 1–30 ms.

Advantages of a TOF mass filter include its simplicity and ruggedness and a virtually unlimited mass range. Additionally, virtually all ions produced in the ionization chamber enter the TOF mass filter and traverse the drift tube. However, TOF mass filters suffer from limited resolution, related to the relatively large distribution in flight times among identical ions (resulting from the physical width of the plug of ions entering the mass analyzer). In order to illustrate the TOF MS system, view Video S.13.6.

Video S.13.7 illustrates how a pulsed accelerator plate/slit acts as a gate for a reflective TOF mass filter system. (The system shown illustrates the analysis of metal isotopes with an inductively coupled plasma (ICP) but the reflective TOF works the same for organic analytes.

13.5.4.1 Ion Mobility Mass Spectrometry

If you have been in an airport recently, you have seen or your luggage has been analyzed by an ion mobility spectrometer (IMS). Although originally developed by Earl W. McDaniel of Georgia Institute of Technology in the 1950s, IMS systems have gained popularity recently due to their versatility—having been designed for specific classes of compounds, they have excellent detection limits, and they can be manufactured to be light-weight and mobile.

The basic design is similar to the TOF mass filter. Important differences are that they use an easier ionization source, they can be operated at atmospheric pressure and therefore do not necessarily require pressurized gases or high vacuum pumps, and as a result of their atmospheric pressure sample introduction, they have good detection limits. Samples are introduced at atmospheric pressure and ionized by corona discharge, atmospheric pressure photoionization (APPI), ESI, or a radioactive source such as a small piece of ^{63}Ni or ^{241}Am, similar to those used in ionization smoke detectors or GC electron capture detectors. The ionized analytes are then introduced to the drift tube by a gate valve similar to the one described earlier in this section for TOF mass filters. However, the IMS drift tube is different in that it can be operated at atmospheric pressure and is in a countercurrent environment. The analytes travel from left to right in the one-meter drift tube due to a 10–30 kV potential difference between the inlet and exit. As the analytes are mobile due to the potential they travel through a buffer gas that is passed from right to left in the drift tube (and atmospheric gases are commonly used). Separation of different analytes is achieved due to each ion having a different mass, charge, size, and shape (the ion mobility). As the ions are electrically drawn toward the detector, the ion's cross-sectional area strikes buffer gases and its velocity is reduced based on its size and shape. Larger ions will collide with more buffer gas and be impeded, travel slower, and arrive at the detector after longer times in the drift tube. Detectors for IMS are usually relatively simple Faraday cups but better detection limits can be obtained with an EM.

The most common use of IMS is for volatile organic molecules. IMS has been expanded for use in gas, liquid, and supercritical fluid chromatography.

13.5.5 Double-focusing Systems

The magnetic sector MS described earlier is referred to as a single-focusing instrument since it only uses the magnetic component to separate ion mass to charge ratios. This can be improved by adding a second electrostatic-field-based mass filter and is referred to as double focusing. A magnetic field instrument focuses on the distribution of translational energies imparted on the ions leaving the ionization source as a means of separation. But in doing so, the magnetic sector instruments broaden the range of kinetic energies of the ions, resulting in a loss of resolution. If we combine both separation techniques by passing the ions separately through an electrostatic field (to focus the KE of the ion packet) and magnetic field (to focus the translational energies of the ion packet), we will greatly improve our resolution. In fact, by doing this we can measure ion masses to within a few parts per million (precision) which results in a resolution of ~2500. Compare this to the unit resolutions (28 versus 29 Da) discussed at the beginning of this section (under resolution). On the downside, these instruments can be costly.

13.5.6 Fourier Transform Ion Cyclotron–Mass Spectrometry: (by Nicole James)

Developed by Alan G. Marshall and Melvin B. Comisarow at the University of British Columbia, the use of FT–ICR MS first began in 1974 with approximately 235 instruments in use by 1998. FT–ICR MS has higher mass resolution and accuracy than any other MS system and can detect multiple mass to charge ratio ions simultaneously. However, FT–ICR MS can be prohibitively expensive at $1–2 million for a standard instrument.

The general steps of an FT–ICR MS experiment are: (1) ion formation outside of the detector, (2) ion focusing and accumulation, (3) transportation of ions into a Penning trap, (4) selection of ions based on mass to charge ratio and ejection of these ions from the Penning trap, (5) excitation, (6) detection, (7) fast Fourier transform (FT) of the digital time-domain signal, and (8) conversion of frequency to mass to charge ratio.

13.5.6.1 Ion–Cyclotron Motion by Nicole James

If a moving ion is exposed to a uniform magnetic field, it is subject to a force dependent on the mass, charge, and velocity of the ion. If an ion does not collide with another particle and hit off its natural course, the magnetic field will bend the ion's path into a circular orbit (Figure 13.22).

The motion of the ion can be described by the equation below, where w is the unperturbed ion cyclotron frequency, B_0 is the magnetic field in Tesla, q is the charge in Coulombs, and m is the mass in micrograms.

$$w = \frac{qB_0}{m} \tag{13.20}$$

This equation can be rearranged into the following equation where v is the velocity and z is the charge of the ion in units of elemental charge (e.g. +1, +2, etc.).

$$v = \frac{w}{2\pi} = \frac{1.5356 \times 10^7 B_0}{m/z} \tag{13.21}$$

It is important to note that the above equation is dependent on only the mass to charge ratio of the ion and not its velocity. This makes ion–cyclotron resonance especially useful in mass spectrometry, as one does not need to focus translational energy—which requires longer experiment times, larger apparatus, and more powerful electronics—in order to obtain high-accuracy results.

The radius of the circle an ion makes when exposed to the magnetic field can be found by the equation below, where r is radius in meters, and T is the temperature in Kelvin:

$$r = \frac{1.3365 \times 10^6}{zB_0}\sqrt{mT} \tag{13.22}$$

One can see from the above equation that an ion with a mass of 100 amu and a charge of +1 in a magnetic field of 1 Tesla at room temperature (298 K) would have a radius of 0.2 mm; the same ion with triple the magnetic field (3 T) at room temperature would have a radius of 0.077 mm. Thus, ions can be easily confined to a relatively small orbit by a reasonable magnetic field; this is called ion trapping and is vital to ICR–MS because the longer (approximately 1 s) experiment times require one to be able to retain the ions in a designated space. Additionally, a 3 T magnetic field

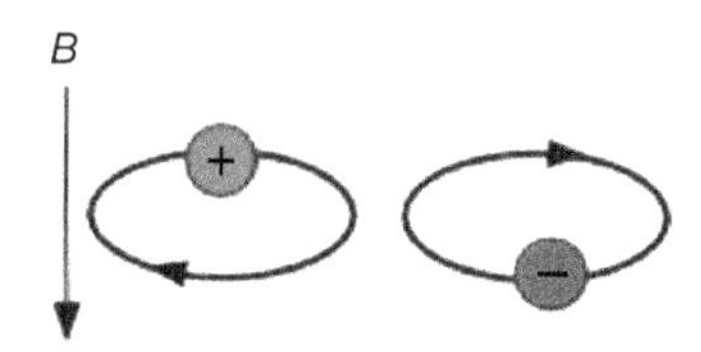

Figure 13.22 Illustration of a particle in a magnetic field (Source: Dunnivant and Ginsbach (Authors)).

is easily attainable for commercially available electronics. The largest FT–ICR MS built as of 2010 can attain a magnetic field of 15 T, allowing one to confine an ion with an m/z value of 60 000.

13.5.6.2 Ion–Cyclotron Excitation and Detection

A number of ions at a specific mass to charge ratio spinning in an ion–cyclotron orbit does not, itself, generate an observable electric signal because (1) the ions were randomly placed (i.e. incoherent; ions are spread throughout the radio of orbit) as they began orbiting, meaning that an ion at a specific position will have its charge cancelled out by an ion half an orbit away from it, leaving no net electrical current and (2) the radius of the orbits are generally too small to be detectable, even if all ions were in the same phase. Thus, ions must be excited in order to be detected.

Particles in an ion–cyclotron orbit can be excited by applying an oscillating or rotating uniform electric field at or near the frequency of ions of a given mass to charge ratio. This excitation can be used for three purposes: (1) accelerating the ions into a larger orbital radius for detection, (2) accelerating the ions to a larger orbital that is ejected from the ion trap, and (3) increasing the KE of an ion to the point that it further ionizes or reacts with another molecule. For the purposes of this text, excitation in order to accelerate the ions for detection is most significant.

Applying an oscillating or rotating radiofrequency electric field in resonance with (at the same frequency as) a specific m/z value or range applies a force on the ion(s) that continuously enlarges the circular orbit of the ion(s) at one point—in other words, the orbiting ions begin to spiral outward. Ions of different types will spiral outward at different rates. The post-excitation orbit for an ion excited for a period of time, t, is shown in the following equation, where E_0 is the applied electric field and B_0 is the magnetic field:

$$r = \frac{E_0 t}{2B_0} \tag{13.23}$$

The fact that the above equation is independent of the mass to charge ratio of the ion means that all ions can be excited by a radiofrequency electric field to enlarged ion–cyclotron orbits for detection. This simultaneous detection vastly decreases both the time an experiment will take and the amount of analyte required.

When a group of ions with the same mass to charge ratio are excited, they are pushed off-axis due to their spiraling nature. By pushing the ions off-axis, not all ions have a "partner" ion half a cycle away—the ions are considered to be "cohered." A cohered packet of orbiting ions causes a difference in current between opposing detection plates within the ion trap; this differential current can be modeled as an "image" current opposing the current on the detection plates; this image current is proportional to the number of coherent orbiting ions. This is the ICR signal; the ICR signal increases linearly with increasing ion–cyclotron radius after excitation and with increasing ion charge. Throughout most of the frequency range possible on the instrument, the signal to noise ratio (S/N) is proportional to the differential current observed. The number of ions required for an S/N ratio of 3:1 on a standard instrument using standard parameters is approximately 190 ions. Other detection processes have been designed for such high accuracy and detection that they are able to detect a single ion and have been used to corroborate the theory that protons and antiprotons do, in fact, have the same mass (Gabrielse et al., 1990) (refer to Figure 13.23).

13.5.6.3 Penning Trap

The most common ion trap used in FT–ICR MS is the Penning trap, designed in the 1950s by Hans Georg Dehmelt, who named it after Frances Michel Penning for his work on the Penning

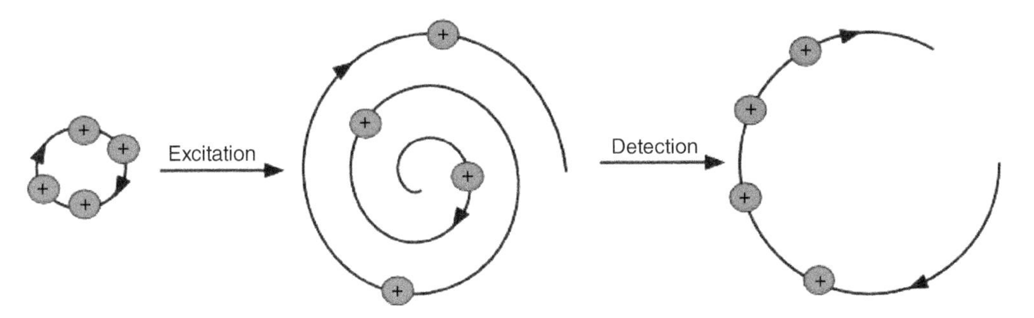

Figure 13.23 Excitation and detection of an ion (Source: Nicole James (Author)).

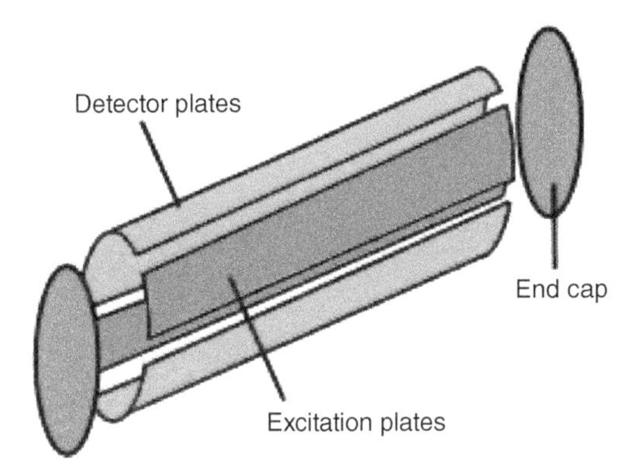

Figure 13.24 Diagram of a Penning trap (Source: Nicole James (Author)).

gauge. The ion–cyclotron motion induced by a radial magnetic field contains ions radially, but it is necessary to add an axial electric field in order to trap the ions axially. Thus, the motion of an ion inside a Penning trap is essentially the combination of three distinct motions: the cyclotron, "magnetron" (a component of ion–cyclotron motion), and the axial motion. The axial containment is accomplished by introducing two "endcap" electrodes. The endcap electrodes are coupled by capacitance, which allows for a nearly perfect rf electric field to be used for the ion–cyclotron excitation without any negative effects on other electronics. Opposing plates with an electric field applied across them within the Penning trap (Figure 13.24) are used as detector plates.

13.5.6.4 Analysis of Results

The signal detected by an experiment is in units of current per time. To extract mass to charge data, one must apply an FT. In general, an FT takes a time-based signal and converts it into a frequency-based plot. Since the initial function is a function of time, it is typically called the *time domain*; the frequency plot is called the frequency domain, or the *frequency domain representation* of the initial function. More specifically, an FT uses the fact that almost any function can be degraded into a sum of sine and cosine waves; each component sine and/or cosine wave represents a periodic component of the data. By finding each component of a sine or cosine wave, one can make a frequency plot by representing a specific sine or cosine wave as a peak on a plot of amplitude versus the frequency of the wave. The sharper the peak, the more "exact" the periodicity is; in most real-life applications, the peak will be somewhat broad—not just a vertical line (see Figure 13.25).

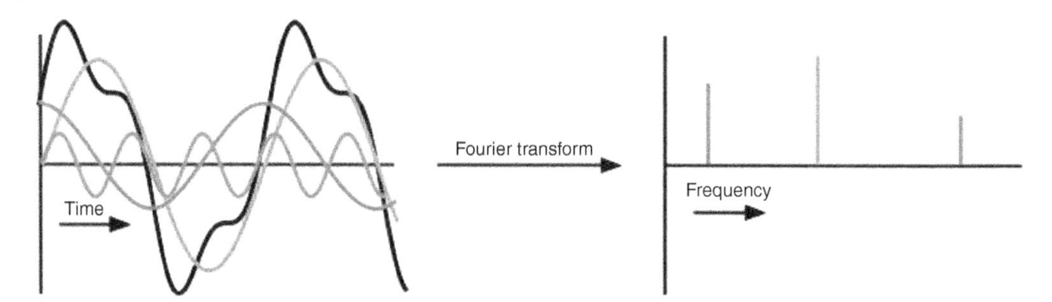

Figure 13.25 Graphic representation of the Fourier-transform process where a time domain signal is transformed to a frequency output (Source: Dunnivant and Ginsbach (Authors)).

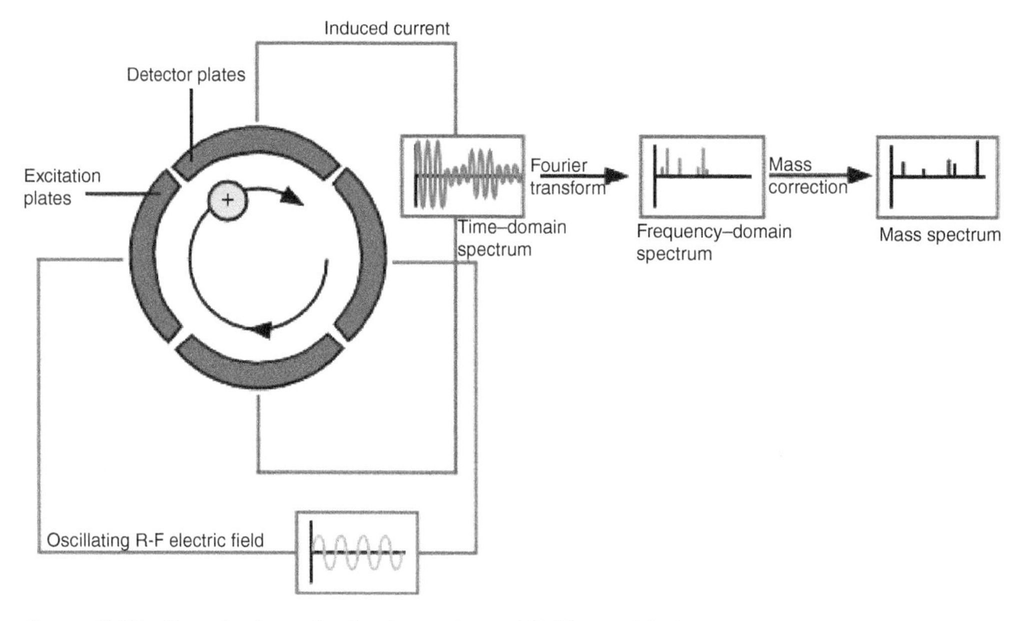

Figure 13.26 Overall schematic of an ion cyclotron MS (Source: Nicole James (Author)).

An FT of the (time-domain) ICR response results in a frequency plot that can be mass-corrected to result in a mass spectrum. Obtaining this mass spectrum with most other types of MS would have required sweeping slowly across the entire range of mass to charge ratios; being able to quickly and simultaneously detect all mass to charge ratios decreases the time, effort, and supplies that must be used to test a sample. In addition, the greatly increased resolution means that FTICR–MS will continue being an extremely powerful instrument (refer to Figure 13.26).

13.5.7 Orbitrap Analyzers (by Nicole James)

Designed in 2005 by Alexander Makarov, the Orbitrap MS features a mass resolution of up to 150 000, high mass accuracy (2–5 ppm, compared with approximately 20 ppm for quadrupole systems), a mass to charge ratio range of 6000 and a dynamic range larger than 1000.

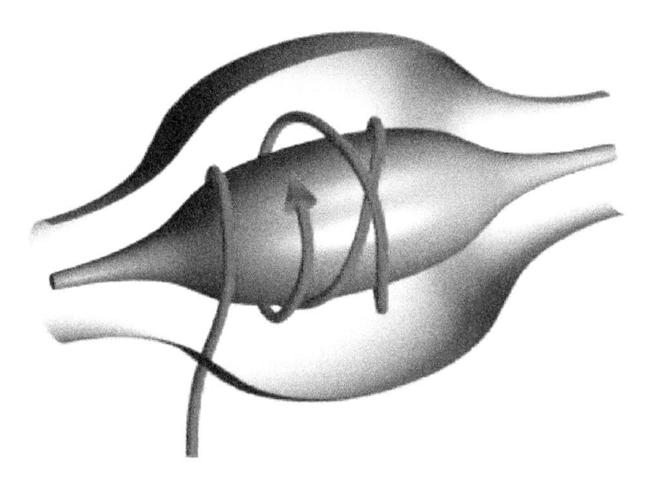

Figure 13.27 The orbitrap (Source: Gbdivers (2020)/Wikimedia Commons/CC BY-SA 3.0 DEED).

The Orbitrap works similarly to an FT ICR-MS: all ions are identified simultaneously by reading an image current of oscillations that are unique to a given mass to charge ratio and an FT is applied to the data to isolate individual signals. However, the Orbitrap requires no magnet, no RF field, and no excitation sequence. Despite this, Orbitrap systems generally cost at least $600 000.

Ions are first ionized by a given source; given the large m/z range, Orbitrap systems are often used to study biological molecules such as proteins, peptides, and oligosaccharides—consequently, one of the most common ionization methods is ESI. The ions are then transported to a storage cell, generally a storage quadrupole, which is kept at a vacuum near 10^{-3} mbar. A series of transfer lenses gradually increases the electric field experienced by the ions until they are at the level of the Orbitrap.

After ions have been transferred into the Orbitrap, the system uses only electrostatic (DC) fields. The Orbitrap itself is composed of an outer "barrel" electrode, an inner "spindle" electrode, and two endcap electrodes. Upon introduction into the Orbitrap, stable ion trajectories will result in orbiting around the center electrode while also oscillating in the z direction. The motion in the z direction can be described as an harmonic oscillator, which is described in Equation 13.24, where ω is oscillation frequency, z is the ion charge, m is the ion mass, and k is the field curvature and illustrated in Figure 13.27.

$$\omega = \frac{zk}{m} \tag{13.24}$$

While the frequency of orbiting the central electrode is also dependent on the ion's mass to charge ratio, this frequency is also dependent on the ion's energy and when it was introduced into the Orbitrap, whereas oscillations in the z direction are independent of energy and any initial parameters. The oscillations in the z direction are read by the image current produced on the endcap electrodes. While all ions of a given mass to charge ratio oscillate in phase for hundreds of thousands of oscillations, small imperfections in the Orbitrap or orbital shape, along with occasional collisions with background gas molecules (despite the 10^{-10} mbar vacuum) can result in the loss or displacement of some ions, ultimately resulting in a slow decrease in the intensity of the signal until it is completely lost in instrument noise. This results in a free induction decay (FID), similar to that which is acquired in NMR analysis. An FT of the FID results in mass spectrum.

13.5.8 Tandem Mass Spectroscopy

Mass spectroscopy is commonly referred to as a confirmatory technique since there is little doubt (error) in the identity of an analyte. To be even more certain of an analyte's identity, two or even three, MSs can be used in series (the output of one MS is the input of another MS). Most often a soft ionization source, such as chemical ionization, is used in the first MS and allows for selection of the molecular ion in the first MS, while harder ionization is used in the second MS to create fragments. A subsequent MS will select a specific ion fragment from the second MS and further fragment it for identification. This technique allows a molecular ion (or ion fragment) to be isolated in the first MS, subsequently fragmented in the second and third MS, and identified based on its final fragment pattern. You should be able to see the confirmatory nature of this technique.

Mass filters of choice for use in tandem include magnetic sector, electrostatic, quadrupole, and ion trap systems. In the absence of HPLC or GC introduction, tandem MS offers many of the same advantages of a single GC–MS or HPLC–MS system but it is much faster since the analyst does not have to wait on the chromatography portion of the analysis. For example, chromatography separations take from minutes to hours prior to entry into an MS, while tandem MS systems (without GC) require only milliseconds. But, of course, this saving in time is considerably more expensive than simple chromatography-based MS systems.

Two popular coupled chromatographic–MS systems are the triple quadrupole (QQQ) and a Q-TOF.

13.5.9 Recent Advances

Additional recent breakthroughs in mass spectrometry include the drastic lowering of detection limits. A new technique referred to as nanostructured initiator MS (NIMS) is being used in research-grade instruments to measure biological metabolites. Utilization of these systems with laser-based systems produces detection limits that are easily at the attomole (10^{-18}) amounts. These systems make the ppb and ppt detection limits discussed in this textbook seem trivial. It is likely that similar detection limits will soon be achieved for ICP-based instruments.

13.6 Ion Detectors

Once the analytes have been ionized, accelerated, and separated in the mass filter, they must be detected. This is most commonly completed with an EM, much like the ones used in optical spectroscopy. In MS systems, the EM is insensitive to ion charge, ion mass, or chemical nature of the ion (as a photomultiplier is relatively insensitive to the wavelength of a photon). EMs for MS systems can be a series of discrete dynodes as in the photomultiplier or they can be continuous in design. Most commonly, continuous EMs are used. Continuous EMs are horn-shaped and are typically made of glass that is heavily doped with lead oxide. When a potential is placed along the length of the horn, electrons are ejected as ions strike the surface. Ions usually strike at the entrance of the horn and the resulting electrons are directed inward (by the shape of the horn), colliding sequentially with the walls and generating more and more electrons with each collision. Electrical potentials across the horn can range from high hundreds of volts to 3000 V. Signal amplifications are in the 10 000-fold range with nanosecond response times. Video S.13.8 illustrates the response of a continuous EM as ions, separated in a mass filter, strike its surface.

At the end of the EM is a Faraday Cup that counts each ion entering the detector zone.

13.7 Three-dimensional Aspects of GC–MS

Typical chromatographic peaks were illustrated in earlier chapters. But as each chromatographic peak enters the MS, it is fragmented and separated into a series of ion fragments. When graphed together on an x, y, and z plots, the x-axis represents time and traces the arrival of each compound at the chromatographic detector and the z-axis represents the total detector response that is related to analyte concentration. The mass to charge spectrum of each chromatographic peak is represented by a series of lines that are parallel to the y-axis and show the arrival of molecular fragments at the MS detector. Again, detector response and concentration are represented by the height of each peak. This is illustrated for one chromatographic peak in Figure 13.28.

13.8 Summary

In this chapter, we illustrated the utility of combining chromatography and MS systems. A variety of possible components provide interesting instruments that can be used to analyze a broad range of analytes. Hard and soft ionization techniques provide for the determination of the molecular weight of the analyte, as well as unique fragmentation patterns for confirmational identification of an unknown chemical structure. More inexpensive instruments, such as quadrupole and TOF MSs, allow only unit resolution of ions while double-focusing instruments yield the determination of differences with resolution of four decimal points in masses. Mass spectrometry, like NMR, is one of the most powerful techniques available to chemists and it is becoming more and more

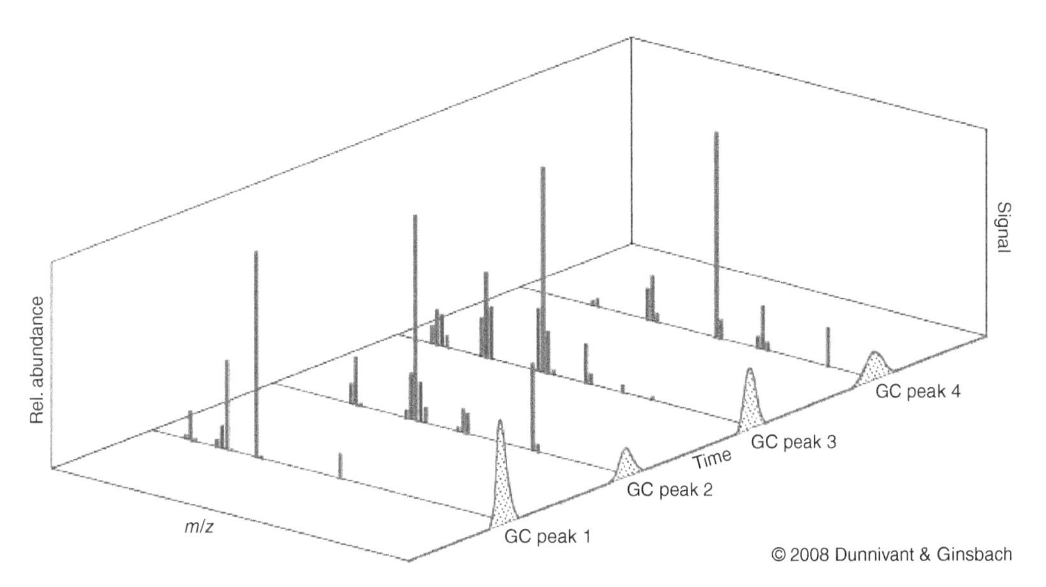

Figure 13.28 The three-dimensional nature of a GC–MS analysis (Source: Dunnivant and Ginsbach (Authors)).

Table 13.2 Summary of mass filter features.

Type of Mass Filter	Resolution	Detection Limit	Approximate Instrument Price (as of 2023)
Routine mass filters coupled with GC, LC, and CE (some not available)			
Ion mobility	50	Low ppm	\$50 000
Single quadrupole (within our chromatography system)	250–500	Low ppb to high ppt	\$50 000–60 000
Ion trap	1000–10 000	ppb	\$400 000–500 000
Time of flight	3000–10 000	High ppt	\$300 000–400 000
Double focusing	10 000–20 000	Mid- to high ppt	\$750 000–1 000 000
FT ion cyclotron	200 000–1 000 000	ppb	\$2 000 000+
New mass filters			
Magnetic sector/Multi-collector with the Mattauch–Herzog geometry	~500	High ppb	\$350 000–400 000
Proton transfer reaction ionization chamber	Depends on type of mass filter	ppt	\$120 000
Orbital trap (electrostatic ion trap)	150 000–200 000	ppb	\$600 000

(Source: Various Company Literature).

important. While most of the instruments presented in the chapter have detection limits in the subparts per million range, extremely lower detection limits (10^{--15} moles) have been obtained in research-grade instruments.

A summary of mass filters and their characteristics is given in Table 13.2.

13.9 Questions

1 Why are most mass filters maintained at a low pressure?

2 List the common ways samples are introduced into an MS system.

3 How can solid samples be introduced into an MS?

4 Draw and explain how the interface between a GC and an MS works.

5 Why do capillary columns, versus packed columns, work best for MS interfaces?

6 Explain the difference between hard and soft ionization in GC–MS.

7 Why does soft ionization reduce the fragmentation of analytes in GC–MS?

8 Write the chemical reactions occurring when methane is used in soft ionization.

9 Draw and explain how the interface between an LC and an MS works.

10 What is the major problem with interfacing LC (ESI) to MS?

11 Explain how MALDI works. What types of samples is it commonly used for? What type of MS is it commonly coupled with?

12 Draw and explain how the interface between a CE (ESI) and an MS works.

13 Explain resolution with respect to mass filters. Give relevant resolution numbers.

14 Draw and explain how a magnetic sector mass filter works.

15 Draw and explain (in detail) how a quadrupole mass filter works.

16 The governing equation of the quadrupole mass filter consists of a six-parameter differential equation. Which two parameters are used to control the mass filter?

17 What is the purpose of the DC voltage in the quadrupole MS?

18 What is the purpose of the AC cycle in the quadrupole MS?

19 How do the low-mass and high-mass filters work to create a stable cation region in the quadrupole MS?

20 Explain the mass scan line in the quadrupole MS figures.

21 What is the purpose of sweeping the DC–AC voltages?

22 Extend the concepts of a linear quadrupole mass filter, explained above, to explain how the quadrupole ion trap mass filter works.

23 How is the mass range of a quadrupole ion trap mass filter extended?

24 Explain the concept of resonance ejection in ion trap mass filters.

25 Draw and explain how a time-of-flight mass filter works.

26 Contrast traditional TOF and ion mobility MS.

27 Draw and explain how a PTR–MS works.

28 Give a brief explanation of how an ion cyclotron works.

29 Draw and explain how a double-focusing mass filter works. What are its advantages?

30 What is tandem mass spectrometry?

31 What types of detectors are used in mass spectrometry?

32 Use the date in Table 5.2 to contrast the various types of mass filters. Which is the most economical? Which has the best mass resolution?

Supporting Information

Additional supporting information may be found online in the Supporting Information section in the HTML rendition of this article.

References

Blakely C.R. and Vestal M.L. (1983). Thermospray interface for liquid chromatography/mass spectrometry. *Anal. Chem.* 55(4), 750.

Gabrielse, G., Fei, X., Orozco, L.A., Tjoelker, R.L., Haas, J., Kalinowsky, H., Trainor, T.A., and Kells, W. (1990). Thousandfold improvement in the measured antiproton mass. *Phys. Rev. Lett.* 65, 1317–1320.

Gbdivers (2020). https://en.wikipedia.org/wiki/Orbitrap

Golhke R.S., McLafferty F., Wiley B., and Harrington D. (1956). *First Demonstration of GC/MS*. Bendix Corporation.

Hansel, A., Jordan, A., Holzinger, R., Prazeller, P., Vogel, W. and Lindinger, W. (1995). Proton transfer reaction mass spectrometry: on-line trace gas analysis at the ppb level. *Int. J. Mass Spect. Ion Process.* 149–150, 609–619.

Miller, P. and Denton, M. (1986). The quadrupole mass filter: basic operating concepts. *J. Chem. Educ.*, 63(7), 617–622.

Steel, C. and Henchman, M. (1998). Understanding the quadrupole mass filter through computer simulation. *J. Chem. Educ.* 75(8), 1049–1054.

Wong, P. and Cooks, R. (1997). Ion trap mass spectrometry. *Curr. Seperations.com Drug Dev.* 16(3).

Yamashita M. and Fenn, J.B. (1984). Electrospray ion source. Another variation on the free-jet theme. *J. Phys. Chem.* 88(20), 4451.

Further Reading

de Hoffmann, E. and Stroobant, V. (2002). *Mass Spectrometry Principles and Applications*, 2e. New York: John Wiley & Sons.

Jonscher, K. and Yates, J. (1997). The quadrupole ion trap mass spectrometer—a small solution to a big challenge. *Anal. Biochem.* 244, 1–15.

Kanu, A.B., Dwivedi, P., Tam, M., Matz, L., and Hill, H.H. (January 2008). Ion mobility-mass spectrometry. *J. Mass Spect.* 43(1), 1–22.

March, R. (1997). An introduction to quadrupole ion trap mass spectrometry. *J. Mass Spect.* 32, 351–369.

Marshall, A.G. and Hendrickson, C.L. (2002). Fourier transform ion cyclotron resonance detection: principles and experimental configurations. *Int. J. Mass. Spect.* 215, 59–75.

Marshall, G., Hendrickson, C.L., and Jackson, G.S. (1998). Fourier transform ion cyclotron resonance mass spectrometry: a primer. *Mass. Spec. Rev.* 17, 1–35.

Skoog, D., Holler, F., and Nieman, T. (1992). *Principles of Instrumental Analysis*, 5th ed., Philadelphia: Saunders College Publishing.
Tarantin, N.I. (1997). Some aspects of the Penning trap theory. *Phys. Res. B* 126, 392–395.
Ubieto-Diaz, M., Rodriguez, D., Lukic, S., Nagy, Sz., Stahl, S., and Blaum, K. (2009). A broad-band FT-ICR Penning trap system for KATRIN. *Int. J. Mass. Spect.* 288, 1–5.
Zolotov, Yu. A. (2006). Ion mobility spectrometry. *J. Anal. Chem.*, 61(6), 519.

14

Fragmentation and Interpretation of Spectra

14.1 Introduction

Before discussing fragmentation and interpretation, it is important to understand the many ways mass spectra are utilized. For the analytical chemist, a mass spectrum is useful for two applications. The first is the relatively simple case when the analyst is looking for a particular compound in a sample and has a reference material to compare spectra. The second occurs when an analyst observes the presence of an unknown and wishes to identify it. The mass spectrum allows an experienced analyst to identify the compound or at a minimum narrow the possibilities down to a few compounds from the millions of potential chemicals. Then, a reference standard can be more easily selected from this knowledge to confirm the identity of this unknown. A similar situation exists for the synthetic chemist except their analytical toolbox is much larger. Sometimes synthetic chemists attempt to synthesize a final product that is known (for example, in an industrial process line). Here, the mass spectrum of the synthesized product is compared to a reference standard. Other times, a synthetic chemist is attempting to make a new compound where a reference standard is impossible to find.

All four problems center on the same difficult task, identifying the structure of a compound under various conditions. There are three main instruments that perform this task for organic compounds: infrared spectroscopy, mass spectroscopy, and nuclear magnetic resonance (NMR). It is very important that both synthetic and analytical chemists are able to choose the best tool for their particular problem. The mass spectrometer has a few advantages over the other analytical methods. Mass spectroscopy, when coupled with either gas or liquid chromatography, can analyze a complex mixture that an NMR or IR could not. An MS is also the only way to determine the molecular mass of a compound. The largest advantage for analytical chemists is that mass spectroscopy can elucidate structural information from a very small amount of a compound (part-per-million quantities).

The MS has a distinctive advantage over IR spectroscopy in that there is more structural information that can be determined, though the information contained in a mass spectrum is more difficult to interpret. The MS has an advantage over NMR in that it can be performed more quickly. However, both IR spectroscopy and mass spectroscopy have a distinct disadvantage when analyzing compounds with multiple functional groups. For these types of compounds and when the analyst has mg quantities of a relatively pure compound, NMR is usually the best analytical tool.

As a result of these advantages and disadvantages, mass spectroscopy is normally utilized to perform three tasks. The first is in analytical chemistry when there is a small concentration of analyte. The second is identifying compounds that contain few functional groups; a common procedure

Essential Methods of Instrumental Analysis, First Edition. Frank M. Dunnivant and Jake W. Ginsbach.

Companion Website: www.wiley.com/go/essentialmethodsofinstrumentalanalysis1e

in industrial synthesis. The third is confirming steps in a complex synthesis of a new product to determine the molecular mass and possibly some structural information. The products in the third example, however, are usually always checked by NMR.

After choosing to use mass spectroscopy, the selection of gas chromatography or liquid chromatography is equally important. Gas chromatography is utilized for volatile and thermally stable compounds (up to 300 °C). Liquid chromatography is usually utilized for all other compounds since it has poorer resolution than GC. As a result, this chapter will focus on interpreting structural information from the types of compounds commonly analyzed with GC–MS. Once a chemist is able to determine the identity of a compound from a mass spectrum, their problem has been solved.

14.2 Creation of the Spectra

As sample molecules exit the GC column and enter into the mass spectrophotometer, they encounter an energy source. For the purposes of this chapter, the source is an electron impact tungsten filament at 70 eV (Section 13.5.2). Energy emitted from the source removes a single electron from a sample molecule. This is the most basic reaction and is illustrated by methanol below (Figure 14.1).

$$e^- + CH_3OH \rightarrow [CH_3OH]^{\cdot+} + 2e^- \tag{14.1}$$

After these products move through the mass spectrometer, the detector is only sensitive to the positively charged molecules and is not sensitive to any neutral or radical molecules. The detector transforms the number of molecules into an electrical signal, and a computer or integrator translates this individual signal peaks into a bar graph. The abundance is plotted as a function of a molecule's mass divided by its charge (m/z) on a bar graph (Figure 14.1). Since almost all of the fragments detected by the GC–MS have only a single positive charge, m/z is also a measurement of the molecules' mass.

There are more bars on the graph than just the mass of the sample molecule. These other peaks are attributed to the cleavage of bonds in the original sample molecule. These fragments allow for the original structure of the sample molecule to be determined by looking at its various components (Section 14.8). Since the energy of the source exceeds the ionization energy of the sample molecule,

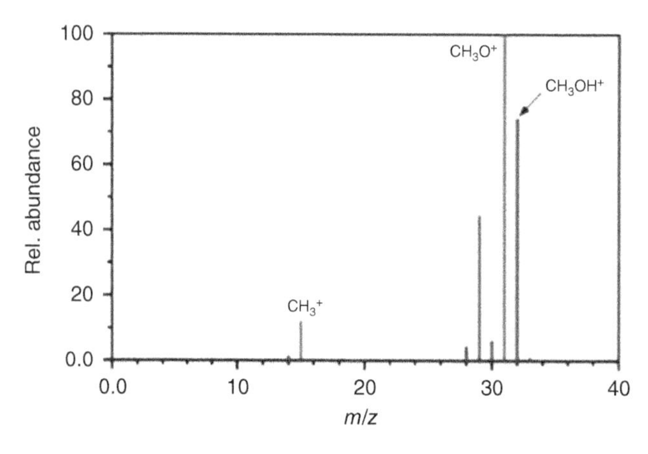

Figure 14.1 Mass spectrum of methanol (Source: NIST/National Institute of Standards and Technology/Public Domain).

the excess energy that is not utilized in the removal of a single electron is distributed over various electronic, vibrational, and rotational degrees of freedom (Section 14.6). Fractionation occurs when this energy exceeds the activation energy of any bond cleavage (Section 14.6). This feature allows the instrument to distinguish between compounds with the same molecular mass and constitutional isomers. The major fragments for methanol (Figure 14.1) can be attributed to the following reactions:

$$[CH_3OH]^{\cdot+} \rightarrow CH_3O^+ + H\cdot \tag{14.2}$$

$$[CH_3OH]^{\cdot+} \rightarrow CH_2O^+ + H_2 \tag{14.3}$$

$$[CH_3OH]^{\cdot+} \rightarrow CH_3{}^+ + \cdot OH \tag{14.4}$$

The two most important peaks in any mass spectrum are the base peak and the molecular ion peak. *The base beak (also referred to as the parent peak) is the largest peak in the spectrum.* In the case of methane, the base peak is the peak at m/z 31 corresponding to the CH_3O^+ fragment. Since the absolute height of any peak is dependent on the concentration of the sample, the other peaks in the spectrum are referenced as a percentage of the base peak and referred to as relative abundance. This normalization of peak heights greatly aids in identification of fragmentation patterns and therefore analyte identification.

The molecular ion peak corresponds to an analyte molecule that has not undergone fragmentation. In Figure 14.1, the molecular ion peak is caused by the $[CH_3OH]^{\cdot+}$ ion and corresponds to m/z 32. The molecular ion peak is often referred to as the M^+ ion. The molecular ion is used as a reference point in identifying the other fragments. For example, the peak corresponding to m/z 15 is referred to as both M–OH and M-17.

14.3 Identifying the Molecular Ion Peak

The molecular ion peak is both an important reference point and is integral in identifying an unknown compound. While it may seem that the molecular ion peak should be the most abundant peak in the spectrum, this is not the case for the majority of compounds. Compounds like alcohols, nitrogen-containing organics, carboxylic acids, esters, and highly branched compounds may completely lack a visible molecular ion. In these cases, it is critical that fragment peaks are not mistakenly identified as the molecular ion peaks in order to avoid misidentification of an analyte. Obtaining a chemical ionization spectrum can assist in correctly identifying the molecular ion.

Even without a CI spectrum of the compound, other rules can assist in ruling out potential masses as the molecular ion. The "nitrogen rule" is one valuable tool for identifying the molecular ion. This rule indicates that if a molecular ion has an odd mass, it must have an odd number of nitrogen and that a molecular ion with an even mass must lack nitrogen atoms or contain an even number of them. Since the majority of organic compounds analyzed with the GC–MS contain either zero or one nitrogen atom, the rule practically states an odd molecular ion is attributed to a single nitrogen and an even molecular ion indicates the sample lacks nitrogen (Figure 14.2). This rule only applies to compounds that contain carbon, hydrogen, nitrogen, oxygen, sulfur, halogens, and a few other less common elements. Since the majority of organic compounds that are analyzed using the GC–MS are made up of these elements, this stipulation is practically ignored.

This rule is a result of nitrogen's unique property. Nitrogen has an even atomic mass but bonds with three other atoms in its most stable form. Other atoms that have even molecular weights like

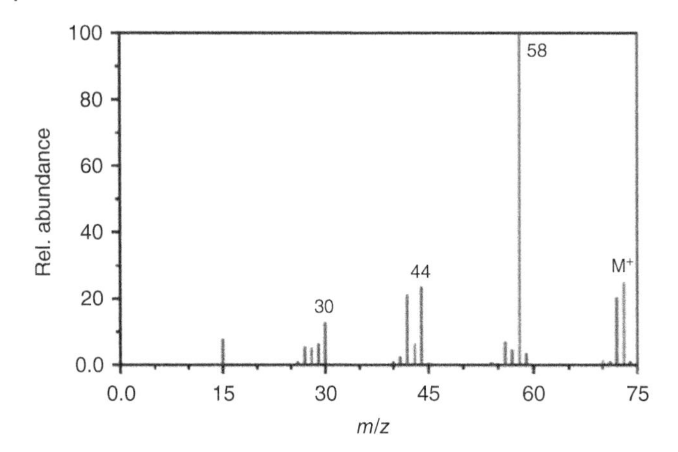

Figure 14.2 The nitrogen rule–the mass spectrum of *N,N*-dimethyl-ethanamine illustrates the presence of an odd molecular ion and even fragments (Source: NIST/National Institute of Standards and Technology/Public Domain).

carbon, oxygen, and sulfur bond with an even number of other atoms. Atoms that bond with an odd number of other atoms like hydrogen, chlorine, bromine, and iodine have odd molecular weights. This rule is invaluable when a chemist knows that a compound lacks nitrogen. This can occur if a sample is prepared from a synthesis whose products and solvents lack nitrogen atoms. In this case, any odd peak cannot be attributed to the molecular ion of the analyzed compound.

Most fractionation excluding rearrangements (Section 14.6) occurs when a single bond is broken. The nitrogen rule indicates that when a molecule with an even mass produces a fragment by breaking a single bond, the fragment will have an odd mass. When the sample mass is odd, fragmentation via a similar pathway will give an even fragment as long as the nitrogen is still contained in the observed fragment. Since this is generally the observed trend (see Stevenson's Rule in Section 14.6), analyzing the major fragments can help determine if the molecular ion should be even or odd. Practically, if the major fragments are mostly odd, the molecular ion is likely even and contains no nitrogen. If the major fragments are even, the molecular ion is likely odd and contains one nitrogen atom as shown in Figure 14.3.

Since molecular ions fragment in predictable ways, the presence of certain fragmentation peaks can suggest that a particular peak is a molecular ion. The observed fragments must be able to be attributed to logical losses. The existence of an M-15 peak from the loss of CH_3, an M-18 peak from the loss of H_2O, or an M-31 from the loss of OCH_3 are a few examples of these logical fragments.

The opposite is true for fragments that are not logical. These peaks suggest that a particular peak is not a molecular ion. Some illogical fragmentation peaks include peaks that are 3–14 mass units away from the peak suggesting that the identified peak is likely not the molecular ion peak. The loss of fragments of mass units 1–3 can result from the loss of up to three hydrogen atoms. From 14 to 18, multiple peaks can be explained by the loss of CH_3, oxygen, a hydroxide ion, or water. The loss of fragments from the 19–25 range is also unlikely except in the case of fluorinated compounds which produce M-19 (loss of F) and M-20 (loss of HF).

The molecular ion is difficult to identify with chemical ionization because there is no definitive test. While these patterns can greatly assist in identifying the molecular ion, they should not

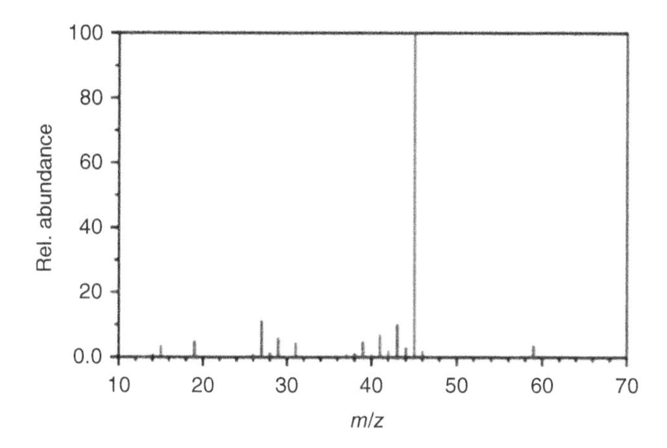

Figure 14.3 The use of the nitrogen rule in determining the molecular ion—should the faint peak at *m/z* 60 be attributed to the presence of C_{13} or is it the molecular ion? The presence of the base peak at 45 in combination with our knowledge about the nitrogen rule suggests that the peak at *m/z* 60 is likely the molecular ion because even molecular ions usually produce odd molecular fragments by breaking single bonds. Given this spectrum is of Isopropyl alcohol, our deduction is correct, although chemical ionization techniques could verify the molecular mass of the sample (Source: NIST/National Institute of Standards and Technology/Public Domain).

be trusted as confirmatory. Complex rearrangements can potentially result in the misidentification of the molecular ion. As a result, it is good practice to double-check with a soft ionization technique such as chemical ionization when in doubt of the identity of the molecular ion.

14.4 Use of the Molecular Ion

Once the identity of the molecular ion has been determined, much can be learned about the compound. One extremely valuable piece of information that can be determined from a high-resolution mass spectrometer is the molecular formula of an unknown analyte. If a molecular ion was identified to be at *m/z* 80 on an instrument with unit resolution, little could be determined about the molecular formula. For example, some of the many possible molecular formulas include $C_4H_4N_2$ (80.0375), C_5H_4O (80.0262), and C_6H_8 (80.0626). A high-resolution instrument measurement of this peak at 80.0372 ± 0.0005 would indicate that the empirical formula is $C_4H_4N_2$. Extensive tables and computer programs are used to perform this technique on a routine basis.

Once the molecular formula is known, it becomes possible to determine the degree of unsaturation. This allows the analyst to know the number of pi bonds and rings that are in their structure. The elements of unsaturation can be computed by using the following equation.

$$\text{Degree of unsaturation} = \#C - \frac{\#H}{2} - \frac{\#X}{2} + \frac{\#N}{2} + 1 \tag{14.5}$$

where H is the number of H atoms, X is the number of halogen atoms (F, Cl, Br, and I), and N is the number of nitrogen atoms in the chemical formula. As Equation 14.1 indicates, the number of oxygen atoms does not affect the degree of unsaturation. By using this equation, the molecular formula $C_4H_4N_2$ has four degrees of unsaturation. The combination of the molecular formula with the degrees of unsaturation is an important tool in identifying a particular compound.

The molecular ion along with other information from IR and NMR spectra can allow the identity of an unknown to be determined. If all three techniques can be utilized on a sample, the strengths of each allow for the easiest identification. Since IR identifies the unknown's functional groups, the mass of these groups is first subtracted from the mass of the molecular ion. This mass frequently represents the mass of the carbon and hydrogen contained in a sample. Taking this number and dividing it by 12 will give the number of carbon atoms and a fraction representing the number of hydrogen atoms. It is important to not blindly trust this method. If the molecular ion minus the mass of the functional groups gives 85, dividing by 12 would give a molecular formula of C_7H. Attempting to create this molecule will quickly indicate that a more logical molecular formula would be C_6H_{13}.

For the example unknown analyte illustrated in Figure 14.4a, the IR spectrum indicates the compound's functional groups. The sharp peak observed at around 1710 cm^{-1} indicates the presence of a carbonyl group. The large round peak centered around 3000 cm^{-1} suggests that the compound is a carboxylic acid. The large peak slightly above 1600 cm^{-1} could indicate that the unknown contains an alkene or even possibly an imine.

After talking mass spectrum of the compound, it is necessary to identify the molecular ion. We can identify the peak at 86 to be the molecular ion using the nitrogen rule (discussed above). Because its major fragments are both odd, 69 and 41, it is reasonable that its molecular ion should be even. The peaks at 85 (loss of H) and 71 (loss of CH_3) can be explained in a logical fashion further confirming the m/z 86 peak as the molecular ion.

From above, the compound's known mass is 86. Thus, we can confirm that the compound is not an imine because it has an even molecular weight indicating that it does not contain an odd number of nitrogen atoms. Now it becomes possible to identify something about the carbon backbone of the atom. By subtracting the mass of the carboxylic acid functional group (COOH), a mass of 41 is obtained. The IR spectrum indicates that the remainder of the molecule is likely only made up of carbon and hydrogen. This allows the analyst to deduce that the rest of the molecule is made up of three more carbon atoms and five hydrogen atoms. From taking these two easy measurements, one is able to determine that this compound's molecular formula is C_3H_5COOH. From this molecular formula, we can determine that the degree of unsaturation is two. One degree of unsaturation is attributed to the acid functional group while the other is a double bond since a three-carbon ring is extremely unlikely.

While both the IR and MS (Figure 14.4b) are able to determine a great deal about a compound's identity, NMR (Figure 14.4c) is necessary to identify this compound. The peak below 12 ppm is a result of the hydrogen in the carboxylic acid group. The doublet peaks 6.3 and 5.7 ppm are split by approximately 1.4 Hz, which indicates that they are geminal protons. The fact that the peaks are doublets indicates that there are only two hydrogen atoms connected to the vinyl group. The presence of a methyl group is indicated by the peak at 2 ppm. As a result, the unknown compound is methacrylic acid (Structure 14.1).

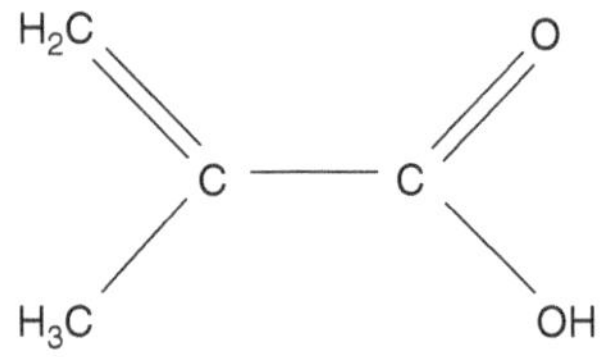

H_2C O
C — C
H_3C OH

Structure 14.1

While utilizing IR and NMR, in combination with the mass spectra, made the identification of this compound relatively simple, these tools are not always available. If the analyte of interest is in a

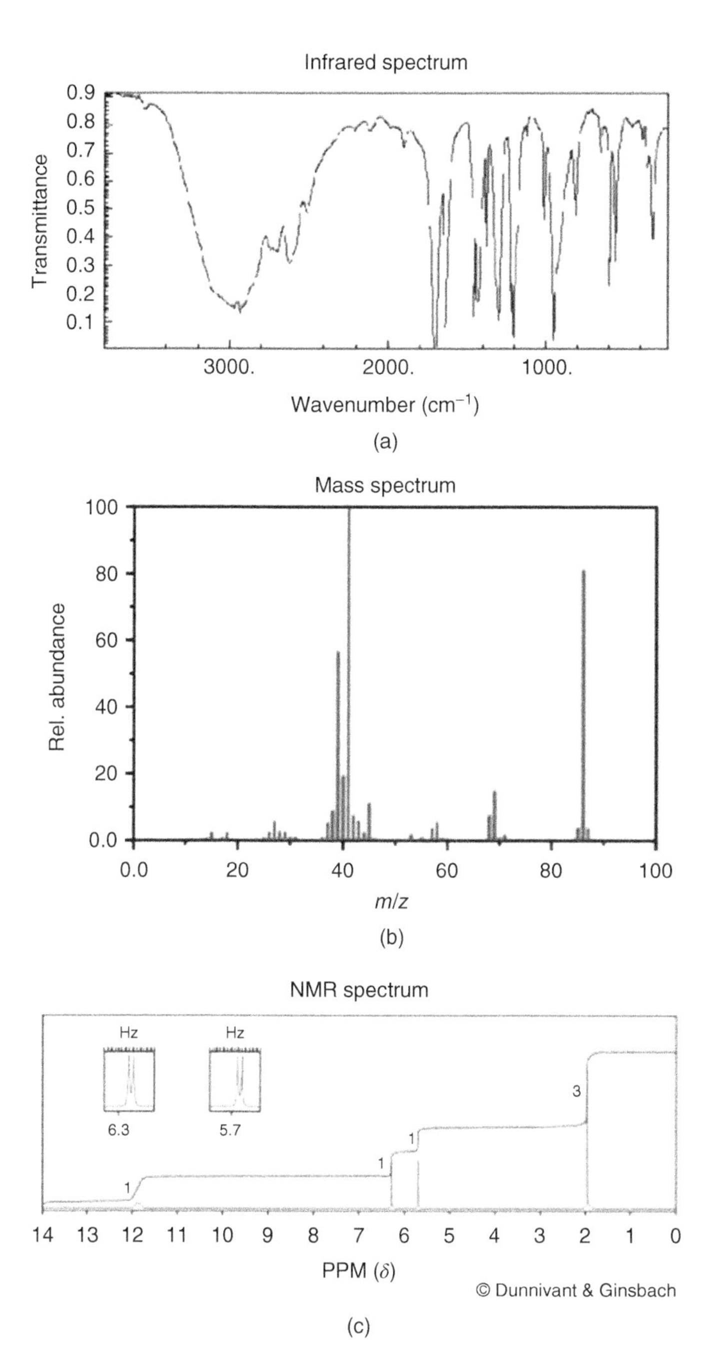

Figure 14.4 (a) IR of an unknown analyte, (b) mass spectrum of an unknown analyte, and (c) NMR of an unknown analyte (Source: NIST/National Institute of Standards and Technology/Public Domain).

complex mixture or there is only a small concentration or quantity (parts per million), both IR and NMR are not effective tools. As a result, it is necessary to be able to identify as much information as possible from the mass spectrum alone. The rest of this chapter will be devoted to such a task by observing common fractionation trends in various types of compounds.

14.5 Identification of Analytes Using Isotopic Ratios

Since the majority of elements have two or more isotopes, the ratio of these isotopes can be a powerful tool in deriving the composition of unknown samples. Prominent peaks will have a smaller peak one mass unit higher than the prominent peak due to the presence of one ^{13}C in some of the sample molecules. Background noise and a lack of resolution in the majority of mass spectrometers prevent the ratio of various isotopes from being an identification technique for all compounds.

However, some isotopes are so prominent that they can easily be observed with a quadrupole mass spectrophotometer with unit resolution. Chlorine, bromine, and sulfur can all be identified by their isotopic ratios. Their exact isotopic ratios are summarized in Table 14.1. Compounds containing chlorine have an M+2 peak that is 25% of the molecular ion (Figure 14.5a). Compounds containing bromine have an M+2 peak that is approximately the same height as an M^+ peak (Figure 14.5b). Compounds containing sulfur have an unusually large M+2 (Figure 14.5c).

Compounds can also contain any combination of multiple chlorine and bromine atoms. These samples will produce distinct peaks due to the various combinations of the isotopes. A compound containing two bromines will have an M+2 peak twice the size of the M+ peak and an M+4 peak the same size as the M+ peak (Figure 14.6a). A compound containing two chlorines will have an M+2 peak that is two-thirds the size of the M^+ peak and an M+4 peak that is 10% of the molecular ion (Figure 14.6b).

Iodine is more difficult to detect because it is one of the few compounds that is monoisotopic. Despite this fact, the large atomic mass of iodine allows for its identification. The combination of a peak at m/z 127 (I^+) and a large gap of 127 mass units between fragments containing iodine and fragments lacking in iodine allows for compound identification (Figure 14.7).

Table 14.1 Isotopic abundances of common elements.

Element	M^+	M+1	M+2
Hydrogen	$^{1}H \rightarrow 100.0\%$		
Carbon	$^{12}C \rightarrow 98.9\%$	$^{13}C \rightarrow 1.1\%$	
Nitrogen	$^{14}N \rightarrow 99.6\%$	$^{15}N \rightarrow 0.4\%$	
Oxygen	$^{16}O \rightarrow 99.8\%$		$^{18}O \rightarrow 0.2\%$
Sulfur	$^{32}S \rightarrow 95.0\%$	$^{33}S \rightarrow 0.8\%$	$^{34}S \rightarrow 4.2\%$
Chlorine	$^{35}Cl \rightarrow 75.5\%$		$^{37}Cl \rightarrow 24.5\%$
Bromine	$^{79}Br \rightarrow 50.5\%$		$^{81}Br \rightarrow 49.5\%$
Iodine	$^{127}I \rightarrow 100.0\%$		

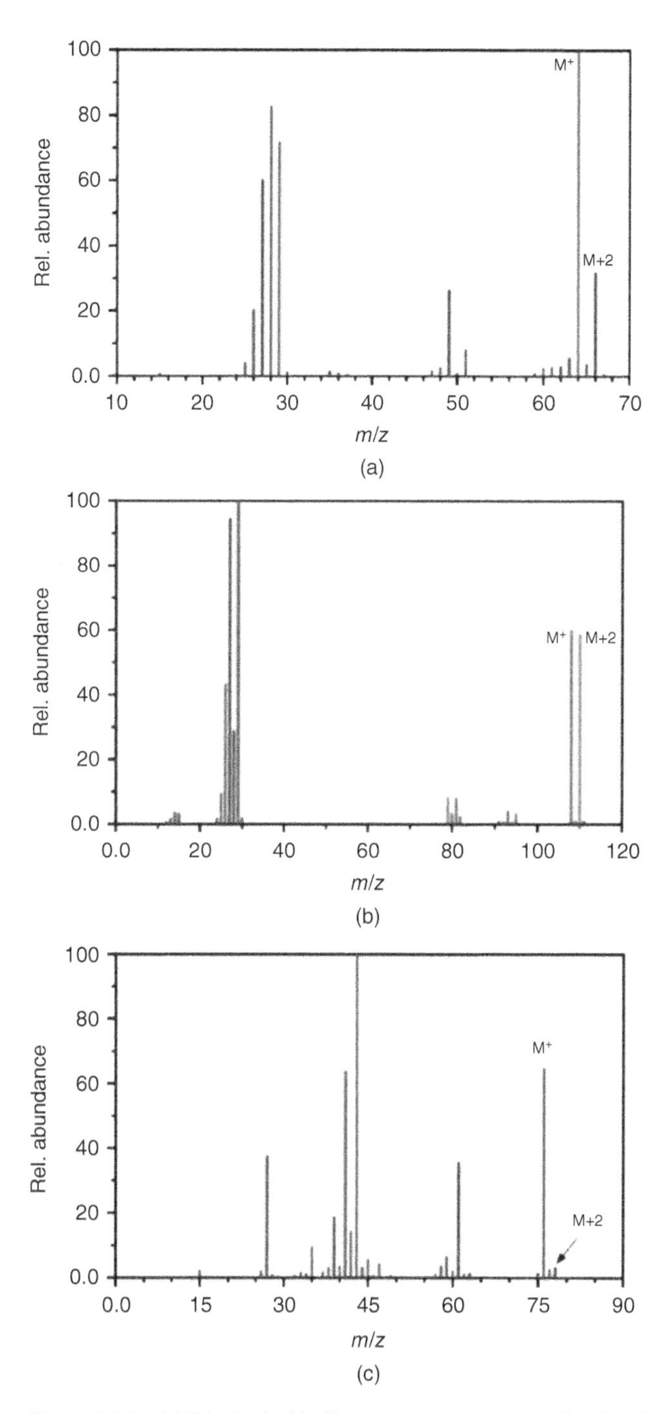

Figure 14.5 (a) Ethyl chloride illustrates the presence of an M+2 peak that is about 25% of the M+ peak, (b) a characteristic M+2 peak that has a similar intensity as the M+ peak, and (c) 2-Propanethiol contains a larger than usual M+2 peak, a pattern observable in sulfur-containing compounds (Source: NIST/National Institute of Standards and Technology/Public Domain).

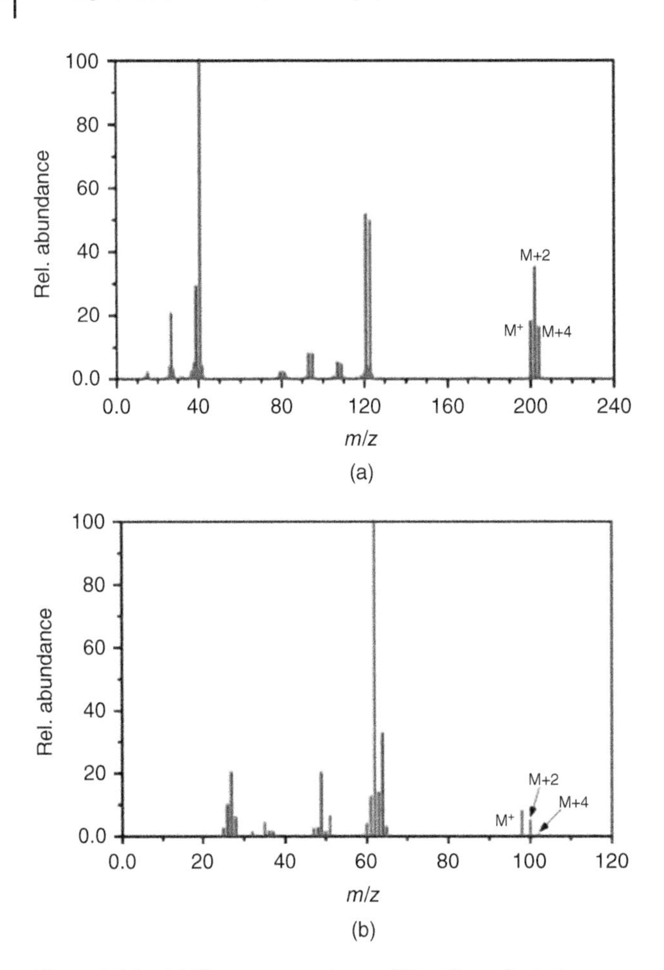

Figure 14.6 (a) The mass spectrum of the dibrominated compound 1,3-dibromopropane and (b) the mass spectrum of the dichlorinated compound 1,2-dichloroethane (Source: NIST/National Institute of Standards and Technology/Public Domain).

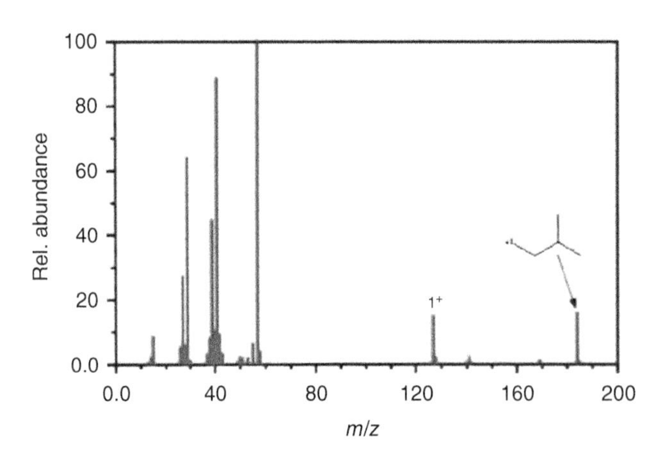

Figure 14.7 Mass spectrum containing iodine (Source: NIST/National Institute of Standards and Technology/Public Domain).

14.6 Fragmentation

While the molecular ion is one of the most important peaks in the spectra, it is also important to gain information from the peaks that are a result of fragmentation. The goal of interpreting mass spectra is to identify the structure of the molecular ion by examining pieces (fragments) of the original molecule. The frequency and size of the fragments are dependent on the structure and bond energy of the sample molecule. This property has resulted in the creation of a unique and reproducible spectrum for a wide variety of compounds.

Before fragmentation can be discussed, it is necessary to develop a new notation because the cation fragments that will be encountered are not present in other branches of chemistry due to their high reactivity. The presence of the vacuum in the instrument prevents collisions with other molecules allowing these reactive cations to exist. The academic convention for notation is to either represent the charge as a delocalized one (Structure 14.2a) or localized it on either a π-bond (Structure 14.2b) or on a heteroatom (Structure 14.2c).

Structure 14.2

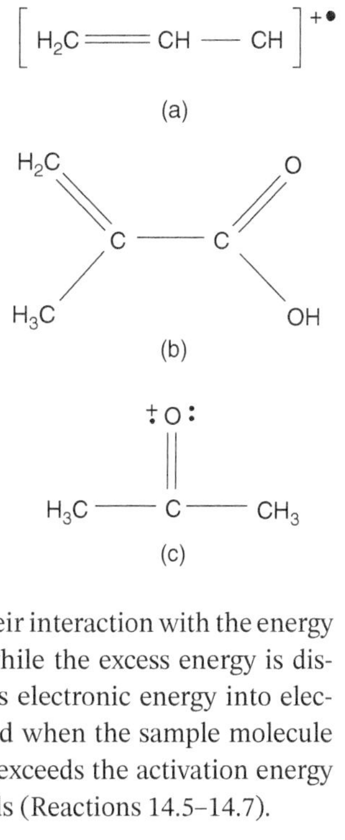

$$\left[H_2C{=}CH{-}CH \right]^{+\bullet}$$

(a)

H_2C, H_3C, C, C, O, OH

(b)

$\overset{+}{\ddot{O}}$:, H_3C, C, CH_3

(c)

The process that creates these observed fragments is the result of their interaction with the energy released from the source. This energy removes a single electron, while the excess energy is distributed over various degrees of freedom. This distribution converts electronic energy into electronic, rotational, and vibration energy. The molecular ion is created when the sample molecule returns to its ground state via a relaxation. Other times this energy exceeds the activation energy of fragmentation and this energy is released via the breaking of bonds (Reactions 14.5–14.7).

$$e^- + R - R' \rightarrow [R - R']^{\cdot +} + 2e^- \quad (14.6)$$

$$[R - R']^{\cdot +} \rightarrow R^{\cdot} + R'^{+} \quad (14.7)$$

$$[R - R']^{\cdot +} \rightarrow R^{+} + R'^{\cdot} \quad (14.8)$$

The fragmentation of a single bond can produce two peaks, one from R^+ and the other from R'^+ since the instrument can only detect the positive ion. According to Stevenson's rule, if two fragments are in competition to produce a cation, the fragment with the lowest ionization energy will be formed more frequently (Figure 14.8).

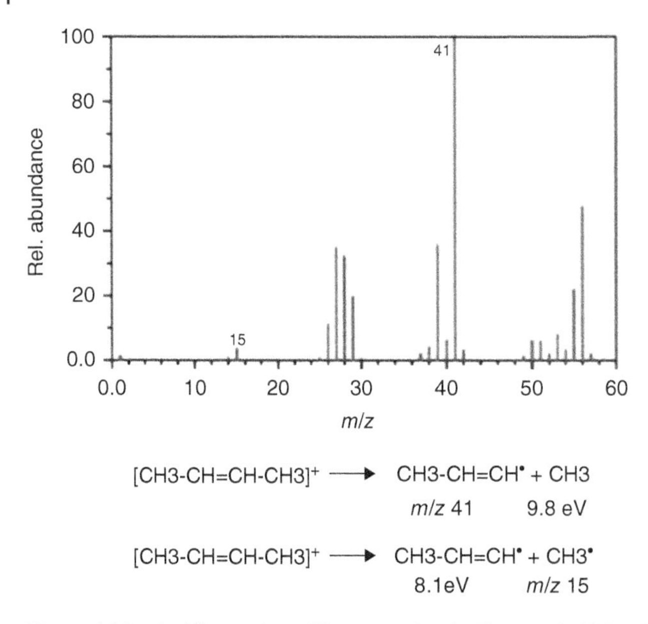

Figure 14.8 An illustration of Stevenson's rule (Source: NIST/National Institute of Standards and Technology/Public Domain).

The fragmentation of a bond can proceed through two pathways, either homolytic or heterolytic cleavage. In the heterolytic cleavage (Reaction 14.8), a pair of electrons move toward the charged site as illustrated by the double-headed arrow producing a cation and a radical (Scheme 14.1).

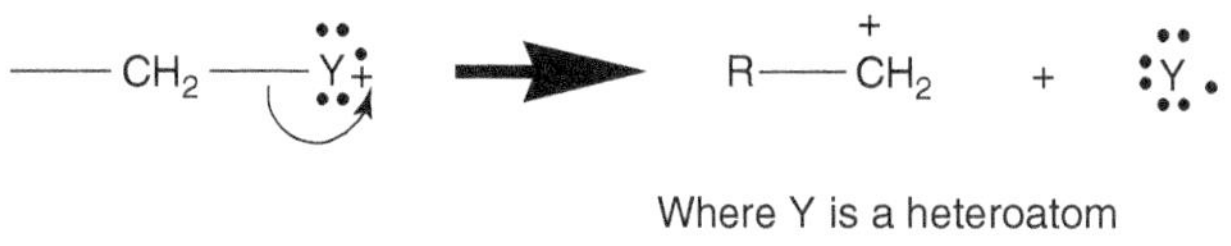

—CH_2—Y^+ ⟶ R—CH_2^+ + Y•

Where Y is a heteroatom

Scheme 14.1

The fragmentation produced by a hemolytic cleavage results from the movement of single electrons (Scheme 14.2).

R—CH_2—CH_2—$O^{•+}$—R ⟶ R—CH_2• + H_2C=O^+—R

Scheme 14.2

For simplicity, usually, only one set of arrows is drawn to illustrate the movement of electrons (Scheme 14.3).

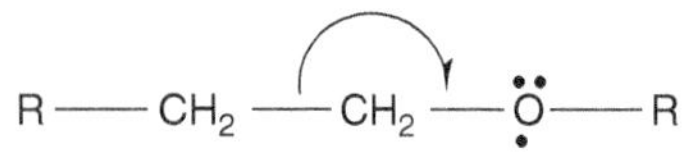

R—CH_2—CH_2—$O^{•}$—R

Scheme 14.3

These fragmentation patterns are usually the result of a functional group contained in the compound. As a result, the bonds that typically break are either located one or two carbon atoms away from the functional group. These carbon atoms are referred to as the α and β atoms (Structure 14.3).

Structure 14.3

$$R - \overset{\beta}{CH_2} - \overset{\alpha}{CH_2} - Y$$

The bond between the functional group Y and the α carbon is called the α bond and the bond between the α and β carbons is the β bond.

14.7 Rearrangements

Some fragments are the result of the cleavage of multiple bonds. The removal of water from alcohol is only one example. The nitrogen rule (Figure 14.3) is helpful in identifying peaks that are produced via a rearrangement. If a molecular ion has an even molecular weight, then generally peaks of even molecular weight were created from a rearrangement. If a molecule has an odd molecular weight, then its rearrangement peaks will also be odd.

One rearrangement is the loss of water from a primary alcohol. The mechanism is illustrated with butanol in Scheme 14.4.

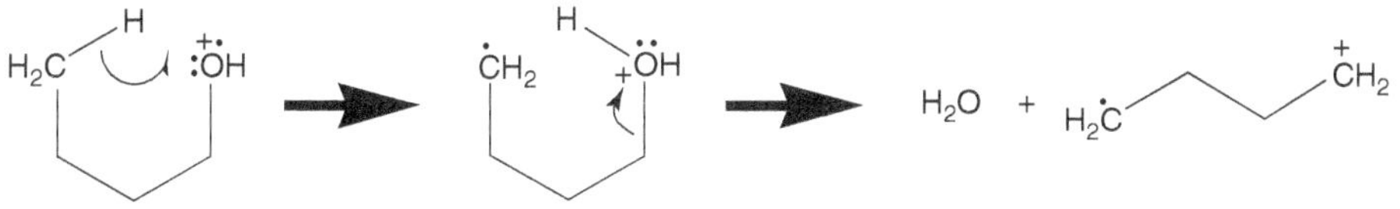

Scheme 14.4

These rearrangements are favored because the low-energy transitions help stabilize the products. Other rearrangements such as the McLafferty rearrangement will be explored in greater detail in the following sections.

14.8 Identification of Compounds

The ability to identify unknown samples is one of the most powerful uses of a mass spectrometer. This, however, requires an understanding of fractionation patterns for commonly encountered compounds. The following trends are only applicable to electron impact with a source at 70 eV. These trends are not comprehensive but are rather a selection of common fragments that are most useful in properly identifying common types of organic chemicals.

The actual likelihood of fragmentation is related to the activation energy of the reaction, the ability for rearrangements to occur, and the stability of the products. Trends that were observed in organic chemistry are helpful in predicting fragmentation patterns. Thinking about the stability of the products as more or less stable cations and radicals is not entirely theoretically accurate but is usually a good, practical way to predict the spectrum of a molecule.

14.9 Fragmentation of Hydrocarbons

There are two types of hydrocarbons that are analyzed with the GC–MS. One is long chain hydrocarbons and the other is the hydrocarbon portion of molecules containing other functional groups. Identifying the structure of these hydrocarbons can be difficult since rearrangements that are not easily explained are frequently observed. It is especially important to utilize reference compounds and GC retention times whenever possible to confirm the identity of the compound.

14.9.1 Fragmentation of Straight Chain Alkanes

Straight-chain alkanes always produce a molecular ion even in long-chain compounds where the molecular ion is usually faint. The base peak in the spectra is usually the peak at m/z 57 corresponding to the C_4H_9 carbocation surrounded by other smaller peaks due to the rearrangement of hydrogen atoms. These groups are separated by 14 mass units resulting from the loss of another CH_2 group. The largest peak in each cluster is caused by the loss of $(CH_2)_nCH_3$ resulting in a fragment of molecular formula C_mH_{2m+1}. The subsequent fragments after the C_4 peak decrease in an exponential fashion to a minimum at M–C_2H_5. The M–CH_3 peak is weak in smaller compounds and absent in long-chain compounds due to the relative instability of the methyl radical. The molecular ion is the unique identifiable peak in straight-chain alkanes longer than eight carbon atoms.

The example illustrated in Figure 14.9 illustrates these common trends discussed above. The prominent peaks at C_mH_{2m+1} combined with the decaying intensity of these peaks indicate this compound is an alkane. The base peak at m/z 57 caused by C_4H_9 is further confirmation that there is a lack of other functional groups causing other prominent fragments. The molecular ion at m/z 170 indicates that this compound is dodecane.

14.9.2 Fragmentation of Branched Alkanes

Branched alkanes have a smaller molecular ion that at times may be absent in highly branched compounds. In larger compounds, branched alkanes contain peaks at C_mH_{2m+1}, similar to straight-chain alkanes. They are distinguished by the lack of smooth exponential decay beginning at the C_3 or C_4 carbon (Figure 14.9). This is caused by the increased frequency of fractionation at

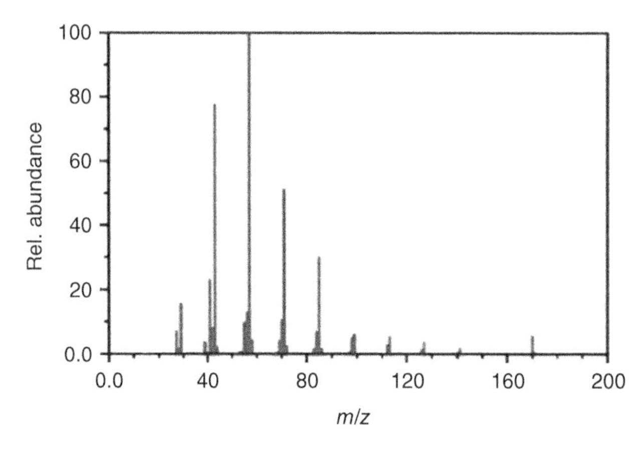

Figure 14.9 Fragmentation of a straight chain alkane (Source: NIST/National Institute of Standards and Technology/Public Domain).

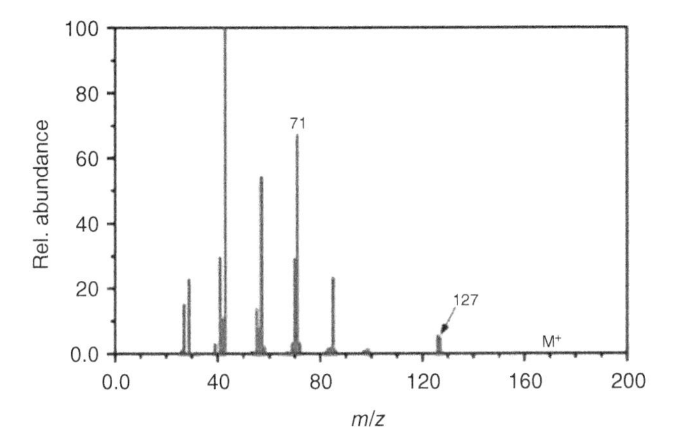

Figure 14.10 MS of fragmentation of a branched alkane (Source: NIST/National Institute of Standards and Technology/Public Domain).

the branch since it results in a secondary rather than a primary carbocation and is hence favored. The loss of the largest alkyl fragment at the branching site is favored because it helps stabilize the radical (Figure 14.10).

This mass spectrum of a C_{12} alkane (determined from the molecular ion by CI at *m*/*z* 170) lacks the exponential decay seen in Figure 14.9 indicating the chain is branched. The intensity of the peak at *m*/*z* 71 indicates a favored C_5 fragment and the fragment at *m*/*z* 127 indicates a favored C_9 fragment suggesting a methyl group on fourth carbon.

The fragmentation at the branching point (Scheme 14.5) is often accompanied by hydrogen rearrangement causing the C_nH_{2n} peak to be more prominent and sometimes larger than C_nH_{2n+1} peak.

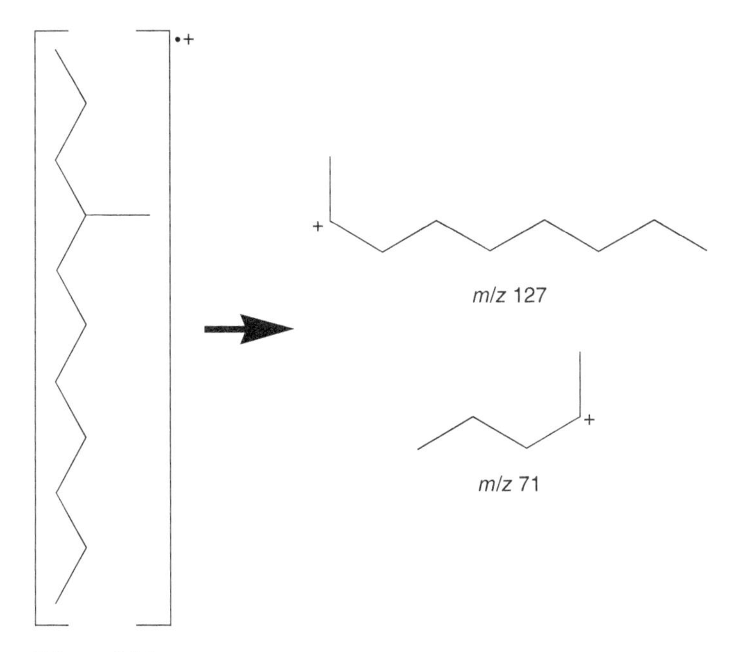

Scheme 14.5

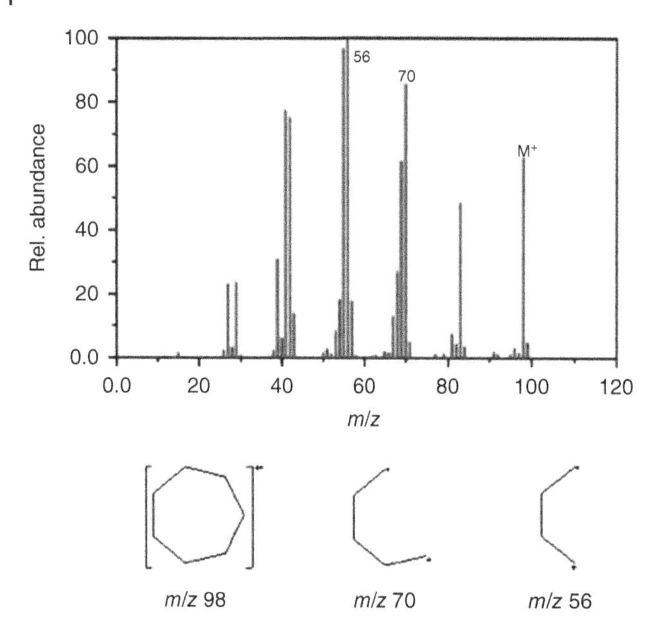

Figure 14.11 Fragmentation of a cyclic alkane (Source: NIST/National Institute of Standards and Technology/Public Domain).

Identifying branched alkanes in organic compounds that contain another functional group is also an important task. The alkane portion of these molecules is usually smaller and is more governed by the stability of the produced radical and cation. Since an ethyl radical is more unstable than a methyl radical, the methyl radical will occur less frequently. Similarly, tertiary carbocations are more stable than secondary, which are more stable than primary. As the alkane portion of any molecule becomes larger, the presence of the C_nH_{2n+1} peaks becomes more prominent.

14.9.3 Fragmentation of Cyclic Alkanes

The ring structure of cyclic alkanes increases the intensity of the molecular ion. Its stability also increases the likelihood that side chains will fragment at the ά bond to the ring. The fragmentation of the cyclic structure is usually most often caused by the loss of more than two carbon atoms. The loss of a methyl radical occurs less frequently because of the instability of the methyl radical in comparison to the neutral ethylene molecule at M-28 or an ethyl radical at M-29 (Figure 14.11).

14.9.4 Fragmentation of Alkenes

The molecular ion of alkenes is usually distinct especially in compounds containing multiple double bonds. Alkene fragments, like alkane fragments are situated in clusters 14 units apart. In alkenes, the C_nH_{2n-1} and C_nH_{2n} peaks are more intense than the C_nH_{2n+1} peak of alkanes.

The presence of double bonds also allows for the production of resonance-stabilized cations. Allylic cleavage results in an allylic cation as shown in Scheme 14.6.

Determining the position of the double bonds in the sample molecule is especially difficult and usually requires reference spectra because of double-bond migration. Cyclic alkenes also undergo a retro-Diels–Alder fragmentation by Scheme 14.7.

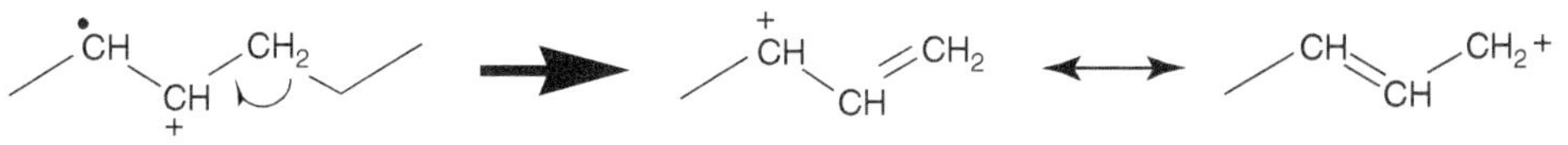

Scheme 14.6

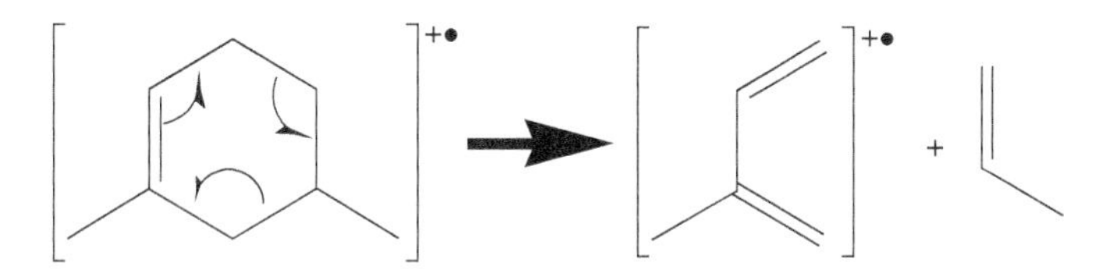

Scheme 14.7

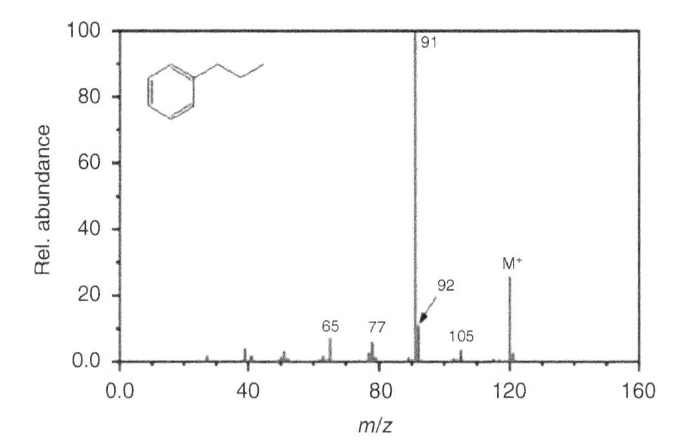

Figure 14.12 Fragmentation of an aromatic (Source: NIST/National Institute of Standards and Technology/Public Domain).

14.9.5 Fragmentation of Aromatics

The presence of an aromatic ring in a compound results in a prominent molecular ion. A common peak at M-1 results from the cleavage of a hydrogen molecule from the benzene ring. Alkyl-substituted benzene rings result in a prominent peak at m/z 91 (Figure 14.12). In most cases, the peak at m/z 91 is the result of a tropylium ion caused by the following rearrangement (Scheme 14.8).

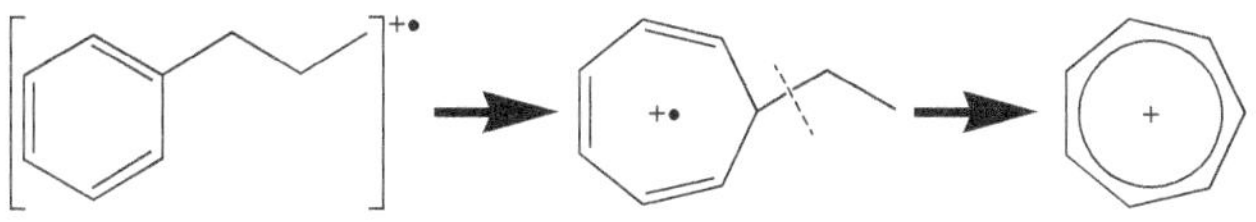

Tropylium ion m/z 91

Scheme 14.8

The peak observed in most aromatic compounds at m/z 65 results from the elimination of an acetylene molecule from the tropylium ion (Scheme 14.9).

Benzene rings with highly branched substituted groups produce fragments larger than m/z 91 by intervals of 14 units. The largest of these peaks will result in highly substituted cations and a large

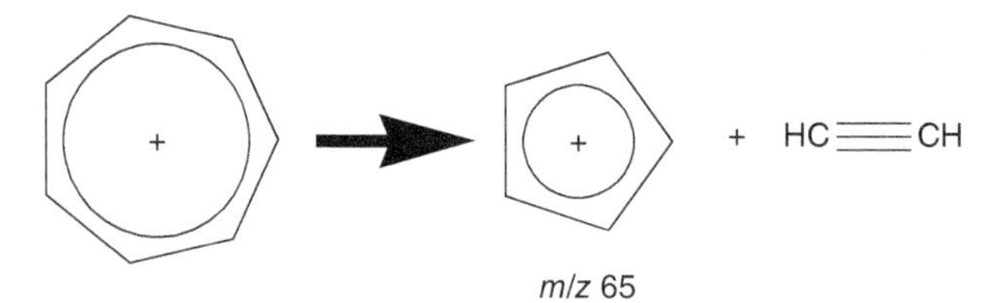

Scheme 14.9

radical, like a simpler branched alkane. The fragment at *m/z* 105 in Figure 14.12 is relatively small since it produces a primary carbocation and an unstable methyl radical. Substituted benzene rings also first undergo ά cleavage followed by hydrogen rearrangement producing a grouping of peaks at *m/z* 77 from $C_6H_5^+$, *m/z* 78 from $C_6H_6^+$, and *m/z* 79 from $C_6H_7^+$.

Side chains with more than two carbon atoms create a peak at *m/z* 92 (Figure 14.12 and Scheme 14.10). Unbranched groups result in a more prevalent peak than do branched groups.

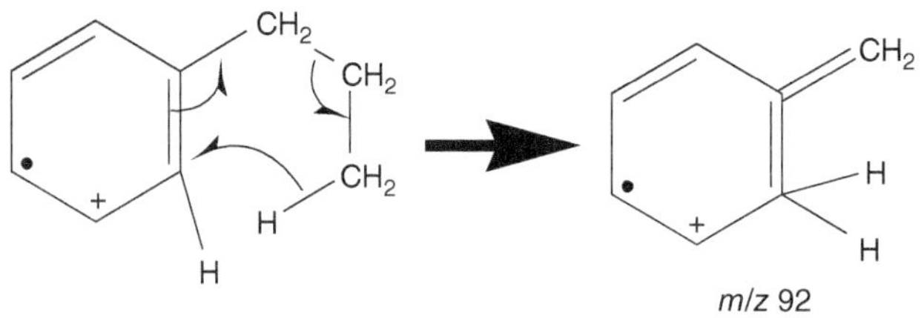

Scheme 14.10

14.10 Fragmentation of Alcohols

The molecular ion of alcohols is usually small and sometimes undetectable, especially in tertiary alcohols. The identification of the molecular ion is complicated by the prevalence of an M-1 peak caused by the loss of a single hydrogen from the α carbon in primary and secondary alcohols.

Alcohols also frequently cleave to give resonance-stabilized cations due to the breaking of the β bond (Scheme 14.11). As a result of this cleavage, primary alcohols show a prominent peak at *m/z* 31 (Figure 14.13a shown later).

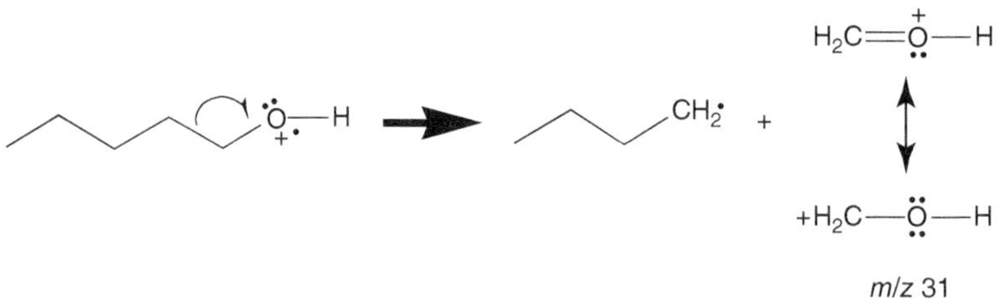

Scheme 14.11

The presence of a *m/z* 31 peak is not a confirmation of a primary alcohol. It is necessary for the peak to be relatively large in comparison to other peaks in the spectrum. This is because secondary

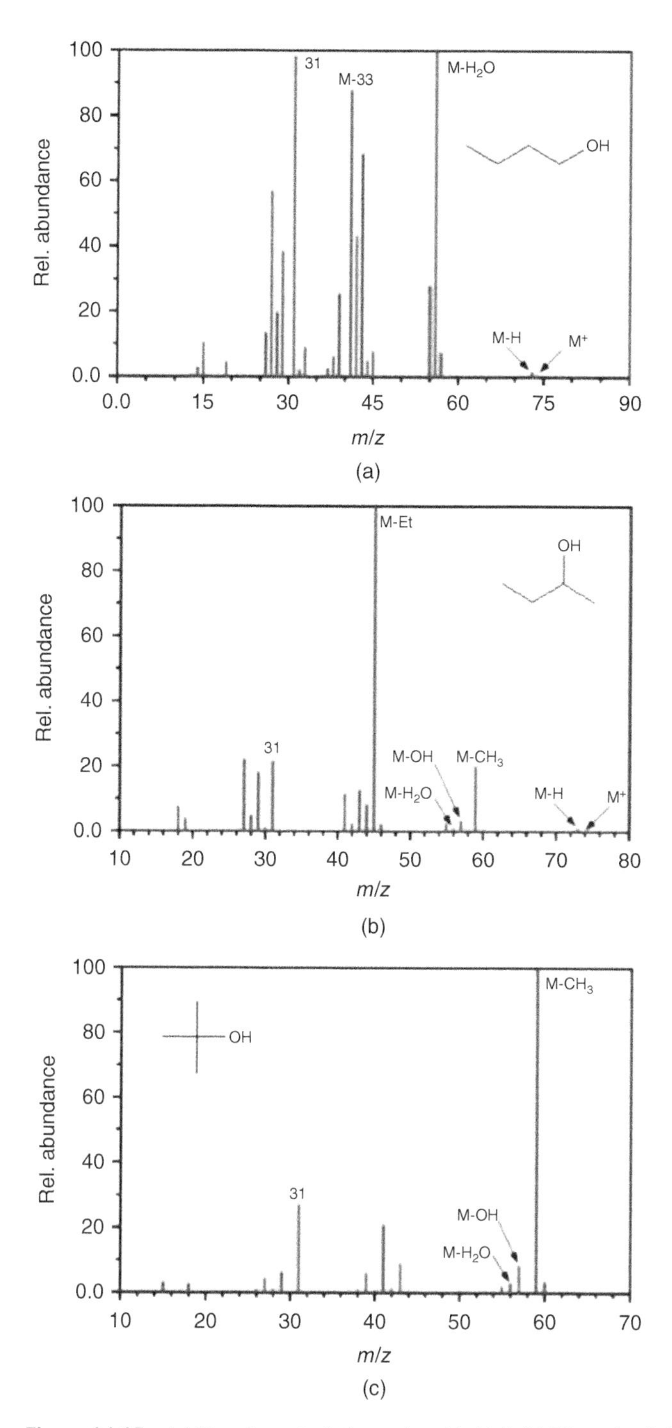

Figure 14.13 (a) The characteristic peak at M−H_2O, M-33, and *m/z* 31, (b) T base peak in the spectrum indicates that this alcohol is not a primary alcohol, and (c) trends common to tertiary alcohols (Source: NIST/National Institute of Standards and Technology/Public Domain).

alcohols and sometimes even tertiary alcohols can undergo a rearrangement resulting in a peak at *m*/*z* 31 (Scheme 14.12).

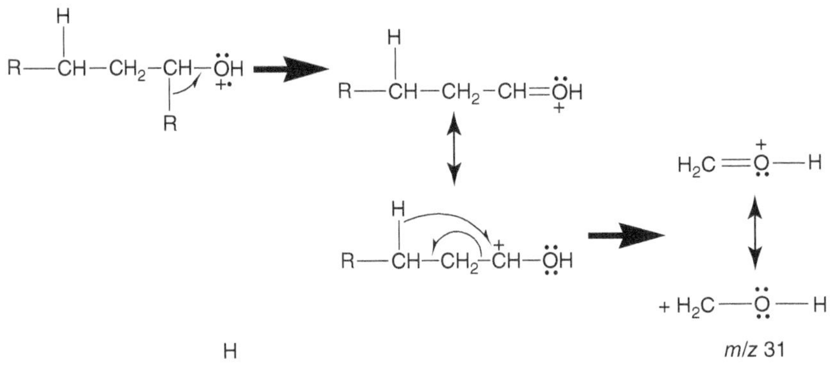

Scheme 14.12

Alcohols also frequently undergo the rearrangement described in Section 2.7 resulting in an M-18 peak from the loss of water. This peak is most easily visible in primary alcohols but can be found in secondary and tertiary alcohols as well. Primary alcohols also can lose both water and an alkene (Scheme 14.13).

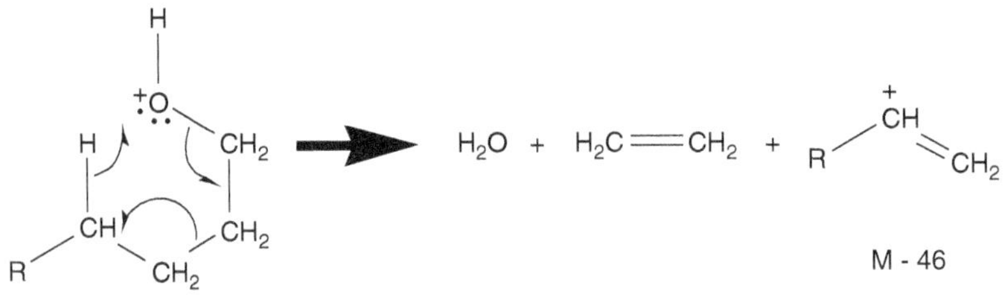

Scheme 14.13

Primary alcohols also produce an M-2 peak caused by R–CH=O+ and M-3 attributed to $R–C{\equiv}O^+$. Alcohols with carbon chains containing methyl groups frequently lose both the methyl group and water at M-33.

Figure 14.13a can be identified as an alcohol because of the characteristic peak at $M–H_2O$, M-33, and *m*/*z* 31. The peak at *m*/*z* 31 can attributed to a primary alcohol because it is one of the larger peaks in the spectrum.

Figure 14.13b has a small peak at *m*/*z* 31 but the base peak in the spectrum indicates that this alcohol is not a primary alcohol. The presence of an M–Et and $M–CH_3$ peak indicates that this four-carbon alcohol (determined from its molecular mass) is the secondary alcohol 2-butanol.

Figure 14.13c illustrates trends common to tertiary alcohols. The spectrum is easily discernable since the single prevalent peak is characterized by $M–CH_3$. The lack of a molecular ion helps confirm the spectrum of a tertiary alcohol.

Cyclic alcohols fragment similar to straight chain alcohols in that they give an M-1 peak from the loss of hydrogen and an M-18 peak from the loss of water. They also create a peak at *m*/*z* 57 via a complex ring cleavage.

Aromatic alcohols, unlike other alcohols, have a prominent molecular ion peak due to the stability of the aromatic group. Phenols usually give a weaker peak at m/z 77 attributed to a rearrangement and can be identified by two peaks at M–CO and M–COH (Scheme 14.14a,b).

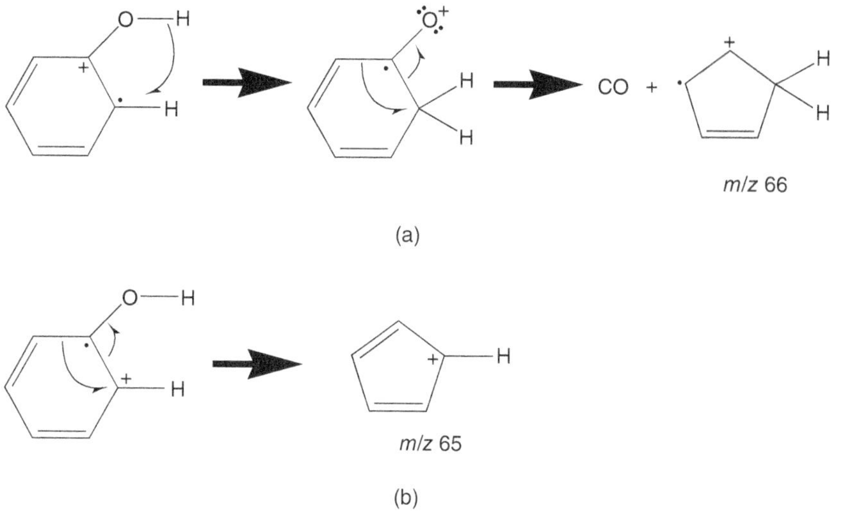

m/z 66

(a)

m/z 65

(b)

Scheme 14.14

14.11 Fragmentation of Ketones and Aldehydes

Both ketones and aldehydes give prominent molecular ion peaks though the M+ peak is more prominent in ketones. The majority of compounds in these categories undergo an important rearrangement, the McLafferty rearrangement (Scheme 14.15).

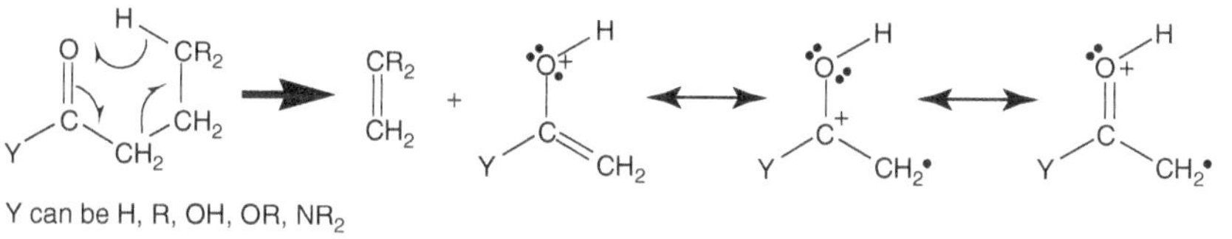

Y can be H, R, OH, OR, NR_2

Scheme 14.15

This rearrangement is mediated by the π systems of the carbonyl group but can occur in other π systems such as in nitriles (Section 14.17). The only ketones and aldehydes that do not undergo this rearrangement lack a three-carbon side chain allowing for the necessary hydrogen donation.

14.11.1 Ketones

One major fragment of ketones is the creation of the resonance-stabilized acylium ion resulting from the cleavage of the ά bond. The base peak in the spectrum is usually caused by the removal of the larger alkyl group since it forms a more stable radical illustrated by 4-Octanone (Scheme 14.16a (top),b (bottom)).

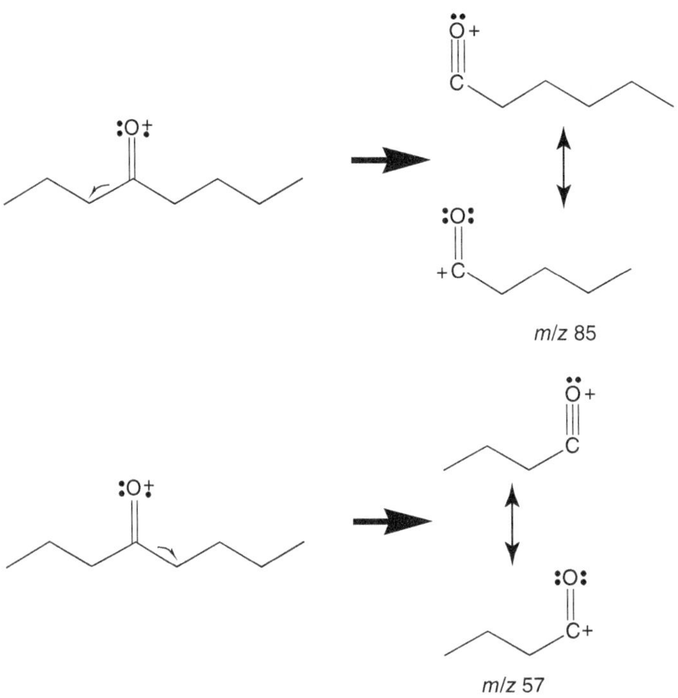

m/*z* 85

m/*z* 57

Scheme 14.16

While ketones undergo a single McLafferty rearrangement described above, they also undergo a subsequent McLafferty rearrangement (Scheme 14.17a,b).

The second rearrangement is mediated by the π system of the alkene group. The ketone functional group is often easily discernable due to the prevalent fragments (Figure 14.14) and rearrangements described above. The configuration of the carbon structure can be difficult to discern. Reduction of the carbonyl group to a methylene group is commonly performed to determine the complete structure of the molecule.

The base peak in Figure 14.14 is the result of a McLafferty rearrangement and an α cleavage (Scheme 14.18).

14.11.2 Fragmentation of Cyclic Ketones

Cyclic ketones' major cleavage is also at the $\acute{\alpha}$ bond. Due to the ring structure, this cleavage will be detected as the molecular ion unless another bond is broken. Saturated cyclic ketones produce a fragment at m/z 55 illustrated by cyclohexanone (Scheme 14.19).

In cyclohexanone, this peak is the base peak. In absence of a reference standard, cyclic ketones are difficult to identify given the difficulty explained earlier in determining the composition of the alkyl portion of the ketone.

14.11.3 Fragmentation of Aromatic Ketones

Aromatic ketones create fragments via almost identical pathways as aliphatic ketones. One prominent peak, and usually the base peak, is the result of the cleavage of the less stable alkyl fragment

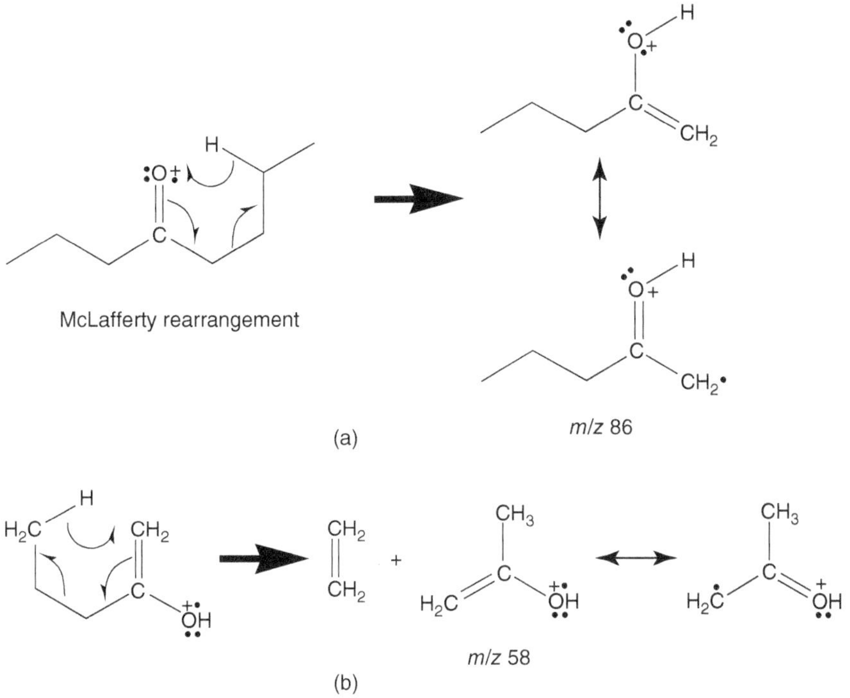

McLafferty rearrangement

(a) m/z 86

(b) m/z 58

Scheme 14.17

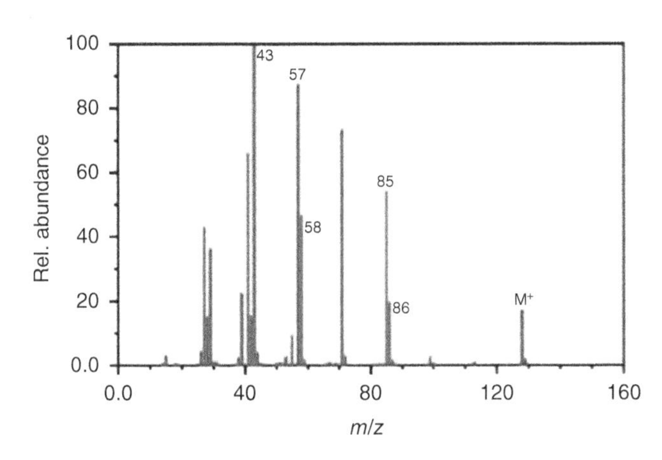

Figure 14.14 Fragmentation of a ketone (Source: NIST/National Institute of Standards and Technology/Public Domain).

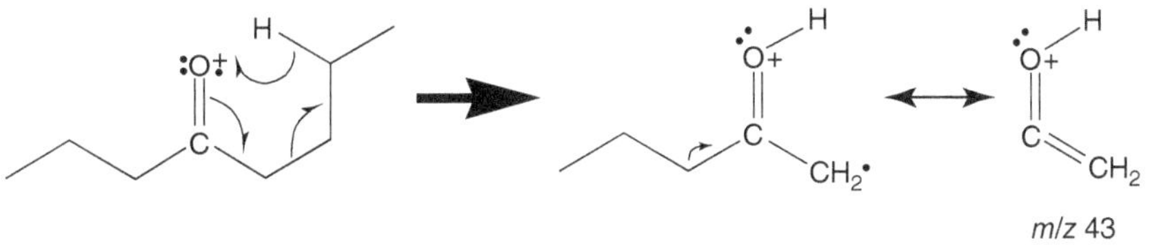

m/z 43

Scheme 14.18

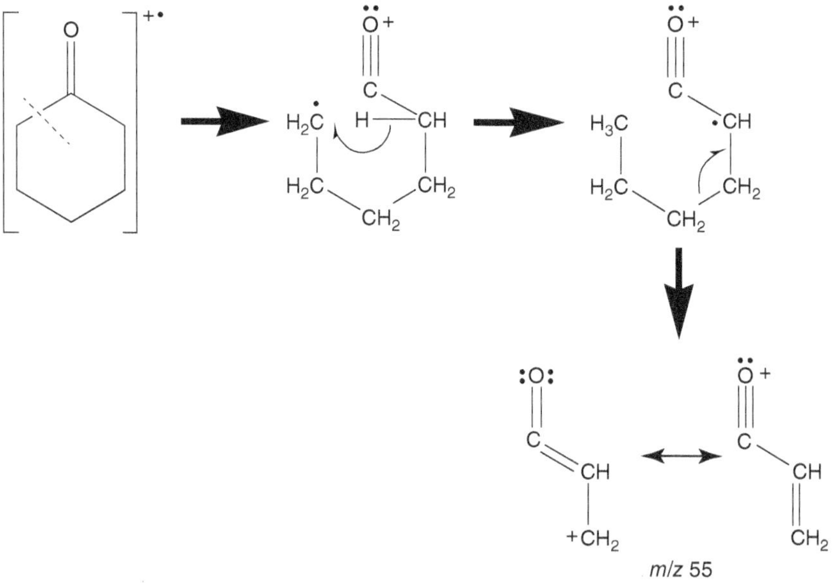

Scheme 14.19

resulting in the ArC≡O fragment located at m/z 105. The alpha cleavage resulting in a benzyl radical is infrequent given the stability of the competing reaction (Scheme 14.20a,b).

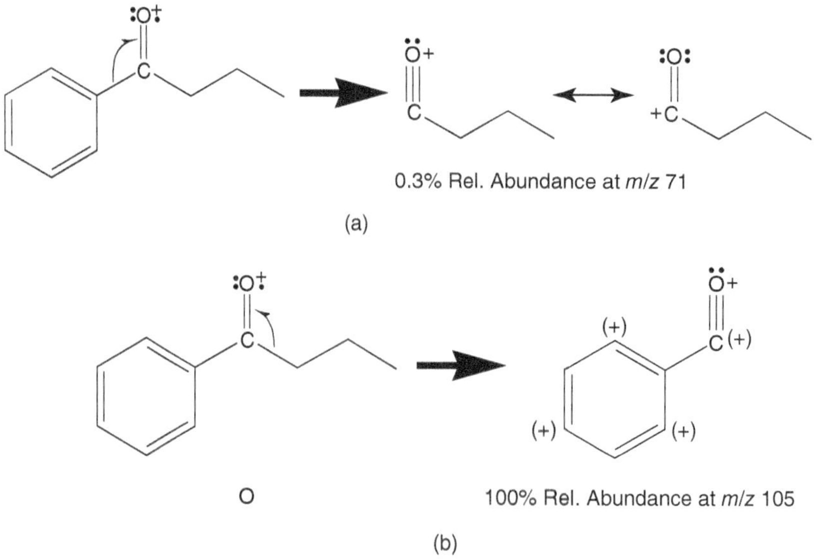

Scheme 14.20

The cleavage of the bond α to the aromatic group results in a peak at m/z 77 (Scheme 14.21).

Further fragmentation results in a peak at m/z 55 after the loss of HC≡CH. Some aromatic ketones undergo the typical McLafferty rearrangement if the other alkyl component contains an abstractable hydrogen atom (Figure 14.15 by Scheme 14.22).

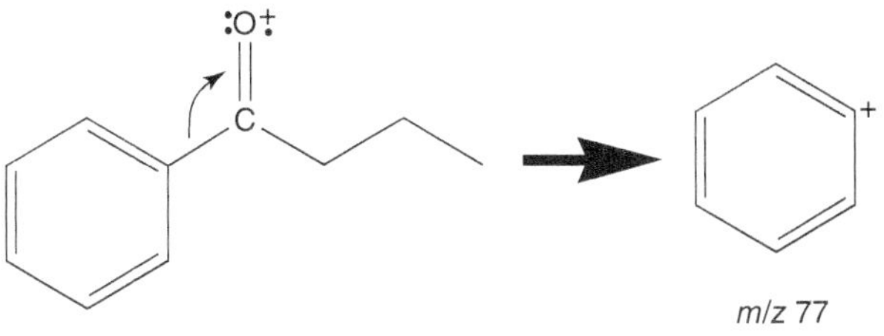

Scheme 14.21

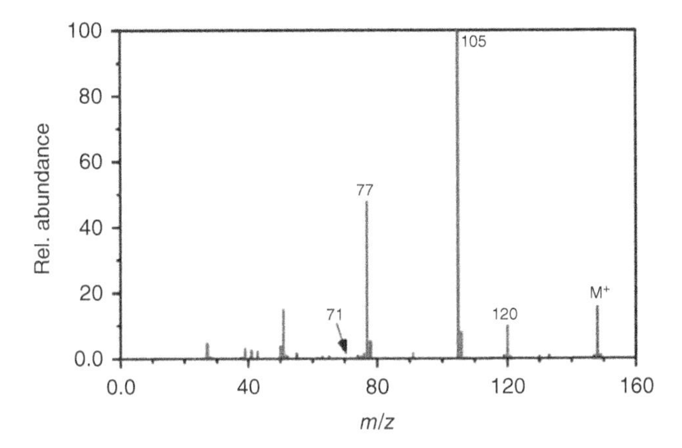

Figure 14.15 Fragmentation of a cyclic ketone (Source: NIST/National Institute of Standards and Technology/Public Domain).

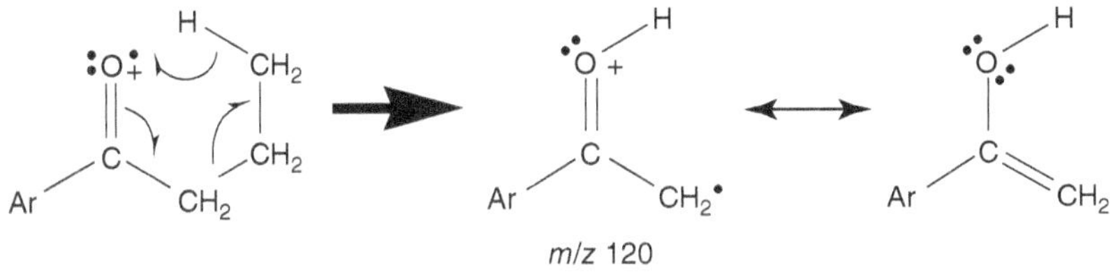

Scheme 14.22

14.11.4 Fragmentation of Aldehydes

The major peaks observed in spectrums of aldehyde are the result of the same ά cleavage as in ketones. This fragmentation results in an M-1 peak and a peak at M–R from the COH^+ ion. The presence of an M-1 peak helps identify the aldehyde but the hydrocarbon rearrangement at C_2H_5 prevents the M–R (m/z 29) peak from being truly useful. Another prominent peak is the McLafferty rearrangement located at m/z 44. The only aldehydes that do not contain this peak are ones that lack the necessary hydrogen atom for this rearrangement.

Straight-chain aldehydes have unique features that help in identification. These compounds will have an M-18 fragment from the loss of water, M-28 from the loss of ethylene, M-43 loss of CH_2=CH–O, and M-44 from the loss of CH_2=CH–OH (Figure 14.16).

The patterns resulting from aromatic ketones are almost identical to those governing aromatic aldehydes. The characteristic molecular ion is accompanied by an M-1 peak from the loss of hydrogen. The ArC≡O fragment loses CO to form the phenyl ion at m/z 77, which further degrades to give a peak at m/z 51.

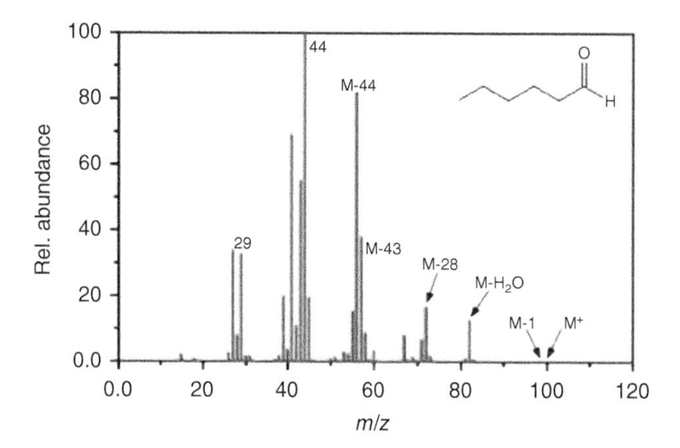

Figure 14.16 Fragmentation of an aldehyde (Source: NIST/National Institute of Standards and Technology/Public Domain).

14.12 Fragmentation of Carboxylic Acids

The molecular ion of straight-chain carboxylic acids is weak but usually present. The prominent and often the base peak results from the McLafferty rearrangement (Scheme 14.23).

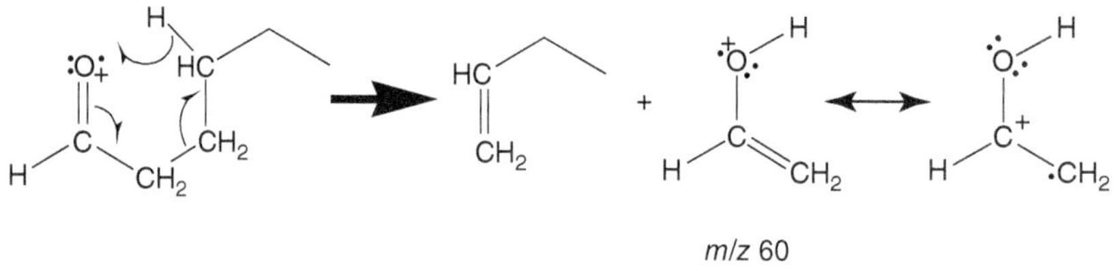

m/z 60

Scheme 14.23

Short-chain carboxylic acids give prevalent peaks at M–OH and M–CO_2H. In larger carboxylic acids, these peaks are less prevalent. Long-chain carboxylic acids are better identified by the fragments at $C_nH_{2n\text{-}1}O_2$ (Figure 14.17). There is also the presence of the hydrocarbon fragment at C_mH_{2m+1} illustrated in Section 14.9.

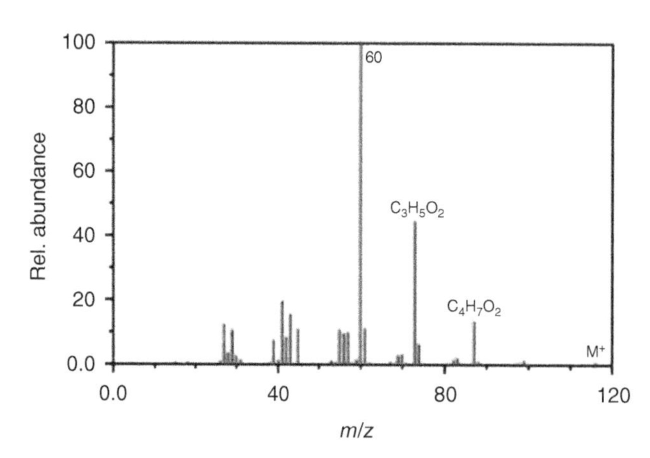

Figure 14.17 Fragmentation of a carboxylic acid (Source: NIST/National Institute of Standards and Technology/Public Domain).

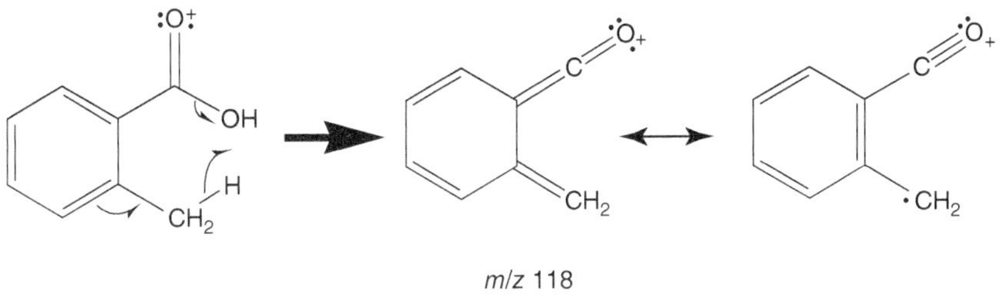

Scheme 14.24

Aromatic acids have a more prominent molecular ion peak but undergo similar fractionation to short-chain hydrocarbons. They produce large peaks at M–OH and M–CO_2H (Scheme 14.24). Aromatic acids can also lose water if an ortho group contains an abstractable hydrogen atom.

14.13 Fragmentation of Ethers

The molecular ion peak is usually weak in ethers. The oxygen atom mediates the major fragment and creates a β cleavage that results in a resonance-stabilized cation (Scheme 14.25). This peak is prominent and sometimes is the base peak.

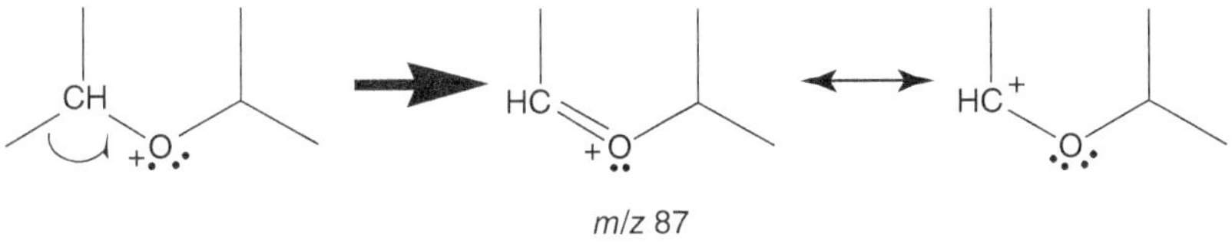

Scheme 14.25

The fragment can also undergo a subsequent rearrangement which typically creates the base peak when the ά carbon is substituted (Scheme 14.26).

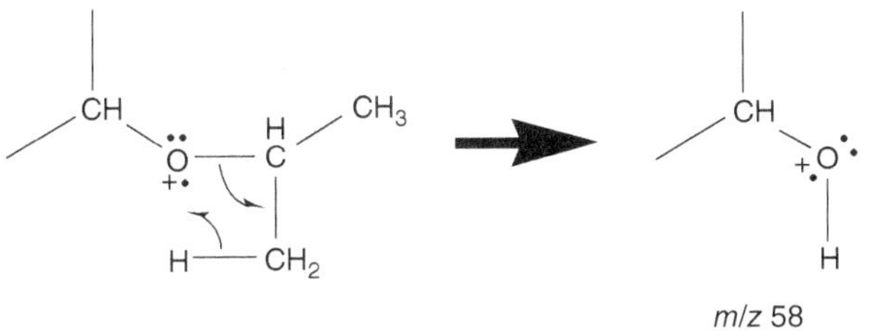

Scheme 14.26

Ethers also produce prominent alkyl fragments when the C—O bond (α bond) is broken and the fragment containing oxygen is a radical (Scheme 14.27).

The base peak in Figure 14.18 is the result of both a β cleavage and the below rearrangement (Scheme 14.28).

Aromatic ethers have a slightly different pattern of fragmentation. They produce prominent molecular ions due to the stability of the benzene ring. The major fractionation occurs at the β bond to the aromatic ring. This fragment can decompose further with the loss of CO (Scheme 14.29).

Aromatic ethers also cleave at the bond ά to the ring to create a peak at m/z 78 and 77 due to hydrogen migration (Scheme 14.30).

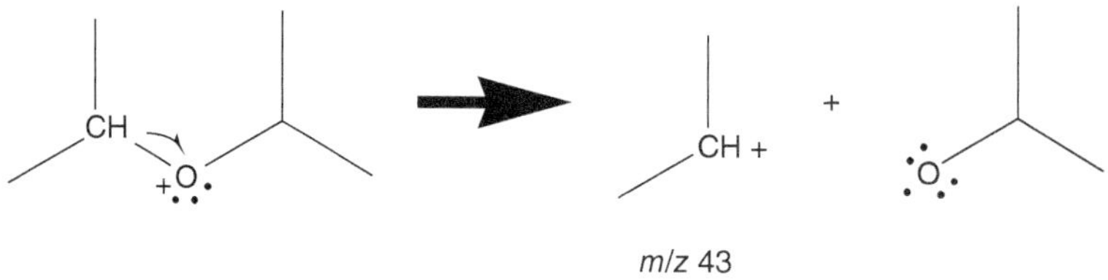

m/z 43

Scheme 14.27

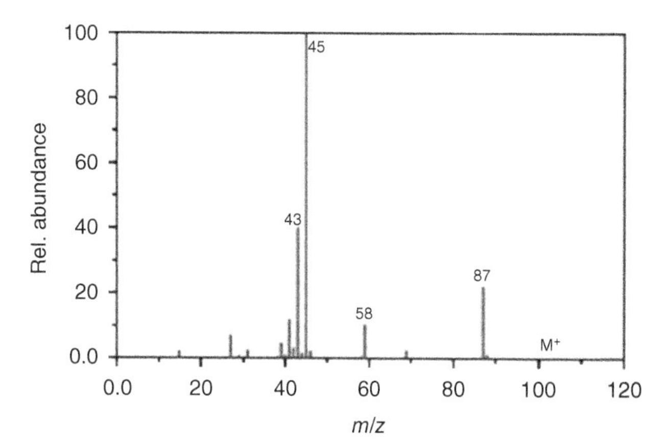

Figure 14.18 Fragmentation of an ether (Source: NIST/National Institute of Standards and Technology/Public Domain).

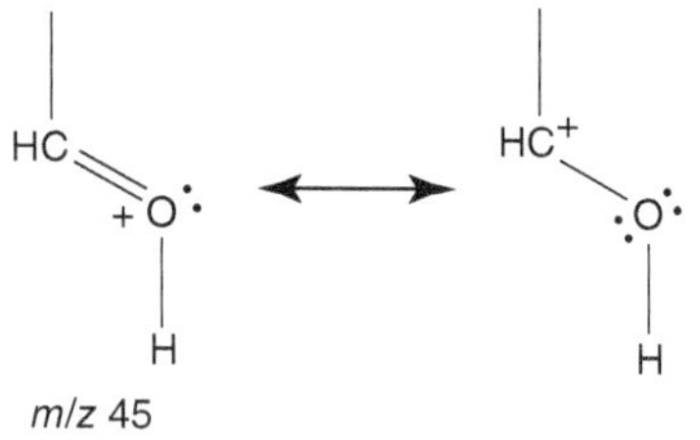

m/z 45

Scheme 14.28

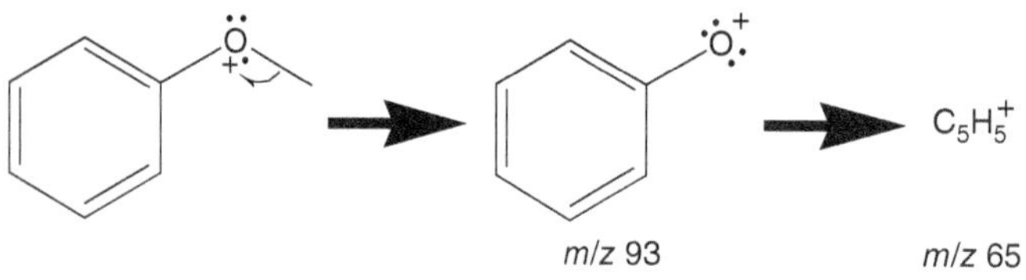

$C_5H_5^+$

m/z 93 m/z 65

Scheme 14.29

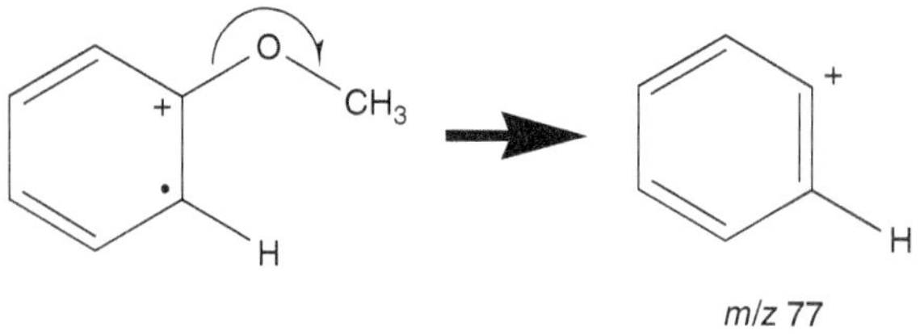

m/z 77

Scheme 14.30

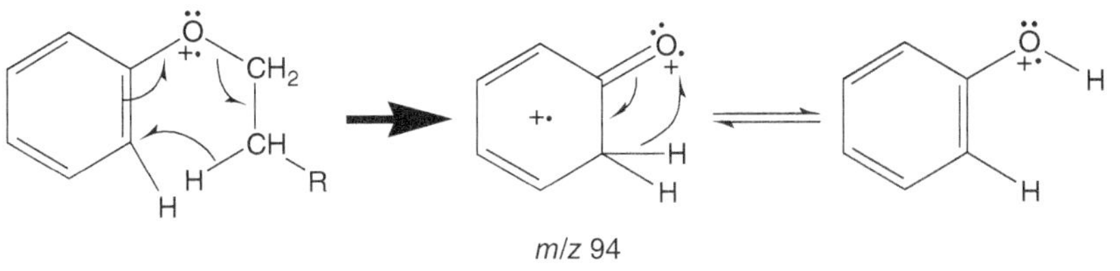

m/z 94

Scheme 14.31

When the alkyl portion of the sample is larger than two carbons, the β is accompanied by hydrogen migration caused by the presence of the aromatic group. This cleavage results in a peak located at m/z 94 (Scheme 14.31).

14.14 Fragmentation of Esters

The molecular ion peak of straight-chain esters is sometimes discernable. A prevalent peak and often the base peak results from the familiar McLafferty rearrangement. The size of the alcohol that formed the ester and the presence of ά substituents can normally be discerned by the mass of these two peaks (Scheme 14.32).

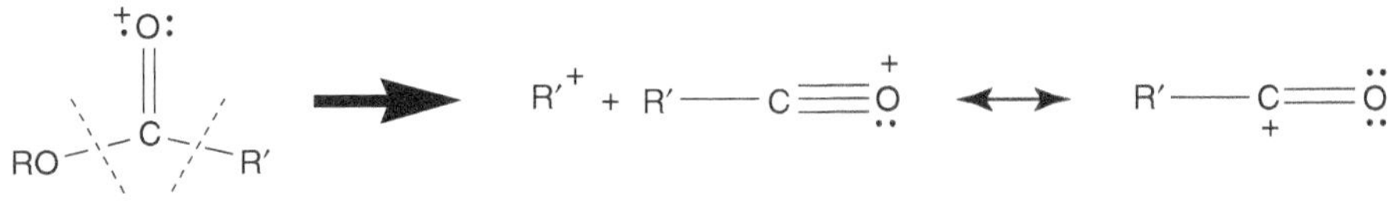

Scheme 14.32

The cleavage of the above bonds results in other fragments. However, these peaks are too small to be of great significance. For example, hexanoic acid methyl ester produces the following fragments (Scheme 14.33a,b).

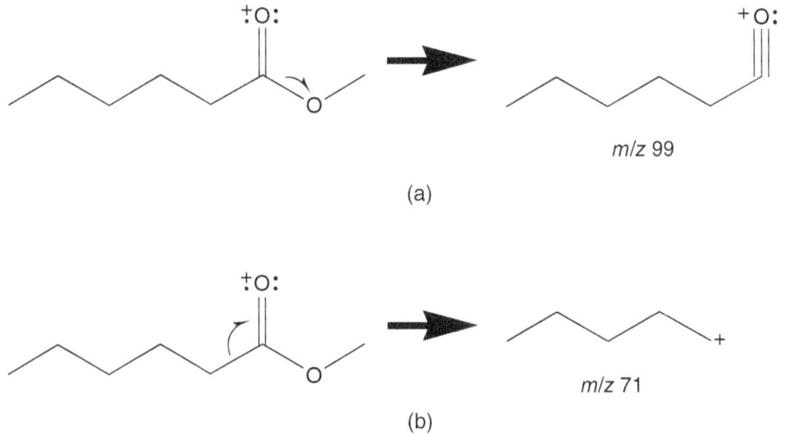

m/z 99

(a)

m/z 71

(b)

Scheme 14.33

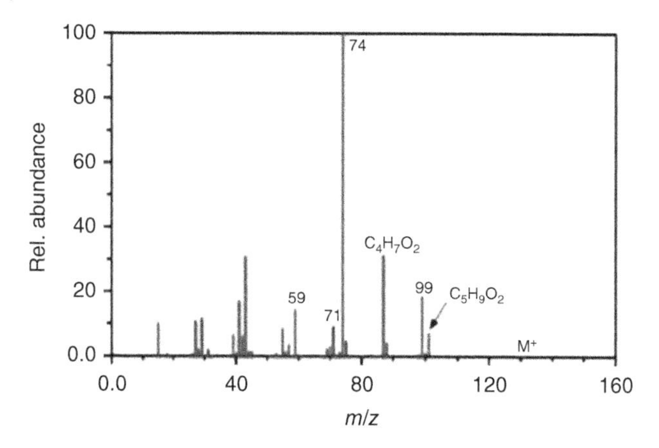

Figure 14.19 Fragmentation of an ester (Source: NIST/National Institute of Standards and Technology/Public Domain).

The resonance-stabilized $R'-C\equiv\overset{+}{O}$ ion gives a discernable peak for almost all esters. The R'^{+} ion is prominent in short-chain esters but is barely visible in esters with more than six carbon atoms. For hexanoic acid methyl ester, the R'^{+} ion is only 9.5% of the base peak.

Until this point, this chapter has covered individual functional groups in isolation. Since esters have both an alcohol and an acid component, fractionation patterns can be observed from both of these types of compounds. The prevalence of the fragments described earlier is dependent on the size of each part of the ester. The increased size of each portion results in a unique rearrangement.

When the acid portion is the major component like in hexanoic acid methyl ester, the fractionation pattern is partially characterized by typical acid peaks. For these straight-chain esters, cleavage of successive carbon atoms gives an alkyl fragment and a fragment containing oxygen. This pattern results in the familiar grouping of fragments spaced 14 units apart with the largest fragment in the cluster resulting from the $C_nH_{2n-1}O_2$ ion (Figure 14.19).

The base peak in Figure 14.19 is a result of the McLafferty rearrangement (Scheme 14.34).

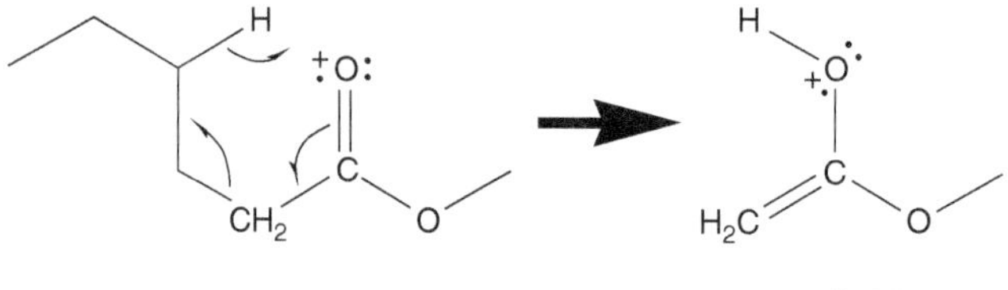

m/z 74

Scheme 14.34

When the alcohol portion of the ester is the prominent portion of the ester, fragments similar to that of an alcohol are observed. These esters will lose a molecule of acid like alcohols loose a molecule of water (Scheme 14.35).

Like alcohols, the prevalence of this rearrangement is so frequent that the molecular ion is normally absent from the spectra. These long-chain alcohols will also lose the alkyl fragment from the alcohol accompanied by two hydrogen migrations (Scheme 14.36).

In aromatic esters, the molecular ion is prominent due to the aromatic group. There are two distinctive types of aromatic esters that have their own unique fractionation patterns. Esters synthesized from aromatic acids mostly undergo ά cleavages (Scheme 14.37).

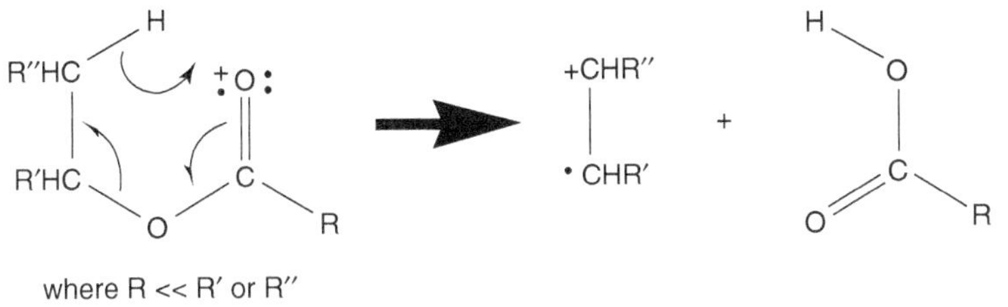

where R << R′ or R″

Scheme 14.35

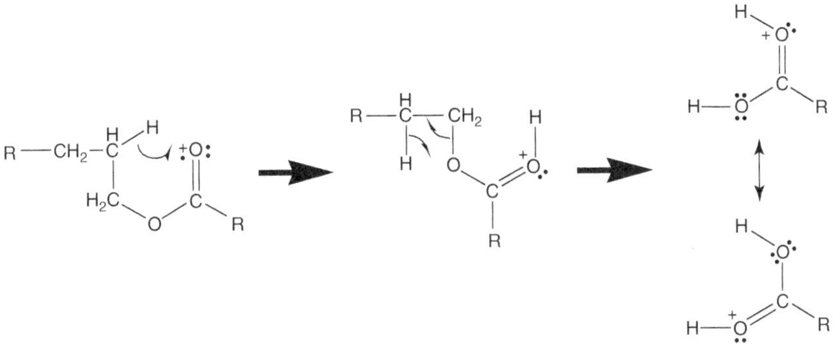

Scheme 14.36

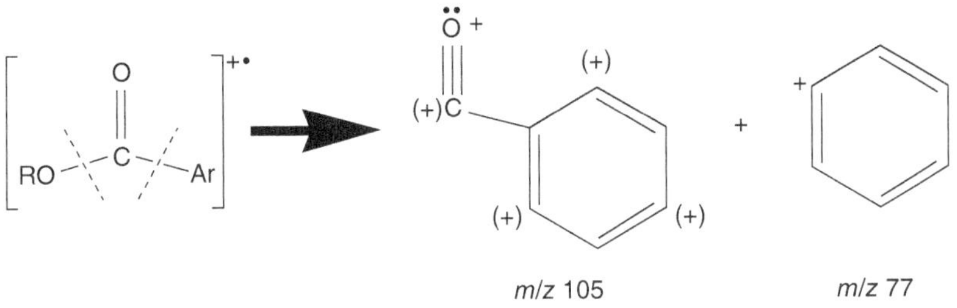

m/z 105

m/z 77

Scheme 14.37

The loss of ·OR results in the base peak because of the multiple resonance forms stabilizing the cation. When the alkyl portion of the alcohol becomes longer, the McLafferty rearrangement and the loss of R with two hydrogen migrations explained above are more favorable. Increasing the size of the R chain will cause the alkyl portion to retain the charge (Scheme 14.38).

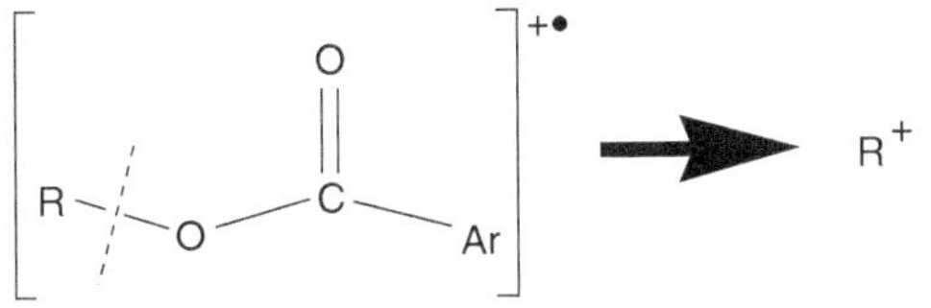

Scheme 14.38

The presence of an aromatic group in the alcohol that formed the ester results in the creation of the $CH_3C{\equiv}O^+$ ion. These esters also undergo a rearrangement that results in the loss of a ketene molecule (Scheme 14.39).

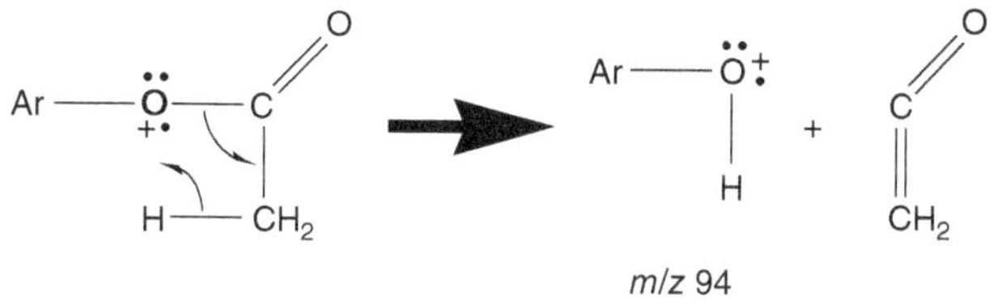

m/z 94

Scheme 14.39

14.15 Fragmentation of Amines

The presence of nitrogen in an amine can be detected by its odd molecular weight and the even fragments that it produces (Section 14.3). Oftentimes the presence of the molecular ion in longer straight-chain amines is not detectable. In these cases, chemical ionization techniques are often used in determining the molecular mass in order to determine the presence of nitrogen.

The base peak in most amines results from the cleavage of the β bond. The loss of the largest branch (R″) is preferred because the larger alkyl fragment stabilizes the produced radical (Scheme 14.40).

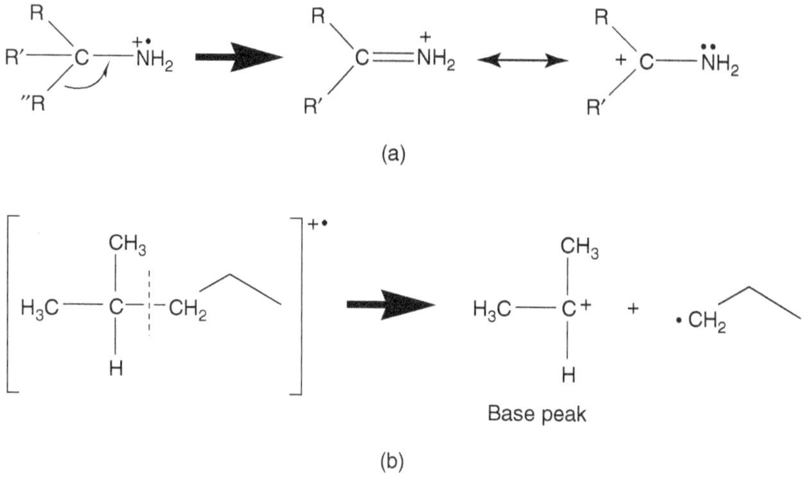

(a)

Base peak

(b)

Scheme 14.40

Like alcohols, if the ά carbon is bonded to a hydrogen atom, an M–H peak is usually visible. In primary amines with an unbranched ά carbon, cleavage of the β bond produces a peak at *m/z* 30 (Figure 14.20a–c). This peak is not conclusive proof of a primary amine because secondary and tertiary amines undergo a rearrangement similar to that of alcohols (Scheme 14.41).

Amines also produce even fragments that cause the cleavage of C—C bonds farther away from the functional group. The fragment containing the nitrogen group usually retains the charge resulting in peaks characterized by $C_nH_{2n+2}N$ spaced at 14 units. There is also the less prevalent hydrocarbon pattern of C_nH_{2n+1}, C_nH_{2n}, and C_nH_{2n-1}.

Cyclic amines produce a discernable molecular ion peak unless the ά carbon is substituted. The loss of hydrogen from the ά carbon is also a prominent peak. The ring is cleaved when

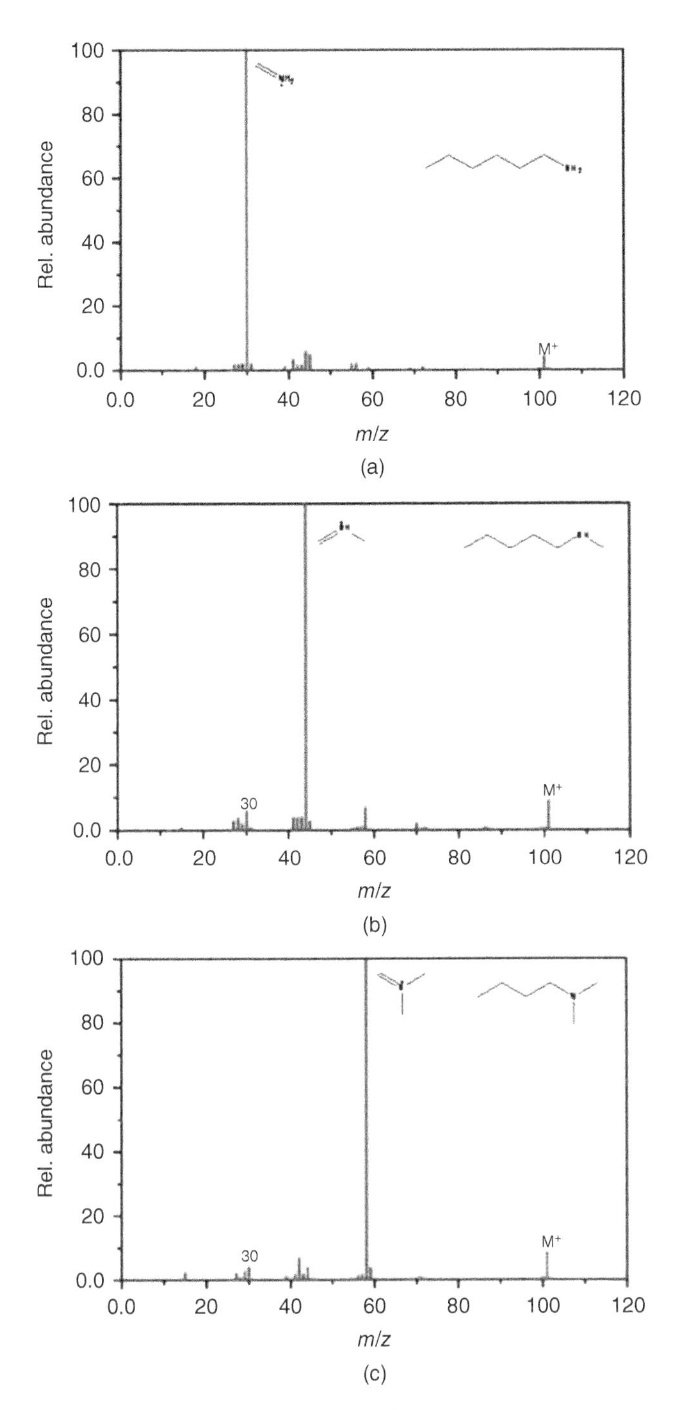

Figure 14.20 (a–c) Fragmentation of three amines—the mass spectrum is easily recognized as an amine due to its odd molecular ion and the presence of even fragments. The base peak in each spectrum, due to b cleavage distinguishes between the primary, secondary, and tertiary amine (Source: NIST/National Institute of Standards and Technology/Public Domain).

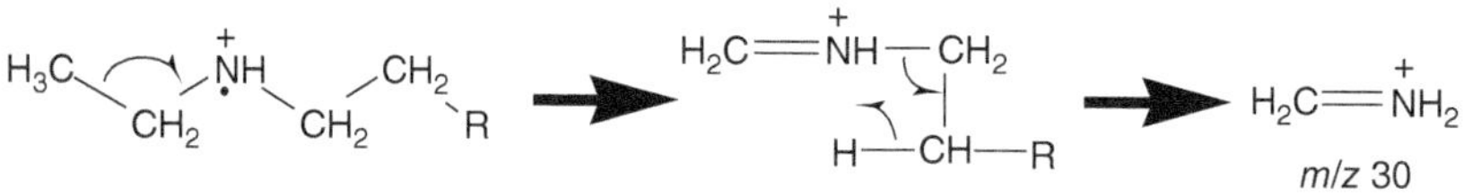

Scheme 14.41

the β bond is broken and subsequent alkene molecules fragment from the remaining ring structure.

The molecular ion of an aromatic amine is expectedly intense. The loss of the hydrogen atom bonded to the nitrogen gives a prominent peak at M-1. Similar to ethers, the loss of H—C≡N: (HCN) from the aniline ion produces peaks at C_5H_6 and C_5H_5. Unlike ethers, however, the heteroatom, not the aromatic group, controls the major pathways of fractionation resulting in β cleavage (Scheme 14.42).

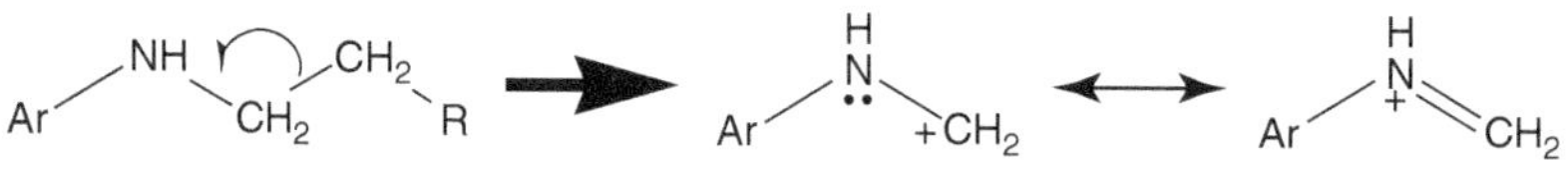

Scheme 14.42

14.16 Fractionation of Amides

The molecular ion of most straight-chain amides is usually discernable, which allows the nitrogen atom to be identified. The fractionation pattern is dependent on the length of the alkyl chain and the degree of substitution of the nitrogen group. Primary amides give a prevalent peak from the McLafferty rearrangement (Scheme 14.43).

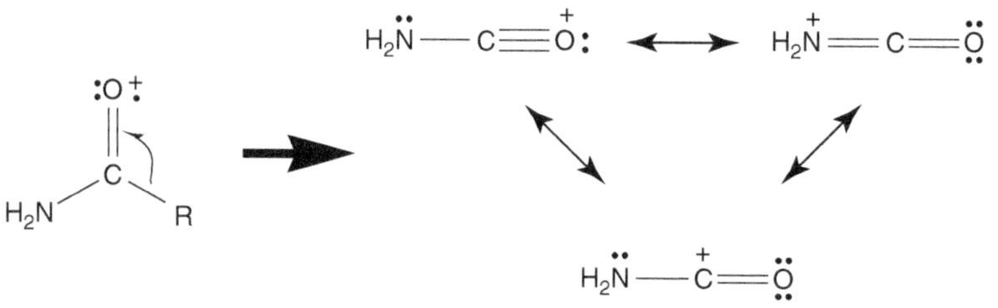

Scheme 14.43

In primary amides that lack an abstractable hydrogen atom for the McLafferty rearrangement, this accounts for the base peak. In the other amides including secondary and tertiary, the base peak is created by the McLafferty rearrangement at *m/z* 59. Primary amines also produce a peak at *m/z* 86 as a result of the following cleavage (Scheme 14.44).

When the alkyl groups bonded to the nitrogen are longer than two carbons, another rearrangement is discernable (Scheme 14.45).

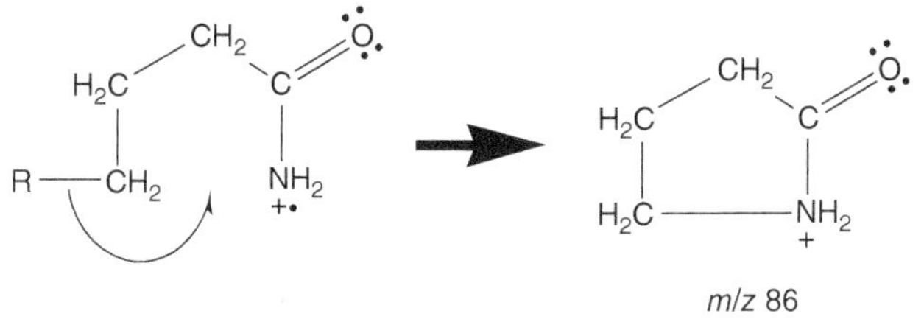

Scheme 14.44

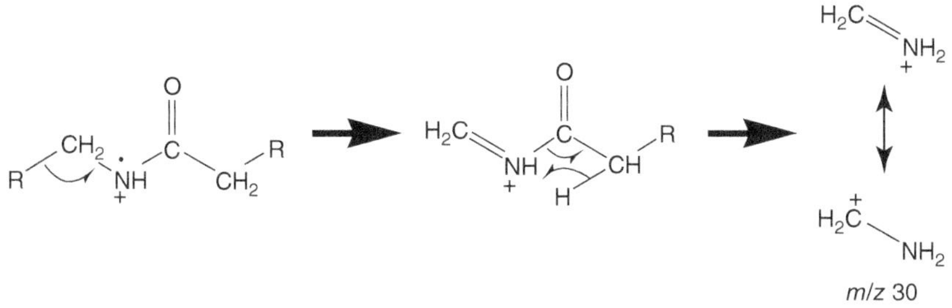

Scheme 14.45

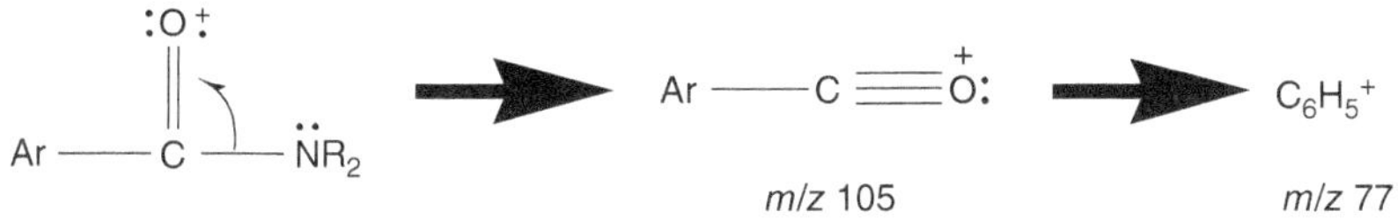

Scheme 14.46

Aromatic amides have a more prominent molecular ion peak. The common fragments are characterized by the loss of NR_2 to form a resonance-stabilized cation followed by the subsequent loss of CO (Scheme 14.46).

14.17 Fragmentation of Nitriles

The presence of nitrogen can usually be identified by the odd molecular weight according to the nitrogen rule (Section 14.3). This identification technique is usually unable to identify a nitrile because these compounds lack a molecular ion. The presence of an M-1 peak complicates the identification of the molecular ion. This peak is formed by a loss of an α hydrogen to form a resonance-stabilized cation (Scheme 14.47).

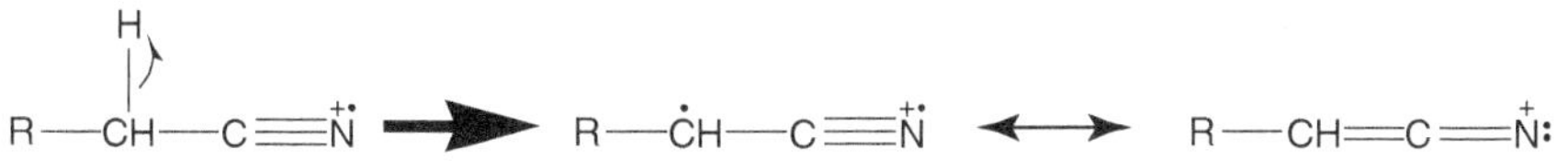

Scheme 14.47

A prominent and frequent base peak is the result of the McLafferty rearrangement at *m/z* 41 in compounds whose α carbon is not branched. This peak, however, is unable to confirm that a compound is a nitrile because hydrocarbon chains frequently form a peak at C_3H_5 (Scheme 14.48).

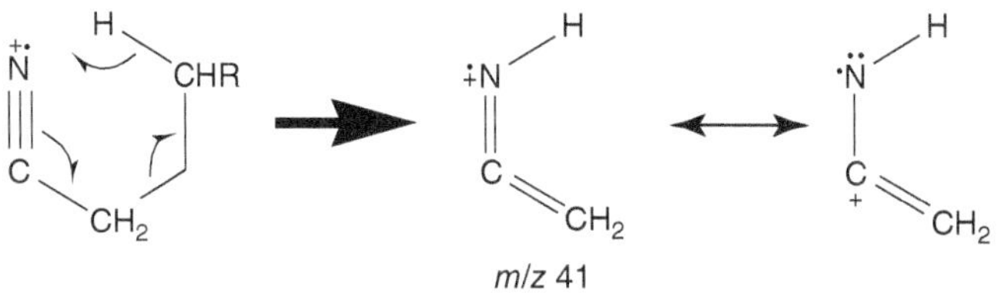

Scheme 14.48

A unique peak at m/z 97 is characteristic of nitriles that contain a straight chain of seven carbons or more (Scheme 14.49).

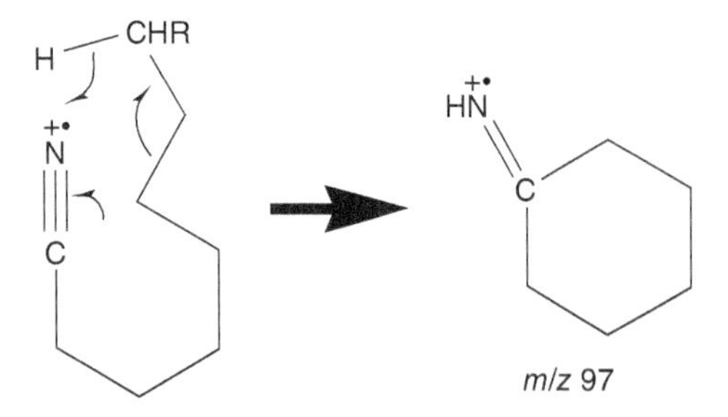

Scheme 14.49

14.18 Reviewing General Principles

After surveying common compounds encountered in organic chemistry and their corresponding spectra, we are able to make some generalizations about fractionation patterns. When dealing with a compound's molecular ion:

1. The molecular ions of ethers, carboxylic acids, aldehydes, and nitrogen-containing molecules such as amides and nitriles can be very faint or potentially absent. The molecular ion of alcohols and branched compounds is almost always undetected.
2. Increasing a compound's size or the branching of the alkyl portion will decrease the intensity of the molecular ion.
3. Cyclic structures, elements of unsaturation, and aromatic groups increase the intensity of the molecular ion.

Unknown compounds, in the absence of a reference standard, can often be more difficult but not impossible to discern given the prevalence of multiple fragments. The majority of compounds, however, will abide by the following rules.

1. Resonance-stabilized cations are favored because they help delocalize the positive charge throughout the molecule.
2. The cleavage of bonds is favored at substituted carbon atoms that produce the most stable cation. As a result, tertiary cations are favored over secondary, which are favored over primary, which are more stable than CH_3^+.
3. The longest chain is eliminated most frequently because the greater number of carbon atoms allows for the delocalization of the radical.

4. The combination of rules two and three can be good predictors of the prevalence of fragments. Achieving a balance between the stability of the radical and the cation produces the most prevalent peaks. The following example (Scheme 14.50a,b) illustrates this point.

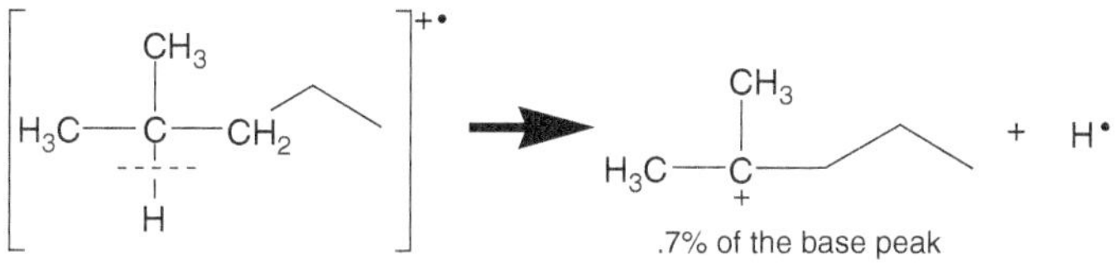

.7% of the base peak

Scheme 14.50

1. The β bond to the heteroatom is frequently broken since the heteroatom's nonbonding electrons allow for resonance forms that stabilize the cation.
2. Rearrangements account for prominent peaks in the spectrum such as the loss of water from an alcohol or the McLafferty rearrangement.

Besides having a general set of guidelines that govern general fractionation, it is also important to be able to identify patterns that are indicative of particular functional groups. As a result, a condensed table of the commonly observed fragmentation patterns is listed in Table 14.2.

Table 14.2 A review of common fragmentation patterns.

Functional Group	Observed Fragments	*m/z* Value
Straight chain alkanes	C_nH_{2n+1}	43, 57, 71, …
	$M-CH_3$	M-15
	$M-CH_2CH_3$	M-29
	$M-CH_2CH_2CH_3$	M-43
Branched alkanes	C_nH_{2n}	Various
Cyclic alkanes	$M-H_2C=CH_2$	M-28
Alkenes	C_nH_{2n-1}	Various
	C_nH_{2n}	Various
Aromatics	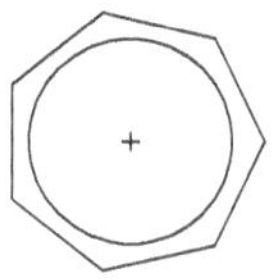	91
	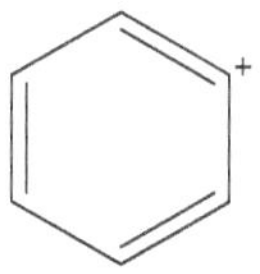	77
	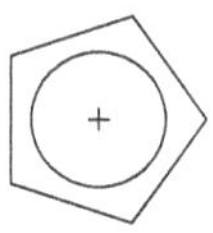	56

(Continued)

Table 14.2 (Continued)

Functional Group	Observed Fragments	*m/z* Value
Alcohols	M–H_2O	M-18
	M–(H_2O & $H_2C{=}CH_2$)	M-46
	M–(CH_3 & H_2O)	M-33
Primary alcohols	CH_2OH	31
Ketones	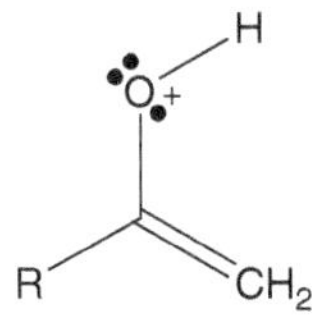 H O^+ R CH_2	43 + R
	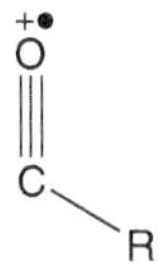 $O^{+\bullet}$ C R	Various
Aldehydes	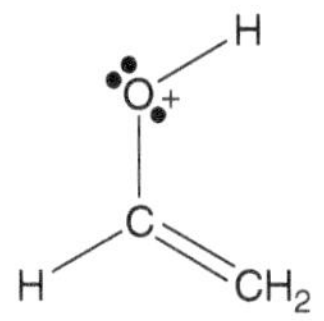 H O^+ C H CH_2	44
	COH	29
	M–H_2O	M-18
	M–$H_2C{=}CH_2$	M-28
	M–$H_2C{=}CH$–OH	M-44
Carboxylic acids	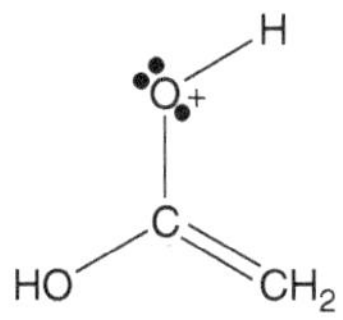 H O^+ C HO CH_2	60
	M–OH	M-17
	M–CO_2H	M-45
	$C_nH_{2n-1}O_2$	73, 87, …
Ethers	β cleavage	Various
	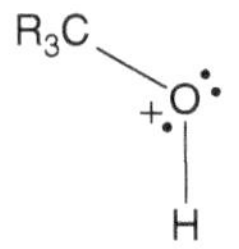 R_3C $O^{+\bullet}$ H	Various

(Continued)

Table 14.2 (Continued)

Functional Group	Observed Fragments	*m/z* Value
Esters	$RO-C(=CH_2)-\ddot{O}^{+\bullet}-H$	74, 88, …
	$H-O-C(=O)-R$	Various
	R'^{+}	Various
	$R'-C\equiv O^{+}$	Various
Amines	$RR'C=\overset{+}{N}H_2$	Various
	$C_nH_{2n+2}N$	58, 72, …
Primary amines	CH_2NH_2	30
Amides	CH_2NH_2	30
	$H_2\ddot{N}-C\equiv\overset{+}{O}:$	44
	$H_2N-C(=CH_2)-\dot{O}^{+\bullet}-H$	59
	$H_2C-CH_2-C(=\ddot{O})-\overset{+}{N}H_2-CH_2$ (ring)	86
Nitriles	M–H	M-1
	$H-\dot{N}^{+}=C=CH_2$	41

14.19 Searchable Databases

The proliferation of databases and the number of compounds that they contain has made the interpretation of spectra less important. These databases cover over 200 000 compounds, the two most commonly used databases are the ones produced by the National Institute of Standards and Technology (NIST) along with the *Wiley Registry of Mass Spectral Data*. Using these databases can be of great assistance when performing routine analysis and sometimes is the only way to positively identify a particular compound.

These databases cannot be the exclusive tool that chemists rely on to interpret data. Like other tools, it is necessary to know when and how to use it. Databases are a perfect tool for performing routine analysis when the analyte and the reference standard have a high percent match. When the quality of the match becomes low, it is necessary to assess the validity of the database match. It is also necessary to understand these fractionation patterns when performing research, especially when synthesizing new compounds that are not contained in the published databases, and for which there is obviously no reference compound. Only through the combination of manual interpretation along with the usage of a library can the composition of an unknown sample truly be discerned.

15

Common Radiochemical Detection Methods in Analytical Chemistry

15.1 Introduction

No analytical text would be complete without the inclusion of radiochemical methods. As you learned in general chemistry, a radioactive element is one that spontaneously degrades to one or more daughter elements with the emission of some form of radiation. From the analytical detection standpoint used in this text, we will only be concerned with alpha, beta, and gamma emissions. The public does not readily know this but we are composed of radioactive elements such as carbon-14 and are exposed to radioactive elements every day from our surroundings and food such as potassium-40 in bananas. In this chapter, we will look at common radiation techniques used to detect radiation in everyday samples.

Radiological methods also follow the previously used term of "there is no zero or detection limit." Most radiologically decay analytes can be detected to extremely low activities by simply counting the detector signal for longer times.

15.2 Common Sources of Radiation

The unit used to measure radiation dose or exposure is the millirem (mrem). Instead of analyte concentration, we refer to radiation activity. A typical American receives a radiation dose of about 0.62 rem (620 millirem) each year. Half of this dose comes from natural sources such as radon in the air and foods, with smaller amounts from cosmic rays and the Earth itself. The other half (0.31 rem or 310 mrem) comes from human-made sources of radiation including medical, commercial, and industrial sources, and even still from background radiation from nuclear weapons testing. Examples are given in Table 15.1.

One interesting aspect of radiation is the way radionuclides decay. Early researchers thought that the decay rate would follow zero-order kinetics, independent of parent element (analyte) concentration or activity since one of each atom decayed by itself. But experimentation showed that all radioactive decay follows first-order kinetics based on the first-order probability of an atom decaying to daughter products. This makes the study of radioactive decay extremely easy.

$$\text{First-order rate law: } \ln A_t = \ln A_o - kt$$

where ln is the natural log, A is the analyte activity, A_0 is the initial analyte activity, k is the first-order rate constant (radioactive decay rate), and t is the time.

The three modes of radioactive decay of interest here are alpha particles (basically a helium nucleus consisting of two protons and two neutrons with low kinetic energy), beta particles

Essential Methods of Instrumental Analysis, First Edition. Frank M. Dunnivant and Jake W. Ginsbach.

Companion Website: www.wiley.com/go/essentialmethodsofinstrumentalanalysis1e

Table 15.1 Radiation dose by source per person living in the United States.

Source	Dose
Natural radiation	
A 5-h jet airplane ride	2.5 mrem/trip (0.5 mrem/hr at 39 000 feet)
Cosmic radiation from outer space	27 mrem/yr (whole body)
Terrestrial radiation	28 mrem/yr (whole body)
Natural radionuclides in the body	35 mrem/yr (whole body)
Radon gas	200 mrem/yr (lung)
Diagnostic medical procedures	
Chest X-ray	8 mrem (whole body)
Dental X-rays (panoramic)	30 mrem (skin)
Dental X-rays (two bitewings)	80 mrem (skin)
Mammogram	138 mrem per image
Computed tomography (CT) head	200 mrem (whole body)
Upper gastrointestinal tract test	244 mrem (X-ray portion; bone marrow)
Barium enema (X-ray portion)	406 mrem (bone marrow)
Thallium heart scan	500 mrem (whole body)
CT abdomen and pelvis	1000 mrem (whole body)
Consumer products	
Building materials	3.5 mrem/yr (whole body)
Luminous watches (H-3 and Pm-137)	0.04–0.1 mrem/yr (whole body)
Tobacco products (30 cigarettes/day)	16 000/yr (bronchial epithelial)

(Source: Data from Tro and Au-Yeung (2011). Department of Health and Human Services; U.S. National Institute of Health).

(composed of high-energy electrons with higher kinetic energy), and gamma rays (a high-energy massless component of the electromagnetic spectrum). Examples of each are shown below.

$$\text{Alpha Decay: } {}^{218}_{84}Po \rightarrow {}^{214}_{82}Pb + {}^{4}_{2}He$$

$$\text{Beta Decay: } {}^{234}_{90}Th \rightarrow {}^{234}_{91}Pa + {}^{0}_{-1}\beta$$

$$\text{Gamma Decay: } {}^{240}_{94}Pu \rightarrow {}^{240}_{94}Pu + {}^{0}_{0}\gamma$$

15.3 Detection of Alpha, Beta, and Gamma Emission

Most everyone has seen an old-fashioned Geiger counter in movies, such as the one shown in Figure 15.1.

The design of these has progressed considerably in recent years being much smaller, easier to use, and only costing tens of dollars. Geiger counters count the total amount of radiation, not distinguishing between alpha, beta, or gamma emissions.

Most analytical instruments are divided into alpha, beta, and gamma detectors.

Figure 15.1 An old model Geiger counter for detected nuclear radiation (Source: Dunnivant and Ginsbach (Book Authors)).

15.3.1 Alpha Detectors

Alpha emissions are more commonly released from heavier mass radionuclides, those with a mass number greater than 150 (or the number of protons greater than 60). Again the alpha is a helium nucleus consisting of two protons and two neutrons containing low kinetic energy, so low that a piece of paper can effectively shield exposure, and eventually capture two electrons and become a helium atom. Alpha emissions do not even penetrate the human skin but are most harmful if ingested or injected.

Modern methods of alpha detection use liquid scintillation counting (LSC), where the sample is directly mixed with a scintillation cocktail where the individual alpha emissions react with a reagent to produce visible light emission events that are counted (an example of luminescence).

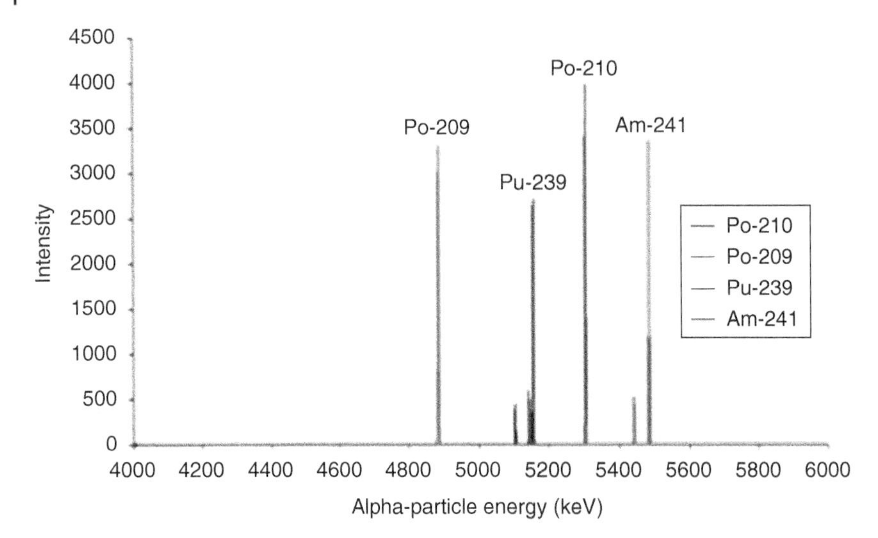

Figure 15.2 A simulated alpha spectra (Source: Wikimedia Commons/https://en.wikipedia.org/wiki/Alpha-particle_spectroscopy#/media/File:Alpha1spec.png (accessed February 06, 2024).

The LSC instrument records the amount of light energy per radioactive decay event. Such alpha spectra obtained by LSC are shown in Figure 15.2.

15.3.2 Beta Detectors

Beta emissions have greater kinetic energies than alpha particles, in the range of greater than 0.2 MeV. Beta decay can be of the *beta minus-* or *beta plus*-type emissions. Geiger counters are sensitive to them but more analytical grade detections again involve the use of liquid scintillation counters that can discriminate distinct kinetic energies. Common used include the measurement of tritium and carbon-14 activities.

15.3.3 Gamma Detectors

Gamma emissions contain considerably more kinetic energy, tens of eV to MeV, and have considerable penetration power. It takes several centimeters of lead to stop exposure to a gamma source. Radium and potassium radionuclides are two common gamma emitters. Instruments are considerably more complex and expensive, and consist of a supercooled semiconducting germanium gamma-ray spectrometer that correlates the energy of the gamma emission to specific radionuclide emissions (isotopes). Emitted gamma rays react in the detector to produce isotope-specific results such as those shown in Figure 15.2 for beta rays but with considerably more energy and as shown in Figure 15.3.

Most instruments are stationary lab-based but the petroleum has used portable down-borehole probes for decades to evaluate crude oil reserves and exploration. The US Department of Energy also uses these down-hole instruments to evaluate groundwater contamination at its many cold-war laboratory sites. Such an instrument is shown in Figure 15.4.

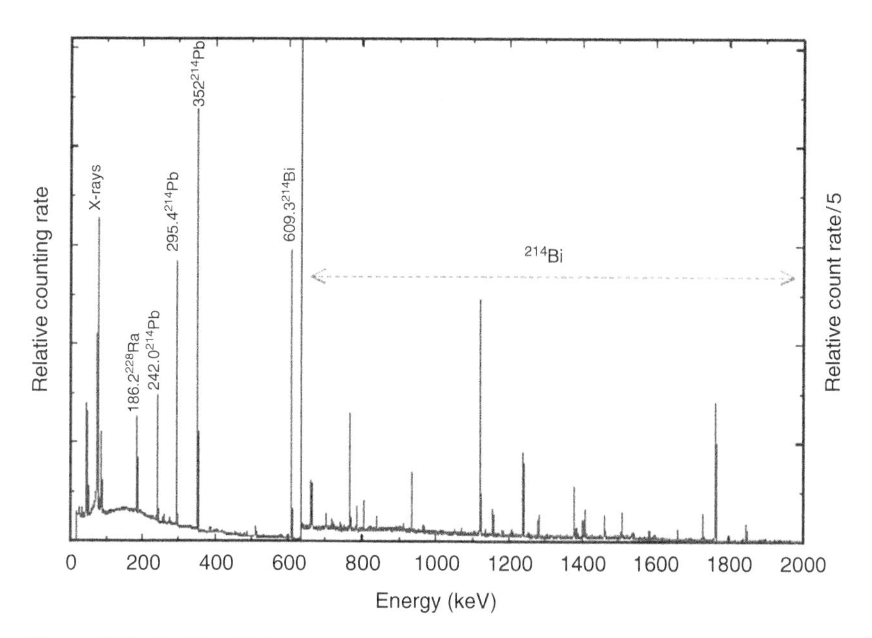

Figure 15.3 Radium-226 gamma-ray spectrum from high-purity germanium (HPGe) detector (Source: NIST, 2021/National Institute of Standards and Technology/Public Domain).

Figure 15.4 A mobile down-hole, isotope-specific gamma-ray detector. The down-hole, liquid N_2-cooled Ge detectors are shown on the inside, left wall of the van (Source: U.S. Department of Energy/Public Domain).

15.4 Case Studies

15.4.1 The dating of the formation of the Earth was measured by the alpha and beta decays of U-238 to Pb-206. An excellent and clear example can be found on Wikipedia at https://en.wikipedia.org/wiki/Uranium%E2%80%93lead_dating.

15.4.2 Radiocarbon dating of the oldest human footprints in the Americas dating seeds preserved at White Sands National Park in New Mexico, United States. https://www.science.org/doi/10.1126/science.adk3075.

Reference

Tro, N.J. and Au-Yeung, H.Y. (2011). *Introductory Chemistry*. Prentice Hall.

16

Instrumental Laboratory Experiments with Results

16.1 Introduction

This chapter contains laboratory experiments that have been developed since 2015. Each experiment has been researched by students, contains sample students' results, and is written by students.

16.2 A Typical Glassware Inventory for an Instrumental Methods Course

Table 16.1 contains a glassware and supplies list for the experiments covered in this chapter. The authors recommend a professional-type lab notebook such as the ones published by *X* and *X* that allows for the subject titles illustrated in Section 16.3 to be documented.

16.3 Maintaining a Legally Defensible Laboratory Notebook

Chemistry is a laboratory science and an important part of this course will be to familiarize the student with proper laboratory techniques and recording of their work in laboratory notebooks. The practices we use here may seem rather detailed or "picky," but these are minor compared to the practices used in some industrial sectors.

The student should keep an elaborate and highly organized account of each lab exercise in their notebook. At the beginning of each lab, the instructor should look for a "Things to do" list, possibly a detailed procedure, a dilutions table, etc., in the student's lab notebook. This tells the instructor that the student came prepared to the lab.

Organization of your notebook; Recommended Title Sections for every lab:

- Title (in designated area)
- Purpose
- Chemicals and solution used (brand, lot number, who made them)
- Calculations
- Cited Standard Operating Procedure (SOP) with modifications and/or detailed procedures
- Data, hand written, spreadsheets, and plots to be completed during each lab (spreadsheets and plots should be neatly taped in your lab notebook)
- File name of Instrument Method, Sequence, and Data files
- Conclusions, what you found in this lab

Essential Methods of Instrumental Analysis, First Edition. Frank M. Dunnivant and Jake W. Ginsbach.

Companion Website: www.wiley.com/go/essentialmethodsofinstrumentalanalysis1e

Table 16.1 Instrumental glassware desk list

Name: CO____________________ Desk Number: _________

	Supplied	Returned
Top drawer: Item		
Beaker, 50 mL	4	
Beaker, 150 mL	2	
Beaker, 250 mL	2	
Beaker, 400 mL	2	
Beaker, 600 mL	1	
Beaker, 1000 mL	1	
Cylinder, 10 mL graduated	1	
Cylinder, 25 mL graduated	1	
Cylinder, 100 mL graduated	1	
Middle drawer: Item		
Flask, 250 mL Erlenmeyer	4	
Flask, 10 mL volumetric	2	
Flask, 25 mL volumetric	8	
Flask, 50 mL volumetric	8	
Flask, 100 mL volumetric	8	
Flask, 250 mL volumetric	1	
Flask, 500 mL volumetric	1	
Flask, 1000 mL volumetric	1	
Funnel, 65 mL filtering	2	
~~Pipet, 1 ml~~	~~2~~	
Pipet, 2 mL	2	
Pipet, 3 mL	2	
Pipet, 4 mL	2	
Pipet, 5 mL	2	
Pipet, 10 mL	2	
Pipet, 25 mL	1	
Pipet bulb	1	
Sharpie	1	
Round bottom flask, 250 mL	2	
Tape	1	
Scissors	1	

My signature below indicates that (1) I have received the above equipment in good condition; (2) I have read the safety precautions and agree to abide by the rules of the laboratory; and (3) I will check out of the lab desk in person within 5 days of leaving this course. Please return equipment to the drawer listed upon checkout.
Signature:____Date:__________

A few widely accepted details concerning the legal laboratory notebooks follow:

(1) Leave a few pages at the beginning for a table of contents (by experiment), a list of commonly used tables, and a list of commonly used figures, and appendices. Our current notebooks have a place for this. Note that at the end of the day, the student should initial and date each page.
(2) Instead of recopying a handout or printed procedure, the student may reference the SOP in their laboratory notebook.
(3) If someone else collects the data and the student needs to recopy it in their notebook, note where and when the data were originally recorded.
(4) Neatness is important, but the student must write in procedures as you plan them (in the lab) and enter data *as you collect it* (not on notebook paper or napkins for later transfer to your official record). *Every lab book must contain all data collected in the lab, no matter who in your group collected the data.*
(5) Record every detail in your notebook for future reference. Some items that may not be immediately evident, but should be included, are the type, model number, serial number of equipment (serial number only if more than one piece of the equipment is available), type of glassware used in dilutions, standard dilutions, and brand name and lot number of all chemicals used.
(6) Record computer file's path and file name for sequence, method, and data.
(7) No skipped pages are allowed in official (legally defensible) notebooks without proper procedures. Any page that contains more than 1/4 of a blank page at the bottom or top or one side of a page must have a line drawn through it and be signed and dated by you at the end of the day.
(8) All entries must be made with a ballpoint pen (not the alcohol-based, water-soluble pens that are more commonly available). If you make a mistake, draw a single line through the mistake, and write in the correction. Do not obliterate a number that you think is in error as you may find that you actually needed that number.
(9) Remember that the goal of this entire process is to produce a document that the student, or one of their peers, can pick up (even years in the future) and exactly reproduce your procedure and hopefully your results.
(10) Everything that you do in the lab should be recorded in your lab notebook as you do it, including lengthy procedures and calculations as you do them.
(11) Do not remove any pages from your lab notebook.

16.4 Solutions, Weights, and Laboratory Techniques

The objective of this lab is to develop and refine student's calculation and laboratory skills and introduce students to analytical equipment (and see what they learned in *Quantitative Methods of Analysis*).

The scientific method requires the collection of experimental data, and the data must be collected in a manner that insures precision and accuracy. No matter what field of science you work in, you will eventually have to make solutions of specified concentrations. There are two main goals of this lab exercise: (1) to test your ability to determine how to make a solution of specified concentration and (2) to test your accuracy and precision in making these solutions. You will complete this experiment using gravimetric and volumetric techniques.

The stock ~1000 mg/L stock solution made in this lab should be stored in a plastic bottle and can also be used in Experiment 16.5.

Experimental solutions needed:

(1) A FAAS with Ca lamp.
(2) Each pair of students will be supplied with:
1 or 2, 5, 10, 25 mL Class A pipets, 25, 50, 100, 250, 500, and 1000 mL Class A volumetric flasks.
(3) 99.6% pure $CaCO_3$ (CAS number 471-34-1) (dried at 104 °C overnight and stored in a desiccator).
(4) 1% by volume nitric acid for dilution purposes (2 L per student or group).
(5) Your knowledge of general chemistry and quantitative laboratory.
(6) NOTE that you may need to use a piece of glassware more than once.
(7) Also NOTE that the fewer dilutions you make, the more precise your solutions will be.

To the Instructor: Normally, a technician would prepare a set of Ca standards and calibrate the FAAS to determine the Ca concentration. This is not necessary in this case. The stockroom only needs to prepare the end target concentration and monitor the absorbance reading. Given the high stability of the FAAS absorbance readout, the student's sample should read within plus or minus 0.003 absorbance units.

16.4.1 Weighing and Dilution Skills

This lab will test your knowledge of converting from grams to molar units (and vice versa), weighing skills, and dilution skills. The latter two skills will require a high degree of accuracy and precision at each step in the procedure. You should have obtained each of these skills in the pipetting labs earlier in Quantitative Methods of Analysis. Your assignment is to make up a 1.18×10^{-4} M Ca^{2+} solution from known purity $CaCO_3$ salt (this may slightly change each year, so ask your instructor what the purity of the salt is this year before conducting calculations). You must figure out how to accurately do this given the available equipment (listed above). As you make decisions on how to make this solution (how much to weigh out and what dilutions to make), consider the accuracy and precision of each step. Your ability to complete this task will be tested using a flame atomic absorbance spectrophotometer (FAAS) and your accuracy and precision will be reflected in your lab grade!

You will immediately note that you cannot accurately weigh out the small amount that you need to make the solution. So, you must first make a more concentrated solution and then dilute it. The question that you must figure out is "what is the concentration of the initial solution that you must make?," and then, "how should you dilute it to achieve the desired concentration (1.18×10^{-4} M Ca^{2+})?"

Strict guidelines for making your solutions:

(1) It is best to make at least 500 mL of your stock solution since this same solution can be used in the following experiment in this chapter.
(2) Carefully clean all glassware to remove any Ca^{2+-} that may be present (rinsing with 1% nitric acid works well for this).
(3) Weigh *at least* 0.100 g on the balance in order to reduce your error from weighing. Note weights to three significant figures.

(4) Your final volume should be at least 10 mL, but 100 mL is better. You may also want to conserve the volume of distilled water used.
(5) All solutions must be made in 1% HNO_3. You do not have to make the 1% very accurate. (HNO_3 is needed to dissolve the salt in the first solution and keep it dissolved in your dilution.)
(6) Bring a FULL volumetric flask to the instrument room for measurement. MIX WELL.

16.4.2 "The Order in Which You Are to Do Things"

Step 1: First you will each prepare a procedure for making your dilution. You will do this using the guidelines given above and without the help of anyone else. Once you have come up with a plan, you will present it to your instructor. If they concur, you can proceed to step 2. (Suggested grading scheme: This lab is worth 25 points. 2.55 points will be deducted for each incorrect calculation or dilution, i.e. check your guidelines!)
Step 2: Make your diluted solution and bring it to the FAAS.
Step 3: If your solution is incorrect, remake your solution one more time paying particular attention to your lab technique. If your solution is the correct concentration, clean up your lab space and you are finished. (Suggested grading scheme: 2.5 points will be deducted from your grade for each incorrect solution.)

CLEARLY show all calculations for making the solution in your lab notebook using the guidelines given in class.

- For large class sizes, work in pairs, each person will make at least one correct solution.
- Dilute the stock using Class A pipettes to make your 1.18×10^{-4} M solution.
- Bring your data sheet to the AA with our samples.

After you finish at the AA: What is your conc in ppm to three sign figs?

Solutions Data Sheet; BRING THIS SHEET TO THE FAAS WITH YOUR SAMPLE

Name: ______________________________

Calculations: Attempt #1 __________
Attempt #2 __________
Attempt #3 __________

FAAS Results: Measurement #1 __________
Measurement #2 __________
Measurement #3 __________
Measurement #4 __________
Measurement #5 __________
Measurement #6 __________

16.5 Determination of a Surrogate Toxic Metal (Ca) in a Simulated Hazardous Waste Sample by a Variety of Techniques

This is a four- to five-week group experiment that allows students to analyze for one analyte by a variety of instrument calibration techniques, compile their data, conduct statistical tests, and write a mock journal article on their findings. The instructor can choose between a number of techniques but we suggest selecting only four techniques to avoid redundancy in FAAS instrument use.

Purposes: To introduce complex sample matrices

To learn flame atomic absorption spectroscopy techniques for analyzing trace metal solutions

To learn to titrate complex samples using the ethylenediaminetetraacetic acid (EDTA) titration method (possibly a review of *Quantitative Methods of Analysis*)

To learn to use solid-state Ca electrodes (optional)

To learn to write in a scientific and professional manner

16.5.1 Background

The global problem of hazardous waste did not occur overnight. Its presence has been documented as early as the Roman Empire with the use of lead 2000 years ago. Early forms of hazardous waste included the smelting of metal ore and the tanning of animal hides. The Industrial Revolution brought about an on-slot of hazardous waste issues that were not addressed until the 1970s and 1980s. But first, what is "hazardous waste?" Each country has its own definition but there are remarkable similarities between these definitions. One definition that summarizes the problem is from the United Nations Environment Programme from 1985 (LeGrega et al., 1994).

> Hazardous wastes mean waste [solids, sludges, liquids, and containerized gases] other than radioactive [and infectious] wastes which, by reason of their chemical activity or toxic, explosive, corrosive, or other characteristics, cause danger or likely will cause danger to health or the environment, whether alone or when coming into contact with other waste ...

Note that radioactive and infectious wastes are not covered under other programs even though they may well be hazardous in the eyes of citizens affected by these wastes. Classic pollutants that are specifically listed as hazardous waste include waste containing DDT, mercury, and PCBs, just to name a few notable chemicals. These and other chemicals have led to highly publicized disasters such as Love Canal in New York and Times Beach in Missouri. Old abandoned sites such as these fall under the Comprehensive Environmental Response, Compensation, and Liability Act (CERCLA in 1980 and subsequent reauthorizations) commonly known as Superfund, which is designed to clean up abandoned sites. Hazardous wastes being generated today are covered under the Resource Conservation and Recovery Act (RCRA in 1976 and subsequent amendments) that will supposedly prevent more disasters such as Time Beach and Love Canal from occurring. Similar programs are in place in the United Kingdom (The Poisonous Waste Act of 1972) and in Germany (the Solid Waste Laws of 1976).

In the United States (under RCRA), hazardous wastes are further characterized into the major categories of (1) inorganic aqueous, (2) organic aqueous waste, (3) organic liquids, (4) oils, (5) inorganic sludges/solids, and (6) organic sludges/solids. These categories are very important and determine the final resting place or treatment of the waste. For example, some wastes are placed in landfills, but prior to the placement of the waste in the landfill, it must be characterized (i.e. analyzed for the type and quantity (concentration) of toxic compounds). This brings up the focus of this laboratory exercise, the characterization of an inorganic hazardous waste. Actually, due to safety concerns, we will be analyzing simulated hazardous waste, carbonated beverages or laboratory-prepared solutions. These solutions make excellent simulated hazardous wastes because of their complex matrices (viscosity due to the presence of corn sugar, the presence of phosphates that selectively bind to Ca in the FAAS unit, their color, their pH, and their carbonation). In place of measuring a toxic metal, which you could easily and safely do, we will be analyzing for Ca since it is present in every carbonated beverage and this will eliminate the generation of real hazardous waste that will be costly for your college/university to dispose of.

16.5.2 Theory

Many chemicals, especially metals, can be analyzed by more than one technique. The focus of this laboratory exercise is to learn the flame atomic absorption spectroscopy (FAAS) unit, but you will also use the EDTA titration from quantitative analysis, or a calcium electrode. The EDTA titration and solid-state electrode are fairly easy to understand since the students have used titrations in all of your chemistry courses and the calcium electrode is only slightly more complicated than the familiar pH electrode. The FAAS unit will need more explanation. Please refer to Chapter 5.

The calibration of an instrument can be relatively simple. One creates a calibration curve (actually a line) with the instrument response (in this case absorbance) as a function of analyte (Ca) concentration, usually in mg/L or parts per million (ppm). Procedures for external calibration, external calibration with releasing again (again refer to Chapter 5), and standard addition.

Standard addition is a bit more difficult to explain. To begin our discussion, refer to Figure 16.1, which contains the results of a standard addition experiment similar to the one the student will be conducting. Remember that the purpose of this approach is to try to minimize the presence of interfering compounds in the sample matrix or overcome these interferences. We do this by making all of our standards in the sample matrix. First, a set of identical solutions, each containing the same volume of sample, are placed in individual beakers. All, but one of these solutions, have increasing masses of standard (Ca) added to them. Each sample is analyzed the same way on the FAAS unit and the data are plotted as in Figure 16.1. The diamonds on the positive side of the x-axis are standard concentrations that have been added to the sample. These should result in a line well above the origin (0,0) of the plot. The line is extrapolated back to a y value of zero to determine the

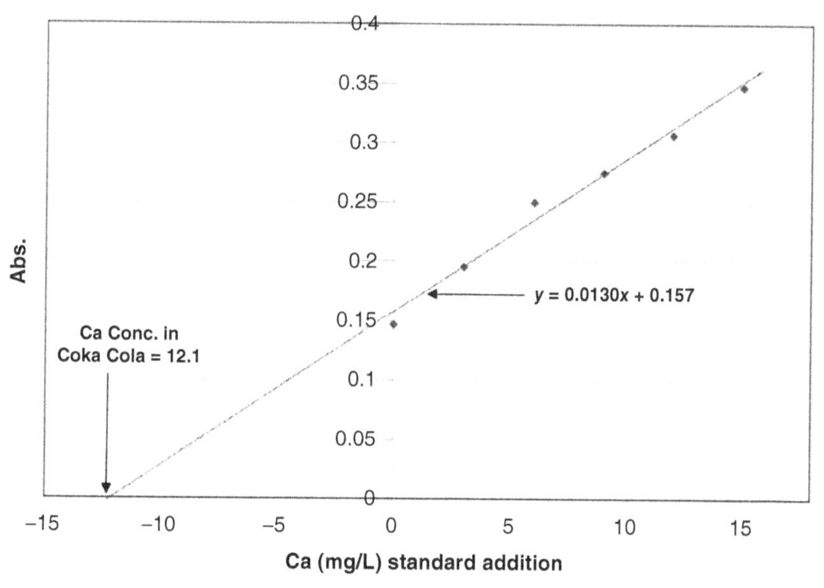

Figure 16.1 An example of the standard addition calibration technique.

concentration of calcium in your diluted sample. Using the dilution factors that you used to make up your sample, you can finally calculate the concentration of calcium in your original sample. Note that the distance from x equal to zero to your highest standard should be of similar or less distance from x equal to zero to your sample concentration. Thus, by using this approach, we have overcome the viscosity effects and most other interferences.

The releasing agent is easier to understand and addresses the fact that calcium will bind to phosphate or sulfate as it dries in the air-acetylene flame and therefore will not be present in its needed form, as a ground-state gaseous atom. We use the periodicity of the elements to overcome this problem. Strontium, an element in the same group as calcium and that acts similarly to calcium, preferentially binds to phosphate in the flame and releases calcium to form free gaseous atoms. You can confirm this by looking at the formation constants for calcium phosphate and strontium phosphate; the strontium phosphate value is much larger. Thus, by adding another metal to each solution (standards and samples) we can overcome the dramatic effect of having phosphate in your samples. Similar approaches are available for other, toxic elements. And remember, if you are involved with the disposal or treatment of hazardous waste, you will want to have an accurate measurement of how much toxin is in your waste sample.

16.5.3 In the Laboratory

The goals of this series of labs are (1) to show that many elements and compounds can be analyzed by more than one technique and (2) to illustrate the nature of one complex sample matrix. This lab is also designed to teach the analysis of a toxic metal in a hazardous waste sample. In an effort to keep the laboratory environmentally friendly and not generate hazardous waste that would need to be disposed of, we will use calcium as our *surrogate* toxic metal. In addition, to avoid exposure of students to real hazardous waste, we will use laboratory-prepared solutions or carbonated beverages that contain complex matrices and interferences in the calcium analysis. However, you should treat the sample as though it was toxic; your professor will observe your laboratory technique and if you improperly handle the sample, a skull and crossbones will be placed at your laboratory station. Some of these techniques we will use are more direct and simple, while others are more involved. It should be no surprise that when a single sample is analyzed by all of these techniques, the resulting concentrations do not always agree. In these labs, you will measure the concentration of Ca in beverages by flame atomic absorption spectrophotometer (using external standards with and without releasing agents, and standard addition techniques as well as matrix modifiers), by EDTA titration (an applied review of quantitative analysis), and using a Ca ion-specific electrode (another review of quantitative analysis) and it will be your task to decide which technique is best for your sample (best by your definition or your professors; you set the criteria).

The order that you do these labs is not important and they will be randomly assigned so that no two groups are doing the same lab at the same time due to time and instrument constraints. To aid in your solution preparation and dilutions, you will be told the actual concentrations of your samples at the beginning of the lab. The suggested laboratory techniques are:

Procedure I: Determination of Ca using Atomic Absorption Spectroscopy and External Standards
Procedure II: Determination of Ca using Atomic Absorption Spectroscopy and External Standards and Releasing Agents
Procedure III: Determination of Ca using Atomic Absorption Spectroscopy using the Standard Addition Technique and a Releasing Agent
Procedure IV: Determination of Ca using Atomic Absorption Spectroscopy using the Standard Addition Technique

Procedure V: Determination of Ca using the EDTA titration
Procedure VI: Determination of Ca using Atomic Absorption Spectroscopy using the Ion-Specific Electrodes
Alternative procedures: FAES or ICP (procedures not given but similar to Procedure I).

If you are using a carbonated sample, the instructor will have the carbonation removed prior to lab since this takes considerable time. Everyone should note that sugars in the carbonated solution will eventually clog the burner head in the FAAS. The other problem with using carbonated store solutions is you will not know the actual Ca concentration in your sample. While use of these commercial solutions can make the experiments interesting, we now use a laboratory-prepared solution following the procedure below.

The surrogate solution preparation follows. In 10.0 L of DI, add:

- 100 mL of concentrated ultrapure nitric acid
- 0.200 g of high purity $CaCO_3$ (critical reagent, must have exact mass)
- ~0.60 g Na_2SO_4–$10H_2O$ (noncritical reagent for Ca interference)
- ~0.25 g NaH_2PO_4–$1H_2O$ (noncritical reagent for Ca interference)
- Note: If all procedures (I–V) are used, each student/group will need approximately 2 L of the sample.

Before beginning any experiments using the FAAS unit, you are expected to learn about the instrument and analytical methods by reading the relevant sections, specifically Chapter 5. Also, some reading may be made available from the instrument manuals.

You may also be given guidelines for instrument startup and shutdown. Follow these closely!

16.5.3.1 Procedure I: Determination of Ca Using Flame Atomic Absorption Spectroscopy and External Standards

The goals of this experiment are (1) refine your ability to make reference standards (Ca), (2) learn to use the atomic absorption spectroscopy system using external standard calibration, (3) determine the linear range for a set of Ca standards, and (4) determine the concentration of Ca in an unknown sample. Analyze the unknown at least 5 times and read the blank sample between each sample. This will allow you to use the spreadsheet described in Chapter 2.

16.5.3.1.1 External Standard Calibration Method This is the normal way of using a calibration curve; you make a set of standards, measure the instruments' response to the standards and unknowns, and make a calibration plot using your linear least squares analysis from Chapter 3 (LLS) (or use the instruments automated calibration software) and use the instrument response to estimate the concentration in your unknown samples. We suggest analyzing your standards from low to high concentration and making a blank measurement between *each* standard. If you have time, repeat this entire process twice. This will give you 15–20 blank measurements that you will need to determine the noise level and your minimum detection limit (equations for these are contained in your spreadsheet included in Chapter 2).

- Make a set of Ca standards (each standard should contain 1% concentrated ultra-pure HNO_3). Calcium concentrations in the final solutions should be 0, 0.5, 1, 5, 10, 15, 25, and 50 mg Ca/L. (Your particular instrument may have different detection limits and linearity ranges than the one used to develop this experiment. Consult your professor for details on your instrument.) Note that some of these standards will be below the detection limit while others may be above the Limit of Linearity (LOL).

- Make five dilutions or samples of the unknown sample.
- Set up the FAAS unit as instructed.
- Analyze the standards and unknown samples on the FAAS unit.
- Plot the data using your LLS spreadsheet (Chapter 2), determine the linear portion of the data, and if the unknown sample signal is in the linear range, determine the concentration of Ca. If the signal of the sample is too high, make the appropriate dilution of the sample in 1% HNO_3, and reanalyze the sample.
- Add all data and analyses to your laboratory notebook and share your group's data with the other groups in your class.

16.5.3.2 Procedure II: Determination of Ca Using Atomic Absorption Spectroscopy and External Standards and Releasing Agent

The goals of this experiment are to: (1) refine your ability to make reference standards (Ca), (2) learn the use of a releasing agent in FAAS, (3) learn to use the atomic absorption spectroscopy system, (4) determine the linear range for a set of Ca standards, and (5) determine the concentration of Ca in an unknown sample (analyze the unknown at least 5 times).

16.5.3.2.1 Procedures This procedure is identical to the one used in other external standard procedures except that you will have to add a releasing agent (Sr) to every solution. The final concentration of Sr in all of your standards and samples should be 100–1000 mg/L. To achieve this, you will have to make a more concentrated Sr solution and add a small, but consistent volume of this concentrated solution to your standards and samples.

Conduct all calculations for dilutions and preparing solutions before you come to lab.

Again, analyze your standards from low to high concentration and make a blank measurement between *each* standard. If you have time, repeat this entire process twice. This will give you 15–20 blank measurements that you will need to determine the noise level and your minimum detection limit (equations for these are contained in your spreadsheet included in Chapter 2).

- Make a set of Ca standards (each standard should contain 1% concentrated ultra-pure HNO_3) and 100–1000 mg/L Sr. Calcium concentrations in the final solutions should be 0, 0.5, 1, 5, 10, 15, 25, and 50 mg Ca/L. (Your particular instrument may have different detection limits and linearity ranges than the one used to develop this experiment. Consult your professor for details on your instrument.) Note that some of these standards will be below the detection limit while others may be above the LOL.
- Make five dilutions or samples of the unknown sample and add Sr to a level of 100–1000 mg/L.
- Set up the FAAS unit as instructed.
- Analyze the standards and unknown samples on the FAAS unit.
- Plot the data using your LLS spreadsheet, determine the linear portion of the data, and if the unknown sample signal is in the linear range, determine the concentration of Ca. If the signal of the sample is too high, make the appropriate dilution of the sample in 1% HNO_3, add 1000 mg/L Sr, and reanalyze the sample.
- Add all data and analyses to your laboratory notebook and share your group's data with the other groups in your class.

16.5.3.3 Procedure III: Determination of Ca Using Atomic Absorption Spectroscopy Using the Standard Addition Technique and a Releasing Agent

The goals of this experiment are to: (1) refine your ability to make reference standards (Ca), (2) learn to use the atomic absorption spectroscopy system, (3) learn the standard addition technique,

(4) learn one technique for overcoming interferences (releasing agents), and (5) determine the concentration of Ca in an unknown sample.

Plan ahead and completely understand this procedure before you come to lab.

Conduct all calculations for dilutions and prepare solutions before you come to lab.

16.5.3.3.1 Procedure Prepare all solutions before using the FAAS unit.

Standard addition calibration method:

Here, we are concerned with viscosity effects of the corn syrup in your hazardous waste sample. We will also evaluate the effect of adding a releasing agent (Sr). You should completely understand why you are adding this before you come to lab.

- Make a stock solution of $Sr(NO_3)$ at a concentration that will serve to meet the requirements below. Check with your professor before you make the solutions to ensure that you have the calculations correct.
- Make a set of standards and samples containing a known amount of Ca (standard) and Sr (at 1000 mg/L in the final solution). Calcium concentrations in the final solutions should be 0, 0.5, 1, 5, 10, 15, 25, and 50 mg Ca/L. (Your particular instrument may have different detection limits and linearity ranges than the one used to develop this experiment. Consult your professor for details on your instrument.) When you make these solutions, we suggest making the samples in 25, 50, or 100 mL volumetric flasks. (For example, if a 25 mL volumetric flasks are used, add 10 mL of sample, a volume of concentrated nitric acid to yield 1% in the final solution, the required volume of Sr solution to obtain 100–1000 mg Sr/L, and fill the remainder with distilled water.) (i.e. to each volumetric flask add (1) an exact and equal volume of sample to each flask based on one of your other experimental results, (2) concentrated HNO_3 to yield 1%, (3) a volume of $SrNO_3$ solution that will give you 1000 mg Sr/L, and (4) fill the flask with distilled water to the mark.) Note that you need your sample concentration (on the negative x-axis) to be within the range of your sample plus standard concentrations (on the positive x-axis).
- Analyze the standards and unknown samples on the FAAS unit.
- Make sure that the data set is linear. If it is not, consult your laboratory instructor before you throw away your solutions.
- Plot the data using your LLS spreadsheet, determine the linear portion of the data, and if the unknown sample signal is in the linear range, determine the concentration of Ca. If the signal of the sample is too high, make the appropriate dilution of the sample in 1% HNO_3, add 100–1000 mg/L Sr, and reanalyze the sample.
- Add all data and analyses to your laboratory notebook and share your group's data with the other groups in your class.

16.5.3.4 Procedure IV: Determination of Ca Using Atomic Absorption Spectroscopy Using the Standard Addition Technique Without Releasing Agent

The goals of this experiment are to: (1) refine your ability to make reference standards (Ca), (2) learn to use the atomic absorption spectroscopy system, (3) learn the standard addition technique, (4) learn one technique for overcoming interferences, and (5) determine the concentration of Ca in an unknown sample.

This procedure is identical to the previous standard addition technique (procedure III) but you will not be using Sr as a releasing agent. Delete all references to it in the other procedure and complete the other procedure as given.

16.5.3.5 Procedure V: Determination of Ca Using the EDTA Titration

The goals of this experiment are to: (1) refine your ability to make reference standards (Ca) and dilutions, (2) review/refine your titration skills, (3) review/learn the details of a complicated EDTA titration, and (4) determine the concentration of Ca in an unknown sample.

Plan ahead and completely outline a procedure before you come to lab. In this procedure, you may use Eriochrome black T or solid hydroxynaphthol blue as a titration indicator. Note: This procedure and indicator has been used for decades but for some unknown reason the indicator does not always change colors appropriately.

16.5.3.5.1 Procedure Use your knowledge of *Quantitative Analysis* to conduct this experiment. Note you may have to dilute your sample (or possibly the EDTA) to dilute the food coloring that may interfere with the endpoint to obtain an acceptable detection limit. It will also be important for you to review exactly what the EDTA titration is measuring as compared to the other procedures in this set of laboratory exercises.

- You may be provided with the known Ca concentration of your sample, Class A volumetric pipettes, and 10, 25, and 50 mL burettes. Given the known Ca concentration of your sample, determine the best combination of volume of sample to be titrated, any dilution, and the appropriate burette to use to maximize your accuracy.
- Pipet a sample of your unknown into a volume flask; the volume is dependent on your above calculations. You will have to determine the initial dilution of the sample and EDTA titrant. The beginning of the procedure will be highly dependent on a trial and error approach and there is not only one correct way of completing this procedure. To each sample (in this step and below) add 3 mL of the pH 10 buffer solution, 30 drops of 50% by weight NaOH, swirl for 2 minutes, and add a small scoop (~0.1 g) of hydroxynaphthol blue (or 6 drops of Eriochrome black T indicator solution). (*Note*: Your sample is naturally acidic, so you may need to add more than 30 drops of NaOH. Check the pH to insure that it is at or above a pH of 10.)
- After you have determined the best dilutions of the sample and EDTA titrant to use, complete at least three sample titrations to find the amount of Ca^{2+} in your unknown sample. Note that you may need to add DI to your flask to give a sufficient volume for your titration.
- Titrate the Ca determinations carefully. After reaching the blue end point, allow each sample to sit for five minutes, with occasional swirling, so that any $Ca(OH)_2$ precipitate can redissolve (if this occurs, the solution will be red/pink in color). Then titrate back to the blue end point. It is always best to perform a blank titration on DI to serve as an endpoint check, but note your sample has a background color.
- Calculate the total Ca concentration in your original sample knowing that 1 mol of Ca^{2+} binds with 1 mol of EDTA.
- Add all data and analyses to your laboratory notebook and share your group's data with the other groups in your class.

Necessary solution for the EDTA titration:

- 3L of $Na_2H_2EDTA{-}2H_2O$ solution (FW = 372.25 g/mol) dry at 80 °C for 1 hour, cool, weigh out ~0.6 g and dissolve in 400 mL DI. Fill to mark in 500 mL vol. Know the exact weight of EDTA for determining the molarity of your standard solution.
- pH 10 buffer: add 142 mL of 28% by wt aqueous NH_3 to 17.5 g of NH_4Cl and dilute to 250 mL with DI.
- 50% wt NaOH.
- 10, 25, and 50 mL burettes.
- Solid hydroxynaphthol blue indicator or Eriochrome black T indicator: dissolve 0.2 g of the solid indicator in 15 mL of triethanolamine plus 5 mL of absolute ethanol.

16.5.3.6 Procedure VI: Determination of Ca Using Atomic Absorption Spectroscopy Using the Ion-specific Electrodes

The goals of this experiment are to: (1) refine your ability to make reference standards (Ca) and dilutions, (2) review/learn the details of ion-specific electrodes, and (3) determine the concentration of Ca in an unknown sample.

Plan ahead and completely outline a procedure before you come to lab. This will involve reading the manual for your Ca electrode. You should also review solid-state electrodes in your *Quantitative Analysis* textbook.

16.5.3.6.1 Procedure Follow the instructions in the electrode manual, and make an external calibration curve to check the slope of the line to ensure that the electrode is functioning properly and for your LLS analysis. You may also choose to complete a set of samples using the standard addition technique.

16.5.3.6.2 Assignment: What Do You Turn In? One of the goals of the chemistry department is not only to teach you proper methods for analyzing samples but to also teach you to effectively disseminate your results. One dissemination method is your lab notebooks and lab reports that you will complete for this and the other labs, but for this lab, you will also do something a little more involved. After completion of all experiments, you are to compile the methods and results and write a journal article suitable for publication in the *Journal of Analytical Chemistry*. The theme of your article will be comparing analytical techniques for Ca analysis of complex aqueous samples. You are upper-level students about to impart on the "real-world," so your professor will treat you as such. You must obtain the "Instructions to Authors" for the Journal from the library or Internet and follow proper scientific writing guidelines (refer to the *ACS Style Manual* on reserve in the library). Remember that in your lab reports, you write down meticulous lab methods, but you will not be able to do this in your journal article (if you did this the article would be 50 pages long!). You must decide the fine line between too little and too much information. The best way, and perhaps the only way, to do this is to review several articles in the journal (perhaps 2 or 3 on AAS, 2 or 3 on titration techniques, and 2 or 3 on ion-specific electrodes). Note that you MUST also do a literature search on your topic and include the results in the introduction. For the introduction, you can introduce the article from a hazardous waste or analytical standpoint. Your article should be no longer than 20 typed double-spaced pages including text, figures, tables, and 20–25 references. In your discussion and conclusions section, *defend* which method(s) is (are) the most accurate for determining Ca in your sample.

Sample description to use in your mock paper:

Description of the waste samples: The industrial process effluents for a multinational enterprise are still under construction. Simulated waste samples were made that were representative of the future process waste streams and designated as Process Waste A is anticipated to contain between 2 and 15 mg/L Be. In an effort to minimize hazardous waste generation and aid in the shipping waste samples to our analytical labs across the country, eight liters of the wastewater were spiked with a surrogate, Ca (at ~8 ppm Ca; the exact Ca concentration of the stockroom-prepared sample should be given to the students) and concentrated nitric acid added to yield a final concentration of 1.00% HNO_3. The wastewater sample was divided into two-liter containers and shipped to each laboratory (student group) for Ca determination. Although the calcium ion is not toxic, it was selected as a surrogate metal for the toxic metal (Be) that would be present in the actual waste stream process. It was assumed that the analytical chemistry between the two metals would be similar.

(Continued)

(Continued)

Labs were asked to evaluate several techniques for approval by EPA and by our industry. The goal of each lab (student group) is to select one SOP (procedure) for analysis and defend it. Primary evaluation criteria are (1) a consistent recovery of Ca in the determination of metal concentration, (2) precision, and (3) cost of instrumentation and speed of analysis given that 50 samples per day will need to be processed at each plant location. Alternative procedures can be evaluated/suggested in your write-up. The student will want to look up beryllium toxicity on the Internet.

16.5.4 To the Instructor

Time requirements: Allow one lab period for each procedure used and one lab period for the laboratory writeup.

Results for this experiment are highly variable and reflect the differences in the experimental techniques used to measure Ca in a complex matrix. Students often have difficulty determining which method is "best," but this will be their job when they enter the real world. I advise them to take one of two possible stances: the best method is the one that (1) yields the highest Ca concentration (in environmental monitoring, you will always want to be conservative in your toxicological predictions); or (2) can be defended to be the best based on the technology of the method (i.e. standard addition and releasing agents). Students often forget to look at exactly what the technique measures. For example, the electrode only measures activity while the EDTA titration method measures all divalent ions in solution at a pH of 10. Some typical results from years of using this exercise are shown below. An additional consideration is that to do with a reproducible procedure that has a low recovery. It is acceptable to have a low recovery as long as the results are reproducible; then you correct the final concentration for the low recovery.

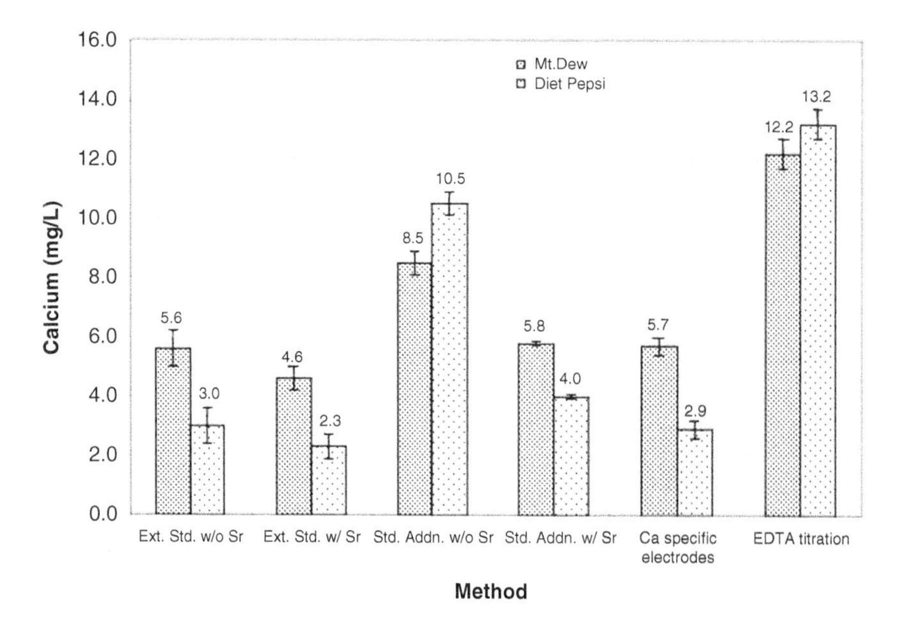

Figure 16.2 Student results for the analysis of Ca in Mountain Dew and Diet Pepsi.

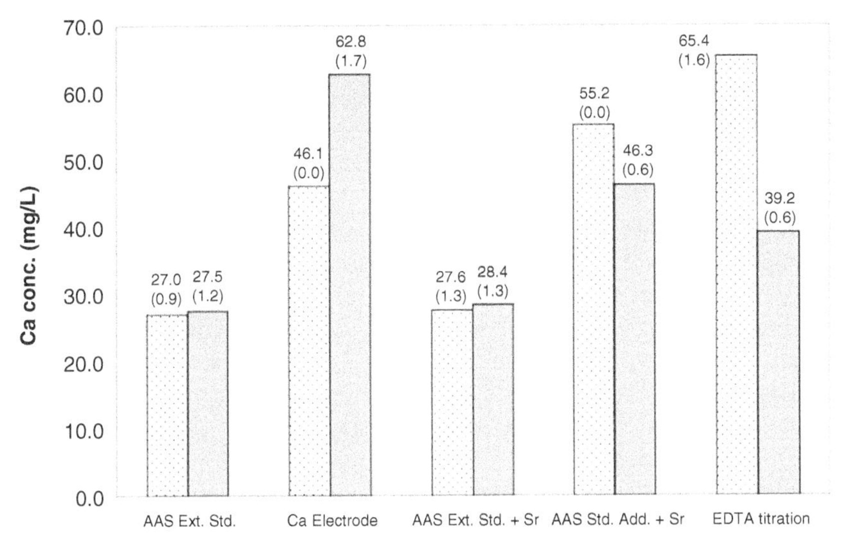

Figure 16.3 Student results for Coke by different techniques.

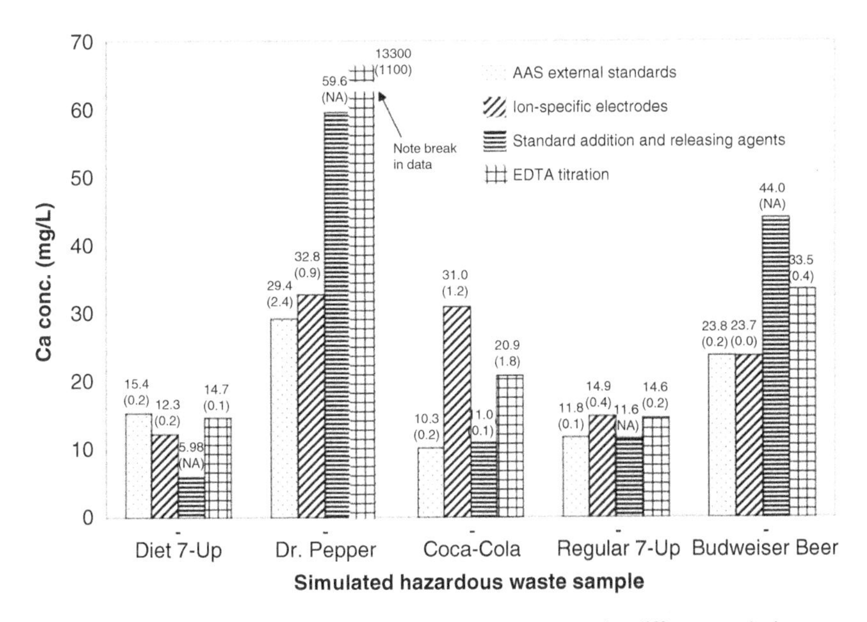

Figure 16.4 Student results for simulated hazardous waste by different techniques.

16.5.5 Student Sample Results

Figure 16.2 shows student results for the analysis of Ca in Mountain Dew and Diet Pepsi.
Figure 16.3 shows student results for Coke by different techniques.
Figure 16.4 shows student results for simulated hazardous waste by different techniques.

16.6 Identification of Components in Liquors and Distilled Spirits

One of the most powerful applications of an MS system is its ability to identify an analyte without a reference compound. In GC, reference compounds are needed to determine the retention time, and the criteria for identification in GC. In MS, the spectrum is the identifier since it can be compared to thousands of reference spectra and a unique match is normally achieved. The laboratory exercises below illustrate the power of MS in identifying unknown analytes in a variety of samples.

Distilled spirits contain a range of flavors that can be identified by gas chromatography–mass spectrometry (GC–MS). Not only can an analyst tell what type of liquor is present (i.e. gin versus whiskey) but they can also compare the presence and abundance of select flavor compounds between different brands of a given type of liquor.

NOTE/WARNING: A few liquors contain nonvolatile components that will coat out the glass liner in the injector port. The injector liner may need to be cleaned or replaced after completing this laboratory exercise to avoid damage to the GC inlet and column.

16.6.1 Experimental Procedures

16.6.1.1 Chemicals and Supplies

Pure samples of a variety of liquors. One approach is to contrast types of liquors (rum versus gin versus whiskey). Another approach is to contrast brands (spiced liquor versus pure liquor).

16.6.1.2 GC–MS Settings

Capillary column	DB-5 Poly(phenylmethyldimethyl) siloxane (5% phenyl) 30 m × 0.25 mm; 0.25 μm phase coating
Injection volume	1.00 μL
Splitless injection	0.50 min
Split flow rate	50 mL/min
Column flow	1.3 mL/min
Linear velocity	42 cm/s
Injector temperature	250 °C
Detector temperature	230 °C
Quadrupole temperature	150 °C
Oven program	Initial temp. at 60.0 °C for 0 min, 2.0–150 °C, hold for 0 min. Post temp. at 280 °C for 10 min
Total run time	45 min

16.6.2 Procedures

Analyze a variety of liquor samples on a GC–MS using the instrumental conditions given above.

16.6.3 Student Results

It is relatively easy to distinguish between most types of liquors. However, pure liquors (unspiced) such as rum and vodka produce similar chromatograms and only contain ethanol, water, and a few trace longer chain alcohols. Other liquors, as shown below, are easily distinguished.

All of the compounds identified in the chromatogram below were conclusively identified by the spectral library (typically 99% confidence/probability) and are known to be present in the liquors based on a scientific literature search or from common information found on company websites or web searches. An interesting project is to Google some of the compounds identified in the spectra and research their origin and why they result in a specific liquor.

Rum: Rum is a fermented beverage made from sugarcane byproducts such as molasses and juice. After fermentation, it is distilled as a clear liquid. Double distillation yields light rums, while single distillation will yield darker rums that were originally thought of as being of lower quantity. From here, the process becomes brand-specific and the initial rum can be aged in a variety of barrels, including oak to impart strong flavors, or filtered through charcoal to remove colors. Spices or color agents are then added.

Figures 16.5–16.7 are chromatograms for Bacardi Gold, Captain Morgan, and Citrus Rum, respectively. Note the lack of compounds in the relatively pure Barcardi Gold rum, only 3-methyl-1-butanol and acetic acid are present in measurable quantities. Barcardi Gold has little presence of the oak flavor compounds such as those found in the other two rum beverages. Captain Morgan's flavor is characterized by additional compounds, most notably oak flavors and vanilla. Citrus Rum, a specifically spiced rum, not surprisingly contains almond, orange, cocoa, fruit, and lemon flavors, as well as extracts from the oak barrel aging process.

As a side note, most of the components shown in these figures are in the ppm range of concentrations.

Whiskey: Whiskey (originally Whisky from its origin from Irish monks) refers to a broad range of alcoholic beverages that are distilled from fermented grain mash and aged (matured) in oak barrels (casks). The age of a whiskey refers to its time in the cask (between fermentation and bottling) and the length of aging greatly affects its chemical makeup and taste from the extraction of wood components from the cask. These components include lacone (3-methyl-4-octanolide), which has a coconut aroma, and numerous phenolic compounds. Grains of choice include barley, malted barley, rye, wheat, and corn, and they may be fermented from single or blends of grains.

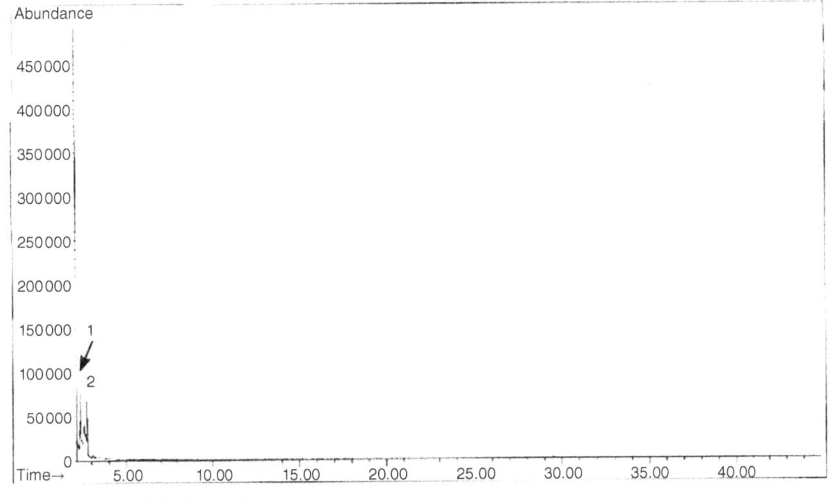

Figure 16.5 GC–MS results for Bacardi gold rum.

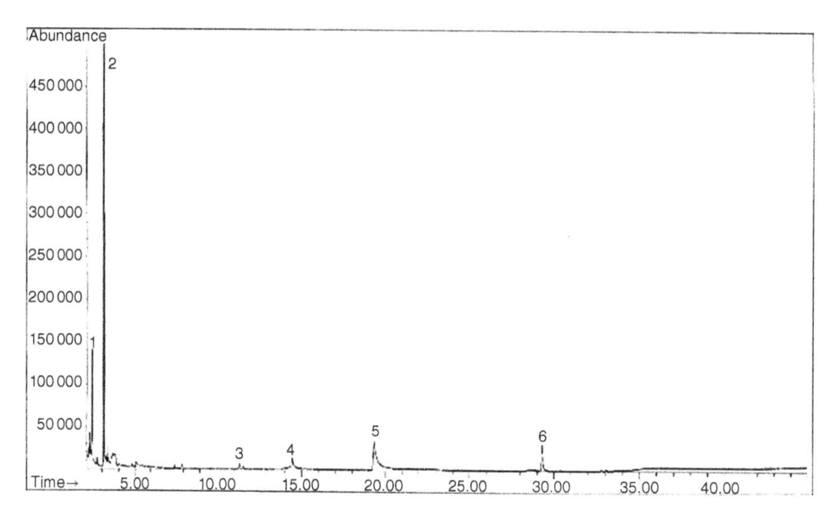

1) Acetic acid
2) 1,2-dihydroxypropane
3) 5-methylbenzene-1,3-diol
4) 2,3-dihydro-3,5-dihydroxy-6-methyl-4H-pyran-4-one (toasty caramel aroma of heated oak)
5) 5-hydroxymethylfurfural (wood flavor)
6) 4-hydroxy-3-methoxy-benzaldehyde (vanilla flavor)

Figure 16.6 GC–MS results for Captain Morgan rum.

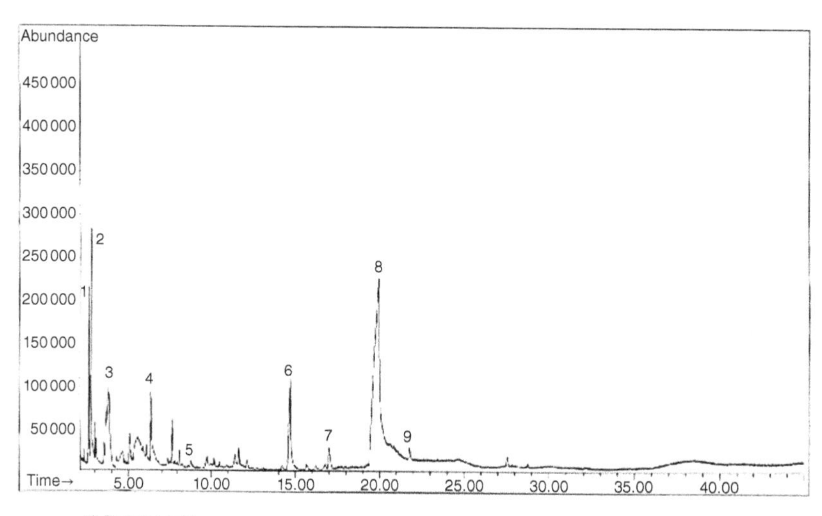

1) Formic acid
2) Acetic acid
3) 2-furaldehyde (wood flavor)
4) Benzaldehyde (almond flavor)
5) Limonene (orange flavor)
6) 2,3-Dihydro-3,5-dihydroxy-6-methyl-4(H)-pyran-4-one (cocoa flavor)
7) α-terpineol (fruit juice flavor)
8) 5-hydroxymethylfurfural (wood flavor)
9) Citral (lemon flavor)

Figure 16.7 GC–MS results for Citrus rum.

Published flavoring chemicals include carbonyl compounds, alcohols, carboxylic acids, and their esters, nitrogen- and sulfur-containing compounds, tannins, and other polyphenolic compounds, terpenes, and oxygen-containing heterocyclic compounds and esters of fatty acids. The nitrogen compounds include pyridines, picolines, and pyrazines. After distillation, the flavoring compounds that are common among different brands of whiskey include fuel oils, which are higher alcohols that are actually mildly toxic and have a strong disagreeable smell and taste in high concentrations. Hence, these are commonly removed by charcoal and linen filtration. Other common flavor agents in whiskey are acetals, such as acetaldehyde diethyl acetal (1,1-diethoxyethane), the principal flavor agent in sherry. The presence of a buttery aroma is due to diketone diacetyl (2,3-butanedione). Some whiskey blends contain specific flavor agents. Use the chromatograms given below to confirm the presence of these known whiskey components. Figures 16.8 and 16.9 show the GC–MS results for Crown Royal and Southern Comfort, respectively.

Cognac: The cognac used here is Grand Marnier, a blend of cognac. Cognacs are brandies produced from specific white grape varieties. Most cognacs are distilled twice and aged in oak barrels. After the review of the liquors above, the student should be able to predict some of the compounds present in cognac. Figure 16.10 shows the GC–MS results for Grand Marnier.

Note that most of the flavor compounds come from the oak barrel or are specifically added.

Peppermint Schnapps: Schnapps is usually a clear, colorless beverage with a light fruit flavor since it is fermented from fruit. The schnapps used in this experiment is infused with peppermint leaf extract or specific flavors (chemicals) found in peppermint. Note the dominant mint flavor compound in the chromatogram. Figure 16.11 shows the GC–MS results for the analysis of Peppermint Schnapps.

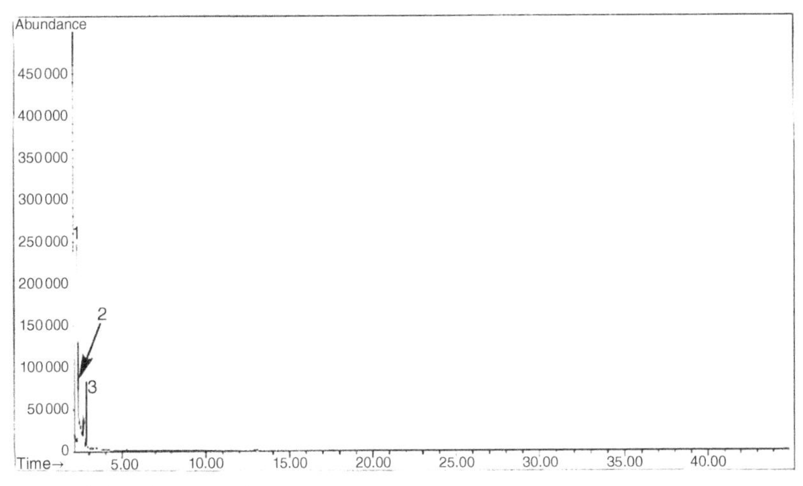

Figure 16.8 The GC–MS results for Crown Royal.

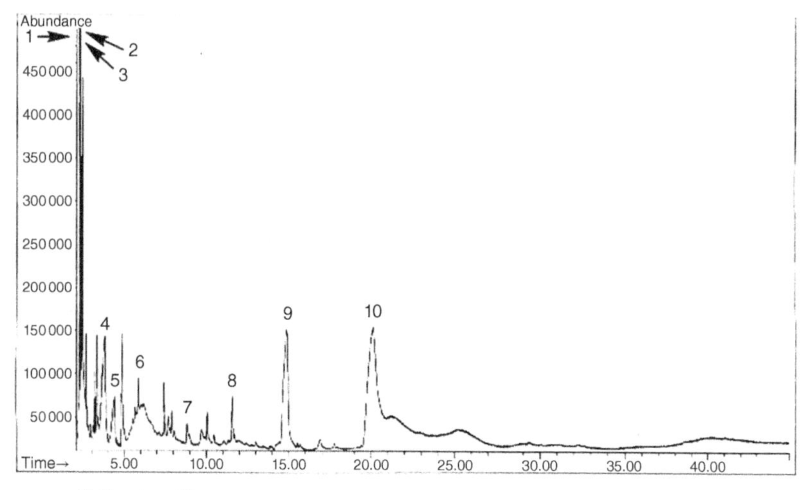

1) Formic acid
2) 3-methly-1-butanol
3) Acetic acid
4) 2-furaldehyde (wood flavor)
5) 2-hydroxymethylfuran (burnt flavor)
6) 1,2-cyclopentanedione (wood flavor)
7) Limonene (orange flavor)
8) Methyl 2-furoate (fruity flavor)
9) 2,3-Dihydro-3,5-dihydroxy-6-methyl-4(H)-pyran-4-one (cocoa flavor)
10) 5-hydroxymethylfurfural (wood flavor)

Figure 16.9 The GC–MS results for Southern Comfort.

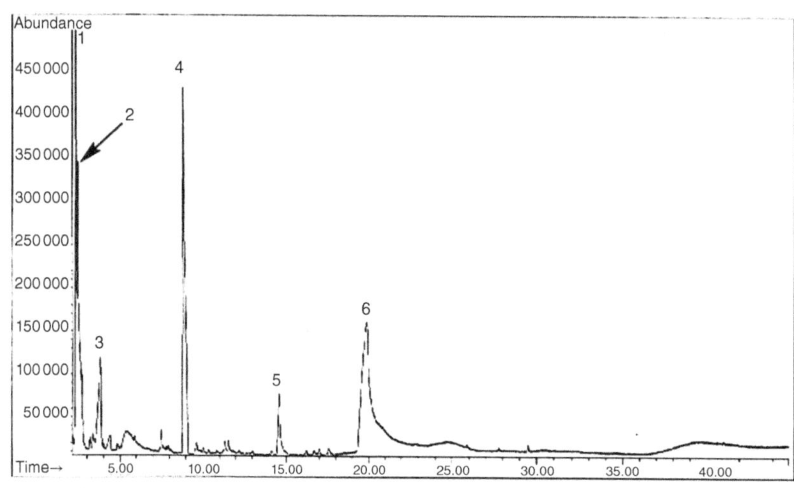

1) 1-pentanol
2) Acetic acid (shoulder peak)
3) 2-furaldehyde (wood flavor)
4) Limonene (orange flavor)
5) 2,3-Dihydro-3,5-dihydroxy-6-methyl-4(H)-pyran-4-one (cocoa flavor)
6) 5-hydroxymethylfurfural (wood flavor)

Figure 16.10 The GC–MS results for Grand Marnier.

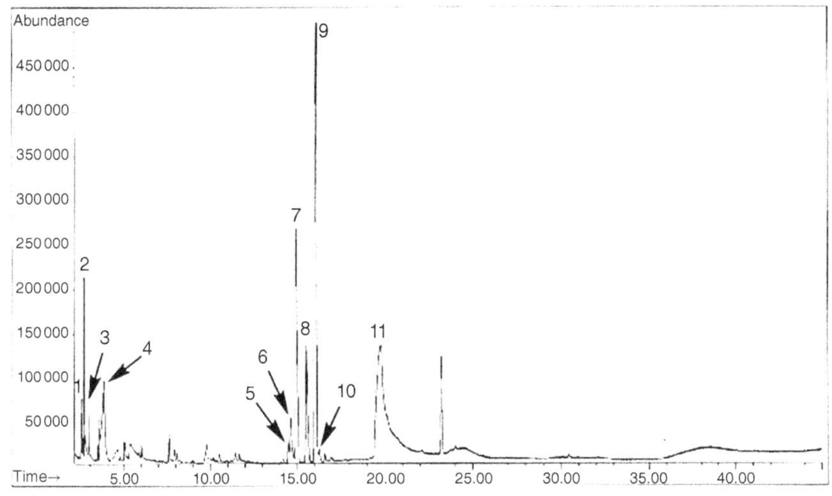

1) Formic acid
2) Acetic acid
3) 1-hydroxy-2-propanone
4) 2-furaldehyde (wood flavor)
5) Isopulegol (mint flavor)
6) 2,3-dihydro-3,5-dihydroxy-6-methyl-4H-pyran-4-one (wood flavor)
7) D-menthone (mint flavor)
8) L-menthone (mint flavor)
9) 5-methyl-2-(1-methylethyl)-cyclohexanol (mint flavor)
10) 4-carvomenthenol (mint flavor)
11) 5-hydroxymethylfurfural (wood flavor)

Figure 16.11 The GC–MS results for Peppermint Schnapps.

16.7 Identification of Fragrances

Fragrances/perfumes provide a "Holy Grail" for GC–MS analysis. As noted in many movies or from a trip to a European fragrance shop (perfumery), a near-infinite variety of combinations of fragrances can be made. In this laboratory exercise, "name brand" fragrances will be compared to their more inexpensive counterparts in an effort to determine if a difference exists in their "fingerprint" based on GC–MS. A fingerprint, in this context, is a characteristic chromatogram of a complex mixture of compounds.

NOTE/WARNING: Most fragrances contain nonvolatile components that will coat out the glass liner in the injector port. The injector liner may need to be cleaned or replaced after completing this laboratory exercise to avoid damage to the GC inlet and column. We suggest not conducting this experiment but using the results, given later, as examples.

Perfumes consist of (1) primary scents at the ppm concentration, (2) modifiers that alter the primary scent to give the perfume a certain desired character, (3) blenders (ingredients that smooth out the transitions of a perfume between different bases; top, middle, and base notes of a fragrance may have separate scents), and (4) fixatives (natural or synthetic substance used to reduce the evaporation rate).

Sources of primary scents include: (1) Plant sources (bark, flowers and blossoms, fruits, leaves and twigs, resins, roots, rhizomes and bulbs, seeds, woods; (2) animal sources (Ambergris which are lumps of oxidized fatty compounds, Castoreum from the odorous sacs of the North American beaver, Civet Musk obtained from the odorous sacs of the animals related to the Mongoose, Honeycombs, Musk originally derived from the musk sacs from the Asian musk deer); and (3) other natural sources (extracts of lichens and seaweed). Synthetic sources of the natural compounds mentioned above are used today, as well as calone, linalool, and coumarin from terpenes, and salicylates (orchid scents) are also used today.

16.7.1 Experimental Procedures

16.7.1.1 Chemicals and Supplies

A variety of perfume samples can be analyzed. In this experiment, Light Blue by Dolce and Gabbana, Shades of Blue by Belcam, Drakkar Noir by Guy Karoche, Classic Match by Belcam, Unforgivable by Sean John, Unjustified by Belcam, and Bring It by Parfums were used.

16.7.1.2 GC–MS Settings

Capillary column	DB-5 poly(phenylmethyldimethyl) siloxane (5% phenyl) 30 m × 0.25 mm; 0.25 μm phase coating
Injection volume	1.00 μL
Split mode of injection	
Split flow rate	131 mL/min
Column flow	1.3 mL/min
Linear velocity	42 cm/s
Injector temperature	250 °C
Detector temperature	230 °C
Quadrupole temperature	150 °C
Oven program	Initial temp. at 40.0 °C for 0 min, 2.0–280 °C, hold for 5 min
Total run time	125 min

16.7.2 Procedures

Analyze a variety of perfume samples on a GC–MS using the instrumental conditions given above.

16.7.3 Results

(1) The relatively expensive "Light Blue" and a generic blend "Shades of Blue": Figure 16.12 contrasts the GC–MS analysis of Light Blue (top figure) and Shades of Blue (bottom figure). Note the presence, absence, or reduced concentrations of the components between the two perfumes. Names and chemical structures for the numbered components are given in Table 16.2.

(2) Drakkar Noir and the generic "Classic Match": Drakkar Noir is a blend of citrus, lavender, spices, and woods. Top notes are citrus, middle notes are woody and herbaceous, and base notes are woodily warmed and spiced with aromatic coriander and juniper berries, strengthened by sandalwood, patchouli, and fir balsam. Figure 16.13 shows the GC–MS comparison analysis of

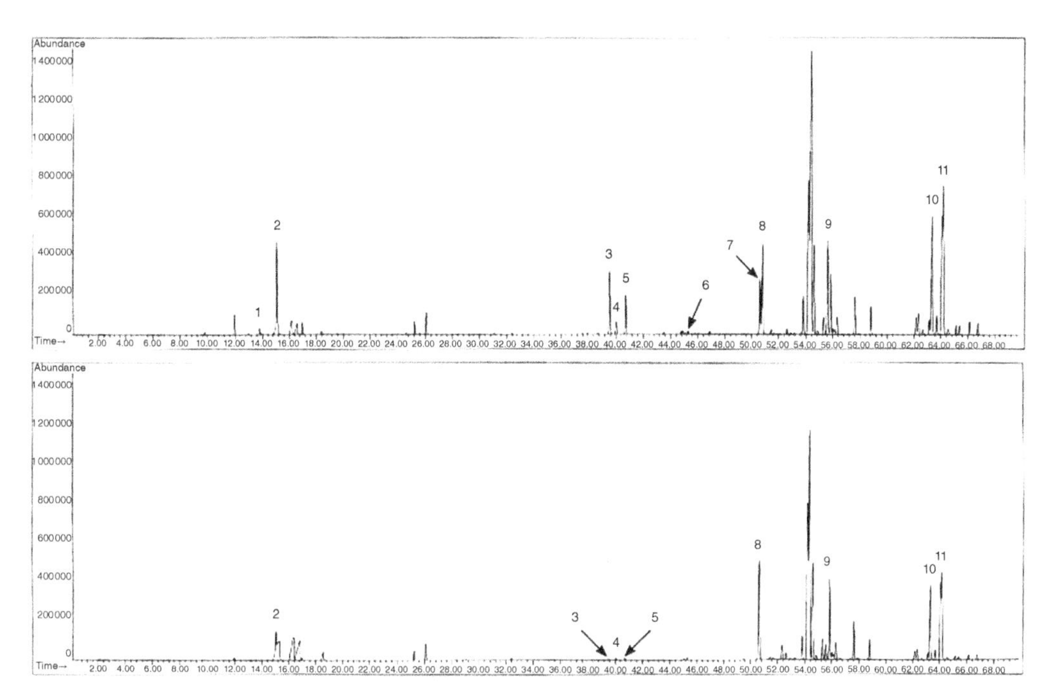

Figure 16.12 Contrasting the GC–MS analysis of light blue (top figure) and shades of blue (bottom figure).

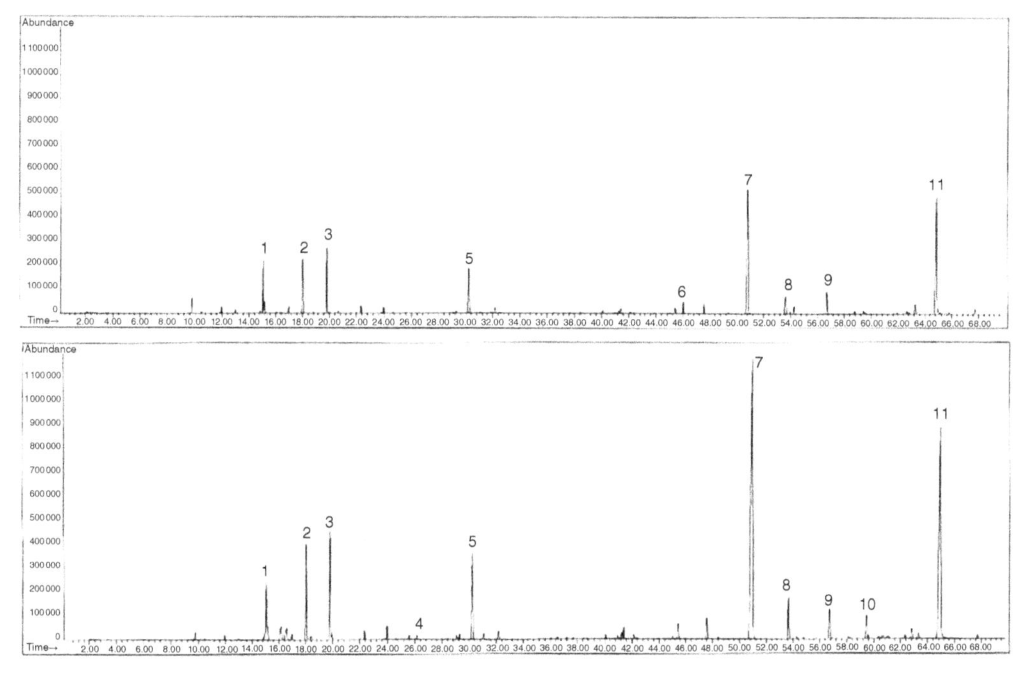

Figure 16.13 The GC–MS comparison analysis of Drakkar Noir (top figure) and Classic Match (bottom figure).

Table 16.2 Labeled components in chromatograms of light blue by Dolce & Gabbana versus shades of blue by Belcam.

	Name	Structure
1.	Diethylene glycol monoethyl ether (preservative)	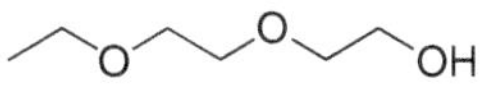 O O OH
2.	Limonene (lemon scent)	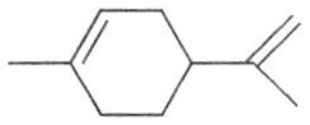
3.	a-Cedrene (wood scent)	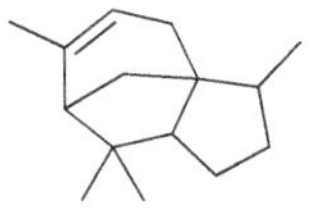
4.	b-Cedrene (wood scent)	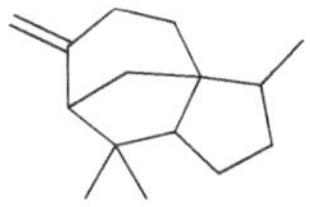
5.	Thujopsene (wood scent)	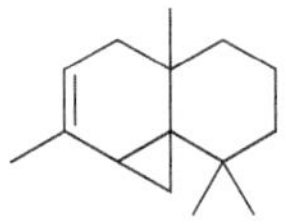
6.	Cuparene (wood scent)	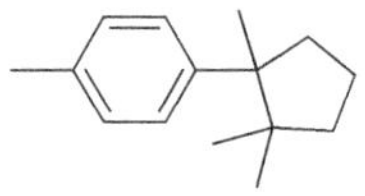
7.	Cedrol (wood scent)	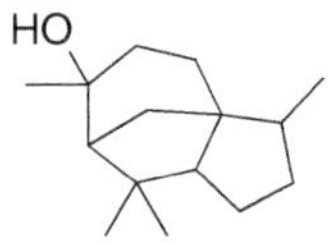 HO
8.	Diethyl phthalate (preservative)	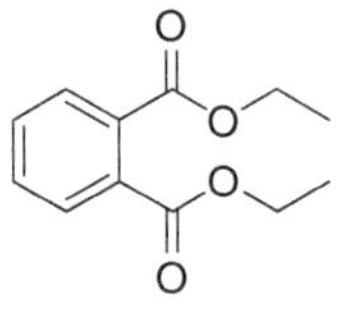 O O O O
9.	Methyl dihydrojasmonate (jasmine)	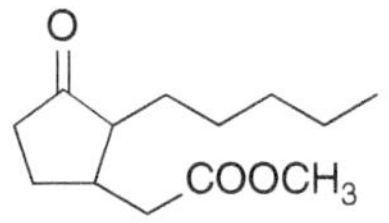 O $COOCH_3$
10.	Isopropyl myristate (skin binder)	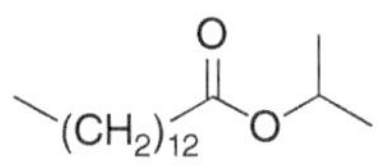 O $(CH_2)_{12}$ O
11.	1,3,4,6,7,8-Hexahydro-4,6,6,7,8,8-Hexamethylcyclopenta-g-2-benzopyran (musk scent)	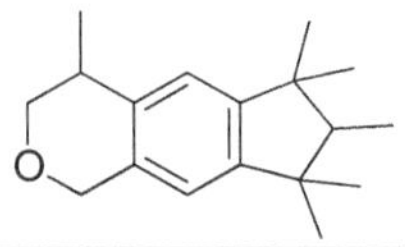 O

Drakkar Noir (top figure) and Classic Match (bottom figure). Compound names in the GC–MS spectra are given in Table 16.3.

(3) Unforgivable by Sean John versus Unjustified by Belcam Inc. versus Bring It by Parfums de Coeur. Comparison GC–MS results are shown in Figure 16.14. Note the presence and peak height differences between the two fragrances.

Table 16.3 Labeled components in chromatograms of Drakkar Noir and Classic Match.

	Name	Structure
1.	Limonene (lemon scent)	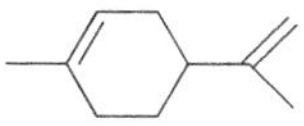
2.	Dihydromyrcenol (lime scent)	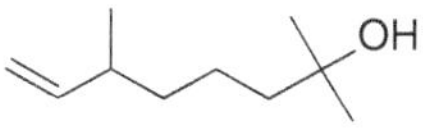
3.	Linalool (spicy floral scent)	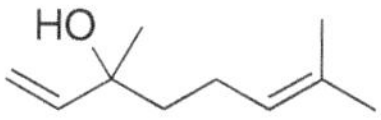
4.	4-Allylanisole (minty sweet scent)	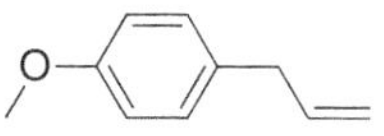
5.	Linalyl acetate (sweet scent)	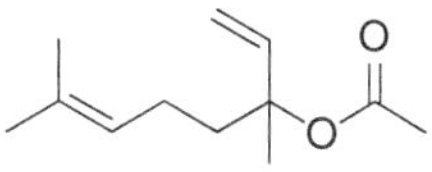
6.	2,6-Ditertbutyl-4-methylphenol (antioxidant)	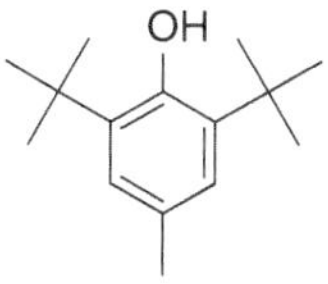
7.	Diethyl phthalate (preservative)	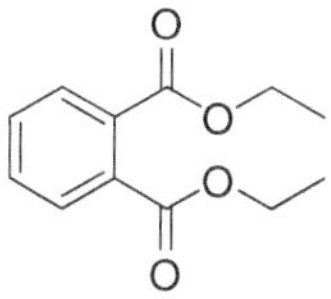
8.	Patchouli alcohol (woody scent)	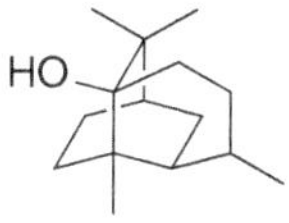
9.	Verymoss (woody scent)	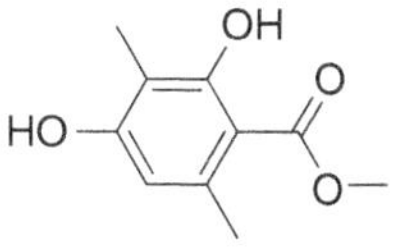
10.	d-Cadinene (woody scent)	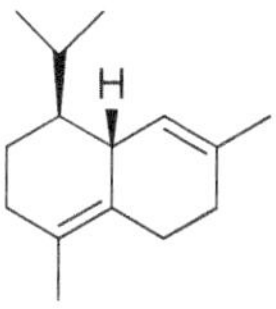
11.	Benzyl salicylate (floral scent)	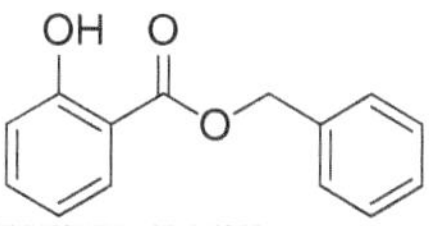

Comparisons of the "real" versus "fake" perfumes show distinct similarities with respect to presence of peaks and their "fingerprint". However, closer inspection of each peak shows differences. These subtle differences change our olfactory perception of their "smell." Note the identification of the components in each table and the type of the compounds present and their purpose.

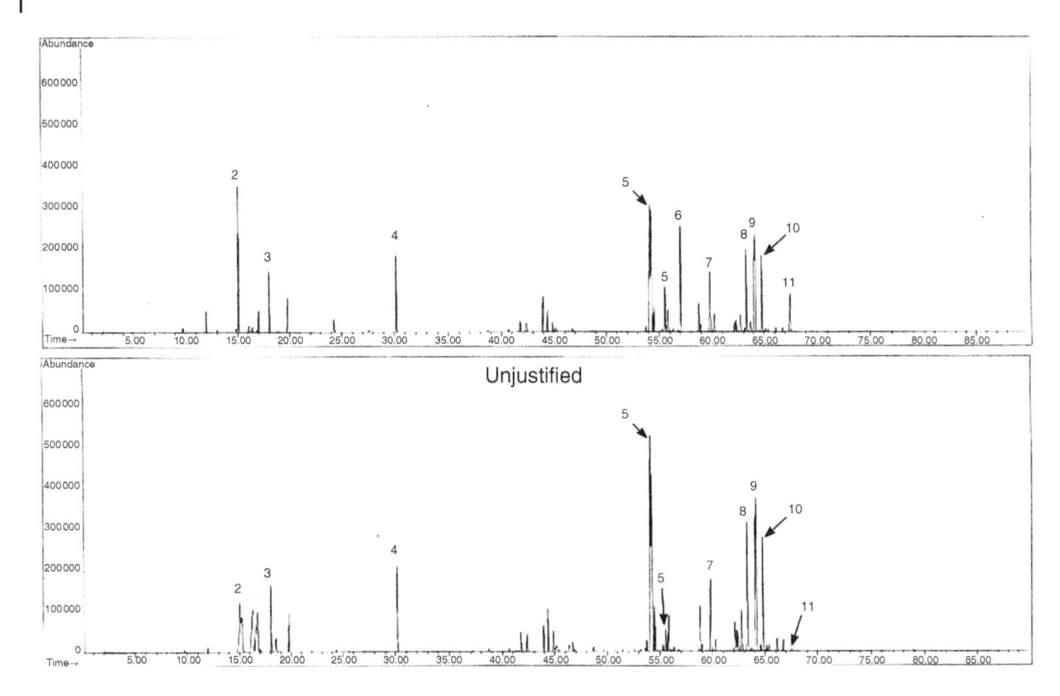

Figure 16.14 Comparison of unforgivable by Sean John (top figure) versus unjustified by Belcam Inc. (bottom figure).

16.8 GC–MS Analysis of Synthetic and Natural Fragrances by Evan Bowman, Annika Mayo, and Aurora Anderson

16.8.1 Summary

In recent years, essential oils have gained increasing popularity compared to their synthetic counterparts. This preference is mainly due to the belief that essential oils are more beneficial than fragrance oils and are safer and better for the body. This paper seeks to understand the difference in composition between synthetically derived and naturally derived fragrances. Various bergamot and lavender essential oils and fragrance oils were compared using gas chromatography and mass spectrometry to analyze similarities and differences in compounds present in the same fragrance.

16.8.2 Introduction

Perfumes and fragrances have been a common part of every household for centuries. It was not until recently, mainly since the 1900s, that chemists were able to start reproducing botanicals synthetically, which makes them much cheaper and accessible (Burger et al., 2019). It removes the hindrance of needing to extract essential oil from hundreds of pounds of natural material to get a few ounces of concentrated oil. It also allowed the isolation of specific compounds that were responsible for the most recognizable scent in people's favorite essential oils such as linalool which smells very similar to lavender (Letizia et al., 2003), and geraniol which has that rosy scent that is in almost every floral perfume (Lapczynski et al., 2008).

While all of this sounds incredibly beneficial to society, it has allowed many big beauty companies to add hundreds of compounds into a single fragrance oil with zero transparency about the actual ingredients. There have been numerous studies on the harmful effects of fragrance oils causing contact dermatitis (Johansen, 2003). Fragrance ingredients are actually one of the most common causes of contact allergic reactions. This was tested by patch testing with a mixture of fragrance ingredients and it was seen that 1.7–4.1% of the general population and 10% with eczema gave a positive patch test reaction. Chemical analysis has revealed that well-known allergens from the fragrance mixture are present in at least 15% of cosmetic products that are applied directly to the skin. Around 2500 different fragrance ingredients are currently used in the composition of perfumes and at least 100 are known to be contact allergens, so they are almost impossible to avoid in today's world (Travassos et al., 2011). While some essential oils can show the potential for also being an allergen, it is not on the same scale as fragrance oils. At the moment, 80 essential oils are identified to cause contact dermatitis (de Groot and Schmidt, 2016).

The other option for using fragrance oils would be to use essential oils. These are volatile and odoriferous oils obtained from plants. The oils are usually present in very small amounts and comprise only a small fraction of the entire plant material. There are numerous ways to extract the oil, but the most common is maceration. This creates an enzymatic reaction that causes the glycosides to undergo a chemical change which releases the essential oil. The essential oil consists mainly of chemicals with a low molecular weight with benzene rings and some esters such as limonene, camphor, α-pinene, and β-pinene (Kumar and Tripathi, November 2011). These oils are then collected through distillation processes such as hydrodistillation or through solvent extraction which pulls the oils into an organic solvent such as hexane or ethyl acetate. The use of different extraction techniques can dramatically affect the number of compounds collected and the quantity of each one (Matulyte et al., 2019). These compounds can then be separated and identified using gas chromatography and then mass spectrometry.

Many consumers as of late show a strong preference for essential oils over fragrance oils. Psychological studies have shown that essential oils have the capability to improve job performance, improve sleep, relieve headaches, and boost mood (Cleveland Clinic, 2022). This is mainly due to their physiological and therapeutic properties such as anti-inflammation and antimicrobial. There is also scientific proof that the extract of *Acorus tatarinowii*, a Chinese herb from the family of Acoraceae, has a variety of biological activities such as hypoglycemic, antioxidant, anti-Alzheimer's disease, antimicrobial, anticonvulsive, and antiepileptic properties (Zhang et al., 2021). While essential oils may not be the cure for everything and cannot fix all ailments, if sourced sustainably and diluted properly, they can be a great way to scent cosmetic products.

The main question this paper seeks to answer is if there is a fundamental difference between synthetic fragrance oils and naturally derived essential oils. There is very little literature that makes a direct comparison between the two because there is already limited knowledge of the compounds that are responsible for their scents. Through using gas chromatography and mass spectrometry, we hope to identify the compounds in various essential oils and their synthetic counterparts to see if there is a fundamental difference in the number and type of compounds within them that are responsible for their scent. The two main fragrances that were written about in the results and discussion were lavender and bergamot because it was theorized that these would have the simplest compositions and the easiest to compare.

16.8.3 Experimental

The students analyzed 13 different essential oils, synthetic fragrance oils, and isolated scent compounds to compare the various molecules that are responsible for their scent. A small sample (~5 mg), approximately a drop, was diluted with hexane and injected into the GC–MS. A variety of different compounds were shown in each sample and they were identified on their GC–MS spectra with their molecular structure.

16.8.4 Materials

- Hexane for dissolving fragrance oils.
- Bergamot- and geranium-distilled essential oil from Eden Botanicals.
- Lavender and Bergamot fragrance oil from P&J Premium Fragrance Oil.
- Geranium essential oils from SVA Organics.
- Lavender essential oil from Majestic Pure Cosmeceuticals.
- Lavender oil from Cliganic Essential Oil.
- Bergamot essential oil from Healing Solutions.
- Patchouli and lavender essential oil from Aftelier Perfumes.
- Linalool and Geraniol isolates from Aftelier Perfumes.

16.8.5 Sample Preparation

Approximately 5 mg of fragrance samples were taken directly from its bottle and diluted with hexane. It was then transferred to a 10 mL volumetric flask and diluted to volume. 1.5 mL GC–MS samples were then made and placed into the GC–MS. The samples were detected by GC–MS equipped with helium gas as the GC carrier gas and a GC column 30 m × 0.25 mm × 0.25 μm film thickness. The max temperature was 280 °C and started at 50 °C with a ramp rate of 3 °C/minute and a 1.2 mL/minute flow rate.

16.8.6 Results and Discussion

Each fragrance was individually analyzed on the GC–MS to determine the compound present in each sample. The resulting data was then assessed and our group identified the major compound groups in each of the bergamot and the lavender fragrances. As a group, we decided not to identify the significant compound groups in any of the Germanium or Patchouli fragrances as those did not constitute any important realizations related to our ultimate goal with these experiments. Is it also worth noting that the max temperature of the GC–MS was 280 °C, and stayed constant throughout the GC–MS runs. Now, to start with the comparison of the Majestic, P&J, and Aftelier (pure) samples as they have similar (but not identical) compound groups.

16.8.7 Comparison of Lavender Samples

Starting with the P&J Lavender samples (Figure 16.15), the major group that sticks out in comparison to the other lavender samples is the acetate compound group at a retention time of 32.33 minutes. Although the peak is only at a 5.89% composition percent, the peak seemed to have a significant impact on the rest of the GC. This acetate compound is a substance primarily used for manufacturing and different inks, not something that would be commonplace in an essential oil. It is also known that this compound is a common skin irritant and when inhaled, can cause respiratory problems at certain concentrations. Therefore, it is important to identify how much of it is within a fragrance used for personal use, for it (along with the alcohol groups) could cause a problematic reaction (Travassos et al., 2011). With the cyclic compounds, and especially with

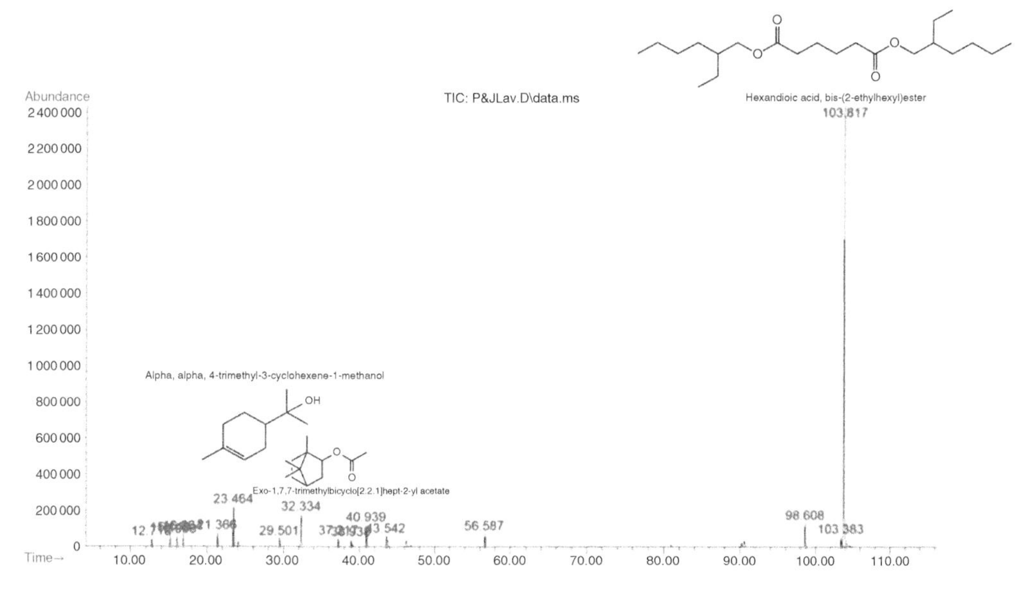

Figure 16.15 GC–MS results for P&J (pure) Lavender fragrance sample, with the significant compound groups, indicated above each peak. Only the five notable peaks were identified and compared between the Aftel Lavender and Majestic Pure Lavender samples. All runs were integrated and their retention times are shown above.

this acetate one in P&J Lavender, the purity of the major lavender compound gets compromised and the peak percent for the repetitive compounds decreases. For the P&J Lavender, though individually, this cyclic acetate compound indicates how much lavender compounds can vary based on the preference of the consumer. The P&J compounds have a particularly sharp and pungent smell, and seeing how most of the esters are odoriferous, it must be that acetate (and the cyclic compounds) is causing that effect.

This leads then to the compounds found in the Afterlier Lavender (Figure 16.16), as there are some similarities in the compounds that are different, but also have some important common compounds as well. In particular, all of the compounds contain hexanedioic acid, bis-(2-ethylhexyl)ester, an important plasticizer that essentially preserves the smell of the

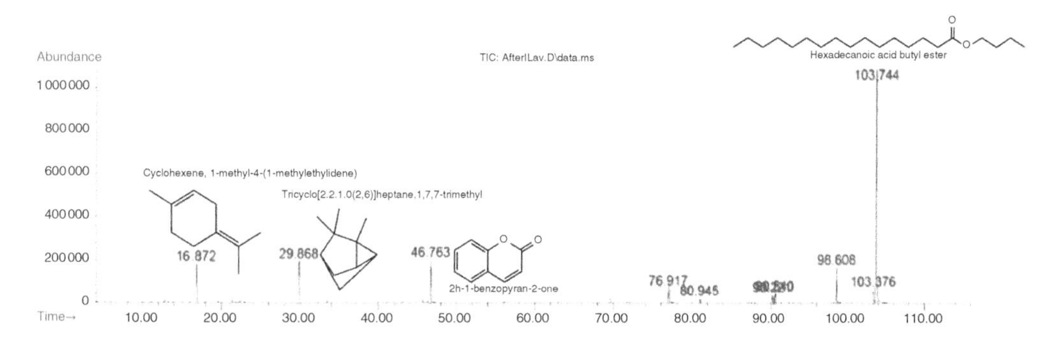

Figure 16.16 GC–MS results of the Afterlier (pure) Lavender fragrance sample, with the significant compound groups, indicated above each peak. Only the five notable peaks were identified and compared between the Aftel Lavender and Majestic Pure Lavender samples. All runs were integrated and their retention times are shown above.

lavender. The retention time for each occurrence of this ester was also very similar, which is to be expected, but what was not expected is the drastic change in peak percent between them. The P&J Lavender sample, which our group predicted has the most synthetic contaminants, had the highest peak percent for that ester at 73.26%. As for the Majestic pure Lavender, the tricyclo[2.2.1.0(2,6)]heptane,1,7,7-trimethyl compound is worth noting as it also comes up in the Afterlier samples as well. However, unlike the acetate compound from P&J, this cyclic causes little to no chemical problems and is just an additive to enhance the pure lavender smell. These conjugated hydrocarbons are seen frequently throughout every lavender sample, which indicates that these essential oil companies prioritize compounds that are easier to extract from nonpolar solvents (Matulyte et al., 2019). This trend can also be followed with the cyclohexene compounds at the lower retention times as well. Especially for the Afterlier sample, the diene compound that was found at 16.87 minutes prompts the question about the different surfactants that were used in these samples. With most essential oils, oil extraction happens by essentially "trapping" scents of different flowers and animals, and although that is usually economically favorable, it also attracts other unwanted, large substances that contaminate the sample after the scent is extracted and processed (Burger et al., 2019). As a result, these surfactants are added to not only decrease the surface tension of the oil but to serve as a hydrophilic head that disallows the absorption of unwanted compounds onto the consumer's skin. The Afterlier sample utilizes these kinds of detergents, and as a result, the cyclohexene1-methyl-4-1methylethyidiene is shown to have a strong peak at 8.51%. So, the use of a surfactant, in this case, was used to increase the peak percent of the other natural compounds (and is also why you do not see any small, unnotable peaks at earlier retention times).

Now we discuss how the Majestic pure Lavender (Figure 16.17) gets included with the other lavender samples. This sample shares the same plasticizer as the P&J lavender sample, although at a much lower peak percent. As these were the samples that our groups predicted to have the most synthetic compounds involved, it comes as no surprise that both have a high peak percent relative to the other compounds. However, it is interesting to see how it has similar compounds in both of the other lavender samples. In particular, the tricyclo[2.2.1.0(2,6)]heptane,1,7,7-trimethyl, which occurs in the Aftelier seems to have a slightly different effect on the overall sample compared to the Majestic pure sample. With the Majestic sample, there is a presence of diene alcohol compound that had a greater peak percent than the alcohol found in the P&J lavender, suggesting that it may have had an effect on how much of the cyclic compound was in each of the samples. Furthermore, the presence of another cyclic compound at 38.959 minutes for the majestic lavender sample could

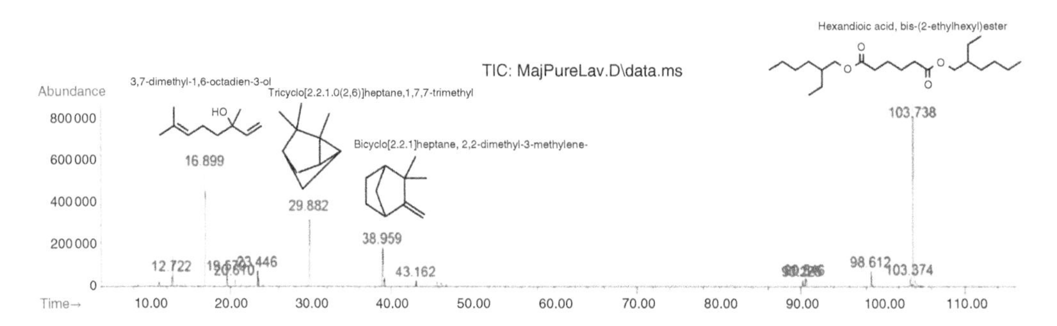

Figure 16.17 GC–MS plot of the Majestic (pure) Lavender fragrance sample, with the significant compound groups, indicated above each peak. Only the five notable peaks were identified and compared between the Aftel Lavender and Majestic Pure Lavender samples. All runs were integrated and their retention times are shown above.

have an adverse effect on the allergens of the sample as well as the severity of the sample reacting with the skin (Tisserand, 2017). It is also worth noting that both of these cyclic compounds have the capability of producing a radical cation, which makes sense as all of the samples are under the presence of an ester (Jjunju et al., 2019). All of these compounds have a ton of similarities with each, but it is fascinating to see how much synthetic elements get introduced into each one and the drawbacks behind it.

There is also something to say about what the linalool GC–MS says and its relation to the other fragrance samples. The main similarity found between the linalool peak and the lavender samples was with the majestic pure lavender sample at retention time 16.89 minutes. These alcohols provide one of the more prominent scents within the sample, suggesting that majestic pure lavender has a stronger overall scent. As for the Bergamot samples, the USDA and Eden samples did not need a plasticizer for preservation, so there was no presence of any esters (or for the most part, alcohols), except for the P&J. With this in mind, the P&J samples are the safest bet to say that out of all of the fragrances, the P&J sample is the most synthetic and was made under the harshest conditions.

The Bergamot samples have some distinct similarities when compared to each other and also some similar compounds from the lavender ones too. Starting with the USDA and Eden Bergamot sample (Figure 16.18), the lack of an ester at a higher retention time indicates that most of the impurities are coming at low retention times and more than likely are cyclic in nature. There also seems to be no presence of linalool or germanium here, further suggesting that these samples were created with natural products that do not need to be synthetically modified. The beta-pinene compound found at 10.31 minutes was also in the Eden Bergamot sample, which comes from the composition of different plants and provides a rustic, wooden smell to the Bergamot. This smell does come out strongly from the sample, and there are various forms of pinene in essential oils to provide a more grounded scent. Studies done with essential oils called *Myristica fragrans* seeds have found plentiful amounts of different conjugated pinenes, suggesting there was an economical advantage to adding this organic compound to the fragrance as well as the ecological benefits of it being straight from plants (Matulyte et al., 2019).

It is also worth pointing out the prevalence of 1-methyl-4-(1-methylethyl) cyclohexane and how both the USDA and Eden Bergamot (Figure 16.19) samples have a similar amount of it. This natural product comes from extracting the juice of aloe plants and is another organic compound that enhances the overall scent. There is also a compound eerily similar to it that was found about

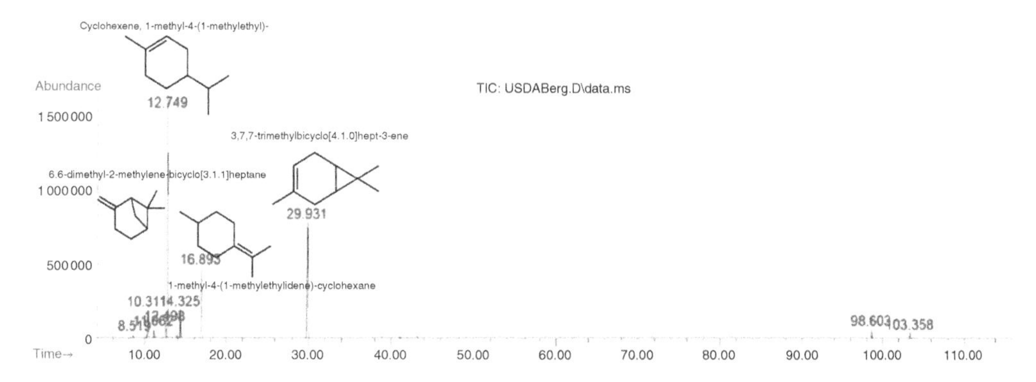

Figure 16.18 GC–MS results of the USDA Bergamot fragrance sample, with the significant compound groups, indicated above each peak. Only the five notable peaks were identified and compared between the P&J Bergamot and the Eden Bergamot samples. All runs were integrated and their retention times are shown above.

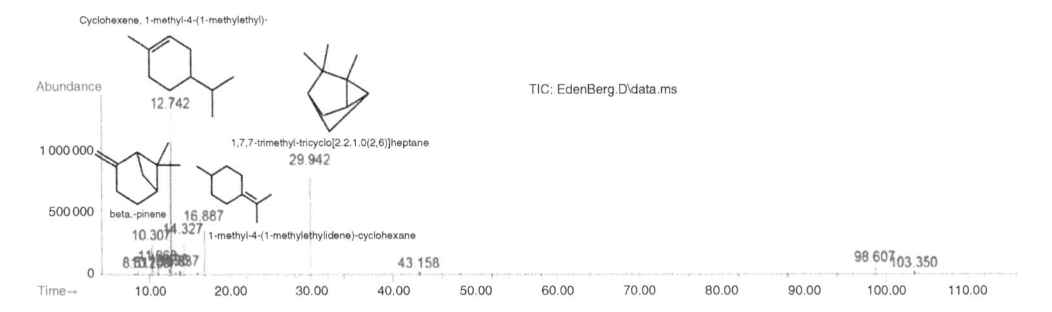

Figure 16.19 GC–MS results of the Eden Bergamot fragrance sample, with the significant compound groups, indicated above each peak. Only the five notable peaks were identified and compared between the USDA Bergamot and the P&J Bergamot samples. All runs were integrated and their retention times are shown above.

four minutes later in both samples, 1-methyl-4-(1-methylethylidene)-cyclohexane. Unlike the previous cyclohexane, this compound is a thick oil that is used as a specific plant metabolite for the sample. Plant metabolites are usually components of NaDES compounds (Natural Deep Eutectic Solvents) and are highly biodegradable at the temperatures which these compounds are usually under. However, their toxicity can fluctuate based on the temperature of the NaDES, so it is best to keep these compounds at a constant temperature when extracting (Burger et al., 2019). That leads into the other Bergamot (Figure 16.20) sample, the P&J, which was unlike any other Bergamot sample found and was mostly synthetic.

Starting with the isopropyl tetradecanoate at retention time 90.23 minutes, there is no other compound like it any of other samples analyzed. Previous GC–MS runs with cleaner fragrances have not found the presence of this compound, suggesting that is not only synthetic, but that is harmful when absorbed through the skin. It turns out that this alcohol is an emollient, or a type of emulsifier that allows for an increase absorption of the sample into the skin. These types of synthetic nonfragrance compounds could attach to potentially harmful allergens that can skin cause a wide

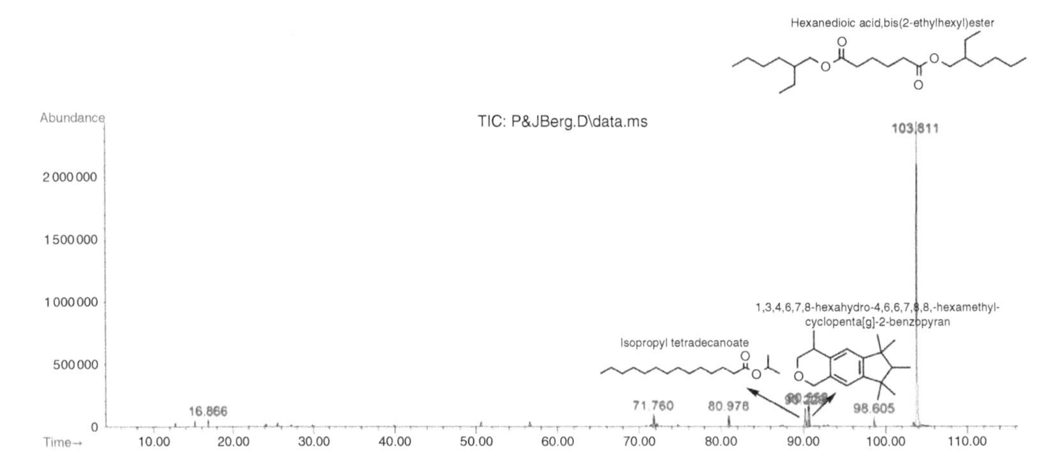

Figure 16.20 GC–MS plot of the P&J Bergamot fragrance sample, with the significant compound groups, indicated above each peak. Only the five notable peaks were identified and compared between the USDA Bergamot and the P&J Bergamot samples. All runs were integrated and their retention times are shown above.

variety of skin issues (Letizia et al., 2003). Now these allergens are usually not notable when it comes to GC–MS, but with companies like P&J that prioritize cost over quality in their fragrances, it's important to analyze these compounds even more closely. Looking at the compounds through an FID would highlight the immediate differences during the analysis between the organic compounds than GC–MS where you have to identify them after the runs.

16.8.8 Conclusion

The results show several essential oils that are marketed as "pure," "natural," and "organic" that are, in fact, synthetically derived fragrances containing compounds including but not limited to plasticizers, inks, and several synthetic and natural organic compounds meant to imitate natural fragrances.

Truly natural essential oils are made up of complex mixtures of volatile organic compounds typically derived from plants. The composition of natural essential oils can vary greatly depending on plant species, the part of the plant extracted, the method of extraction, and the time of harvest of the plant (Price and Price, 2012). Synthetic essential oils, on the other hand, are made in the laboratory using various processes, often by modifying and/or combining simple aromatic compounds to create specific aromas (Tisserand and Young, 2014). We expected that the chromatograms would show more complexity for the more natural oils, and this was confirmed. However, many synthetic compounds had more peaks than initially expected. This could be due to the amount of additives necessary for imitating a given scent; the preservatives needed to reduce volatility and increase packageability; and compounds used to create more appealing textures and colors of oil. While synthetic essential oils offer several advantages, including lower cost and heightened consistency between batches, natural oils are often preferred by the consumer due to their perceived holistic benefits. Synthetic oils can be produced using petrochemicals and other potentially harmful chemicals, which has implications for the safety and sustainability of these supposedly "natural" products (Tisserand and Young, 2014).

The findings were highly limited due to time constraints. In the future, we would recommend full analysis and peak assignments of the gas chromatograms of each oil, as well as the inclusion of more known standard essential oils. Though simply comparing large peaks was effective in seeing the compounds that make up the majority of the oil and was enough to see identifiable differences, identification of each peak could yield information about other potentially dangerous compounds included in synthetic essential oils. Additionally, analysis of all the other oils that we completed GC/MS and FID detections on is likely to have similar findings and could prove to be beneficial in understanding whether differences between synthetic and natural oils are consistently vast among various other essential oil analytes.

Additionally, we found that the oils included in our study largely represented three segments of a spectrum: one entirely synthetic segment, one pure oil segment, and one completely pure fragrance derivative segment (i.e. linalool; Figure 16.21). It may be beneficial to refine the oil in the lab to obtain a high standard of purity and consistency in replications and to compare seemingly natural oils to those produced in the lab. Also, these are far from the only oils that would benefit from this analysis. There are many essential oils on the market that are sold as "food-grade" natural essential oils that are just as affordable as several oils that we obtained and tested. It would be beneficial to complete a similar analysis of these oils, as the common rhetoric among those who use essential oils is that the food-grade ones are healthier and more natural, and less likely to be synthetic replicates of their natural counterparts.

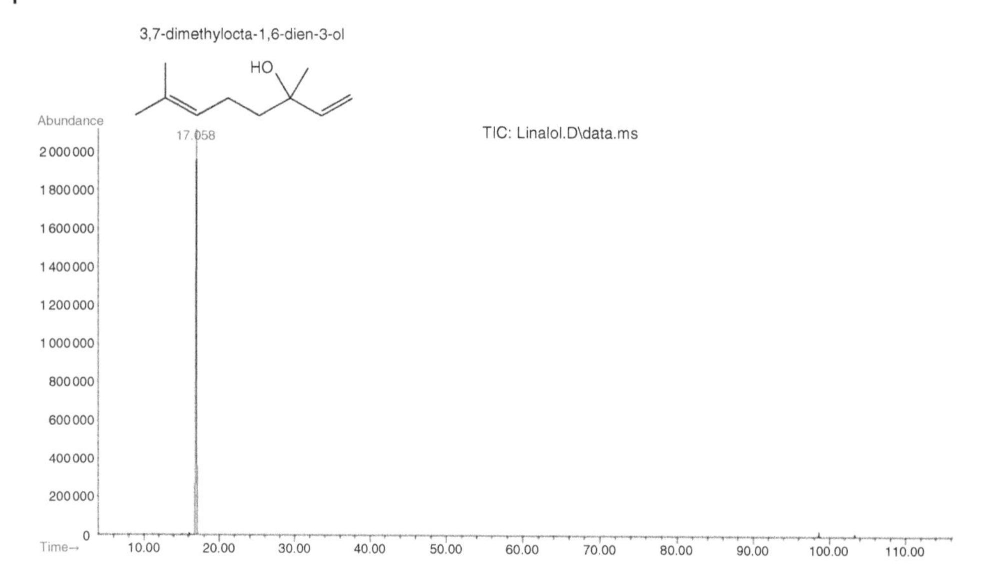

Figure 16.21 GC–MS results of the Linalool fragrance sample, with the significant compound group, indicated above the single peak. This table was used as a comparison tool between the lavender samples and the P&J Bergamot sample.

The findings of this study are highly replicable and show promise of potential implications for advocates of regulation for labeling essential oils, as well as implications for individuals who regularly use these oils in their homes for therapeutic benefits. Further research is needed to yield exactly what these implications may be, as well as to better inform the consumer about potentially deceitful marketing practices of synthetic essential oils.

16.9 SPME–GC–MS Analysis of Wine Headspace by Bailey Arend

For many consumers, the aroma of a wine is nearly as important as the flavor. The wine industry is obviously interested in producing wine with pleasing and abundant aroma. More than 1000 compounds have been identified in the headspace of wine, including alcohols, esters, carbonyls, acids, phenols, lactones, acetals, thiols, terpenols, and many more (Weldegergis et al., 2007; Polaskova et al., 2008). Although human senses can detect surprisingly small concentrations of certain volatile organic compounds in wine headspace, analytical instrumentation provides a more specific and precise way to measure the headspace character of wine.

The complex matrix of wine, as well as the low concentration of some of the volatile compounds presents further obstacles in the characterization of wine aroma. To analyze many of the compounds, sample enrichment techniques must be employed (Weldegergis et al., 2007) liquid–liquid extractions using organic solvents (Ortega-Heras et al., 2002; Andujar-Ortiz et al., 2009) and solid phase extraction (SPE) (Dominguez et al., 2002; Andujar-Ortiz et al., 2009) are both effective for wine analyses, however solid-phase micro extraction (SPME) presents a major advancement in volatile compound analyses.

SPME was first introduced in 1989 by Belardi and Pawliszyn for analysis of organic pollutants in water. The original method involved immersing fiber coated with fused-silica stationary phase directly in the liquid analyte. Analytes are sorbed/adsorbed onto the solid phase, which can then be inserted directly into a gas chromatograph (GC) injector, where the analytes are thermally desorbed and loaded onto the GC column. The newer SPME has drawn much attention for being versatile, yet simple. The technique does not require expensive, high-purity, toxic organic solvents generally associated with instrumental analysis and eliminates many possible sources of error.

SPME was first modified for headspace analysis in 1993. The new method exposed the coated fiber to the sample headspace only, which was found to shorten the extraction time while maintaining detection limits in the ppt range (Zhang and Pawliszyn, 1993). The driving theory behind any SPME is the partition coefficient of the analyte between the coating and the solvent or vapor. The partition coefficient along with the large difference in volume between the coating and headspace volume result in impressive concentration factors. The mass of analyte adsorbed to the coating (n) is given by

$$n = C_o V_1 V_2 K_1 K_2 / (K_1 K_2 V_1 + K_2 V_3 + V_2),$$

where C_o is the original concentration in the liquid phase, and the volumes of the three phases in equilibrium are as follows: V_1 for the coating, V_2 for the liquid phase, and V_3 for the headspace (Zhang and Pawliszyn, 1993). Minimizing the ratio of headspace to sample volume ($V_3 \ll V_2$) and maximizing the affinity of the coating for analyte (large K_1) can boost the amount of adsorbed analyte. This allows direct injection of analyte to the GC without risking instrument damage by injecting concentrated and sugary wine matrix. HS-SPME also avoids many complications of matrix effects, even allowing analysis of solid samples and human blood, as long as the analyte is volatile (Zhang and Pawliszyn, 1993). Tat et al. found that 50/30 μm Divinylbenzene/Carboxen/Polydimethylsiloxane coated fiber gave the most sensitive and repeatable results for the analysis of wine headspace (Tat et al., 2005).

Aside from fiber coatings and volume ratios, other parameters that can affect the sensitivity of HS-SPME are exposure time, temperature, and pH of the sample solution (wine). Exposure time is logically related to the concentration of analyte sorbed to the fiber. Sufficient time must be given for the system to reach equilibrium before the equation above is valid. Temperature governs the fraction of analyte present in the headspace and available for adsorption. Many methods immerse the extraction vial in a heated water bath (Tat et al., 2005), however, care must be taken that the fiber, headspace, and condensed phase are all in thermal equilibrium. The pH values have been adjusted in some studies (Boutou and Chatonnet, 2007) to allow multiple classes of molecules to be in their most analyzable form. For example, to simultaneously adjust for pyrazines (which are best analyzed at neutral to basic pH values, $pK_a \sim 0.50$) and phenols ($pK_a \sim 25$, which are best analyzed at low pH values) Boutou and Chatonnet adjusted all samples to a pH value of 7.

The method of Boutou and Chatonnet was also sufficient for analysis of contaminants that cause off flavors in wine. Compounds such as 2,4,6-trichloroanisole, 2,3,4,6-tetrachloroanisole, 2,4,6-tribromoanisole have olfactory perception thresholds near 10 ng/L and give wine a "barnyard" character (Boutou and Chatonnet, 2007). Wines with these contaminants are referred to as "corked" and are quite undesirable. Analysis of such contaminants can determine the origin of contaminant and improve wine production techniques (Boutou and Chatonnet, 2007).

16.9.1 Materials and Methods

16.9.1.1 Sampling Conditions (Adapted from Tat et al., 2005)

Sample wines and all equipment were stored at room temperature to ensure thermal equilibrium and minimize thermal differences between samples. Thirty-two (32.0) mL of sample wine was pipetted into a 40-mL glass vial equipped with a septum. The septum was pre-punctured with a sharp, hollow needle to avoid contaminating or breaking the fiber by contact with the septum. The extraction fiber was a Supelco 50/30 μm Divinylbenzene/Carboxen/Polydimethylsiloxane Stableflex fiber conditioned at 270 °C in the GC inlet for 1 hour. The fiber was inserted to the vial via septum before being exposed. The fiber was exposed to the headspace while the wine was stirred. The extraction was performed at 25 °C for 15 minutes. When finished, the fiber was immediately inserted into the gas chromatograph injector, where it remained for the entire duration of the temperature program.

16.9.2 Instrumental Parameters (Adapted from Boutou and Chatonnet, 2007)

GC/MS analysis was performed by an Agilent 19091S-433 HP-5MS 5% Phenyl Methyl Siloxoane 30.0 m × 250 μm × 0.25 μm nominal capillary column. The carrier gas was helium (ultrahigh purity, 99.999% passed through hydrocarbon traps) programmed to flow at a constant linear rate of 54.1 mL/minute for the entire run. The injector was ensured to be long enough to allow full insertion of the fiber and operated by manual injection. The injector was operated in splitless mode at 270 °C for the entirety of the each run.

The oven program started at 50 °C for 2.0 minutes, and then increased at 3.0 °C/minute to 190 °C. The temperature was than increased at 50 °C/minute to 320 °C where it was held for 5 minutes. Detection was performed by an Agilent 5975C inert EI/CI MSD quadrupolar mass detector with EI ionization (source temperature 230 °C, quadrupole temp. 150 °C, energy of constant ionization 70). The entire method required a total of 56 minutes.

16.9.3 Results

This study was conducted to demonstrate the ease and versatility of HS–SPME for student chemists. For this reason, no adjustments were made to the wine samples and the recommended heating of the samples was not performed. Although the range of detectable analytes was much smaller than other methods have reported (Tat et al., 2005; Boutou and Chatonnet, 2007), the quick and simple method returned multiple analyte peaks for each wine tested.

It was found to be important to thermally clean the extraction fibers before each exposure. Samples of Black Box Merlot were run without cleaning the fiber and unexpected compounds were observed (shown in Figure 16.22). Many silica compounds were found in the blank runs, which could possibly be attributed to degradation of the fiber coating (shown below in Figure 16.23). If the fiber was "baked out" directly before exposure, the presence of these compounds was minimized (shown below in Figure 16.24).

The method was repeated 5 times on Black Box 2007 California Merlot and the results were found to be reproducible. One advantage of using boxed wine is that the effects of oxidation are eliminated. Samples may be drawn weeks apart, whereas opening a bottled wine introduces oxygen to the entire bottle and could affect the composition of the wine (Simpson, 1978).

While the method was found to be repeatable, only a limited number of identifiable compounds were recovered for each sample. These compounds include 3-methylbutyl acetate, ethyl hexanoate, 2-phenylethanol, diethyl succinate, ethyl octanoate, and ethyl decanoate. The chromatogram for Chardonnay (Buckley's Cove, South Eastern Australia, 2009, Figure 16.25), does not contain 2-phenylethanol and diethyl succinate which were found in all red wine samples. It is possible that the lack of these two compounds could be a signifier of white wine.

It was thought that a higher-quality wine would have a stronger bouquet and yield more volatile compounds, however the chromatogram of Red Table Blend from Walla Walla Village Winery (82%

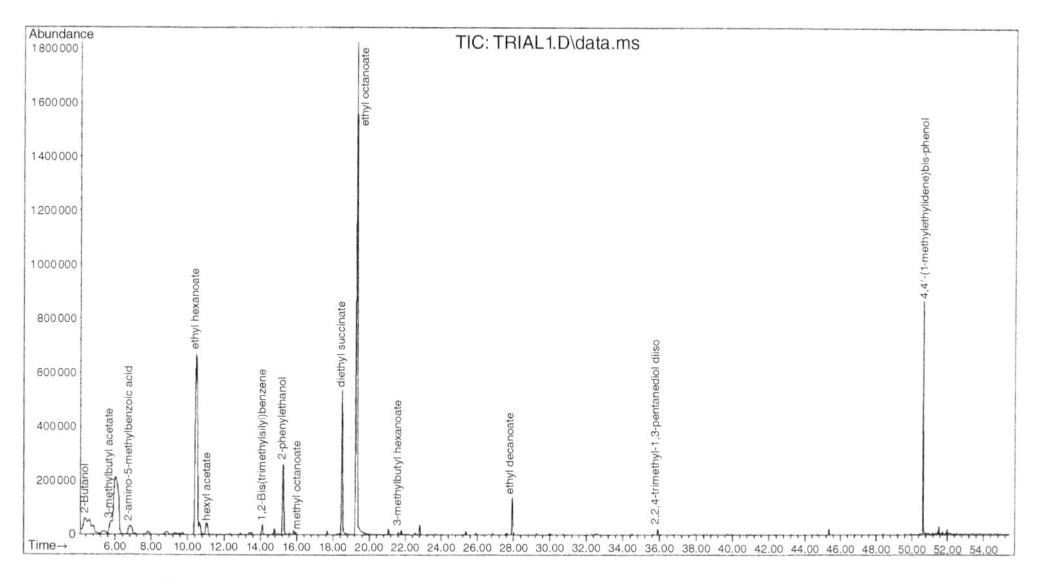

Figure 16.22 The chromatogram for Black Box California Merlot, 2007. No blank was run and unusual peaks were observed.

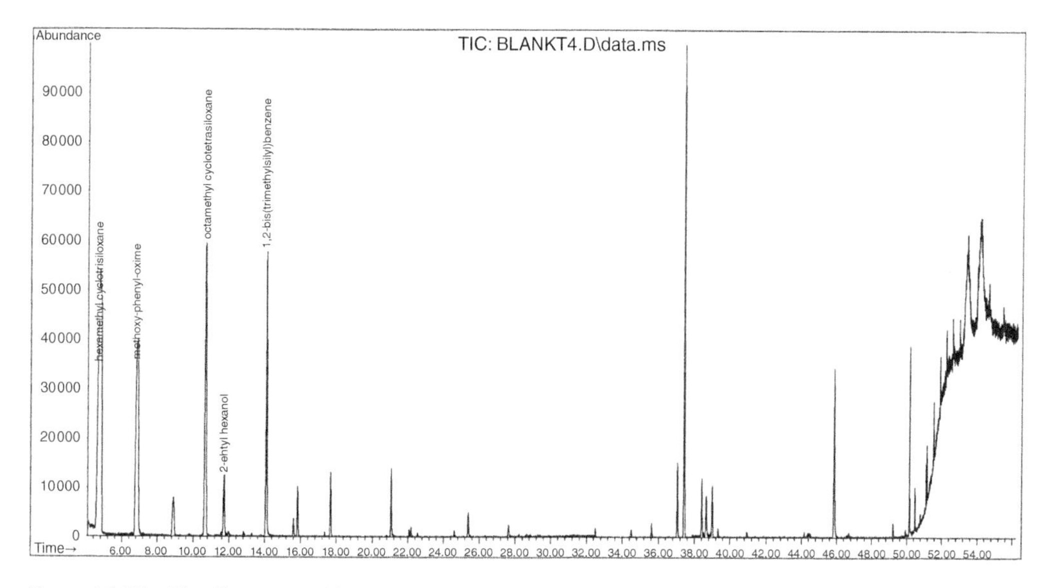

Figure 16.23 The fiber was subjected to an entire run without exposure to any sample. Multiple contaminants were observed, demonstrating the need to run blanks and thermally clean the fiber between samples.

Cabernet Sauvignon, 9% Merlot and 9% Cabernet Franc, Figure 16.26) did not show additional peaks. This method was also insufficient to characterize contaminants in a sample of corked wine from Foundry Vineyards in Walla Walla (2005 Red Wine) (shown in Figure 16.27). It is likely that heating of the samples or adjustment of the other parameters, such as pH, would increase the range of compounds detectable by this method.

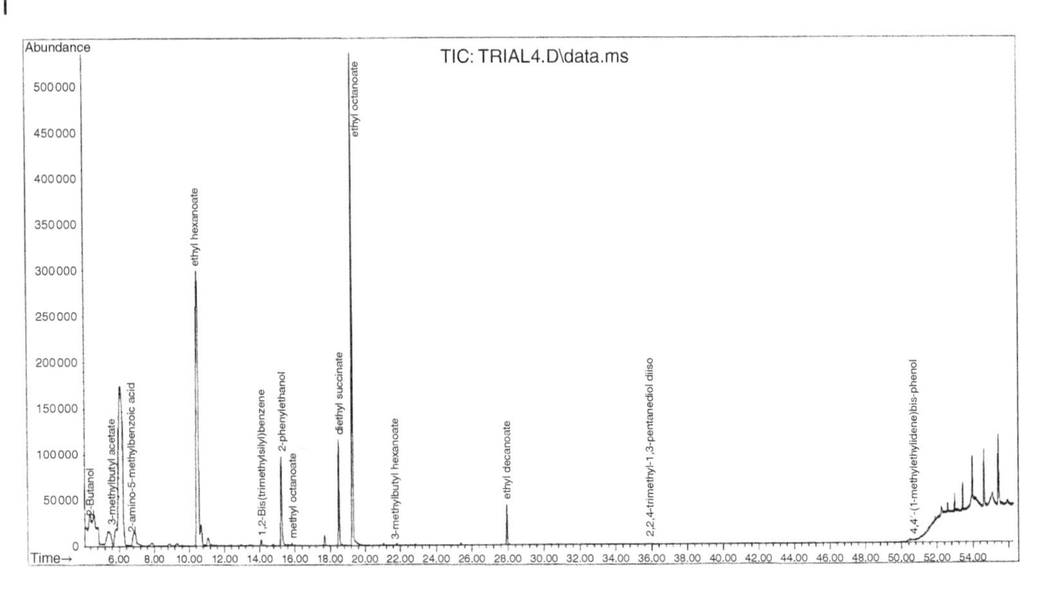

Figure 16.24 The chromatogram for Black Box California Merlot, 2007. The fiber was exposed to the sample immediately after subjecting it to a blank run. Fewer contaminants were observed.

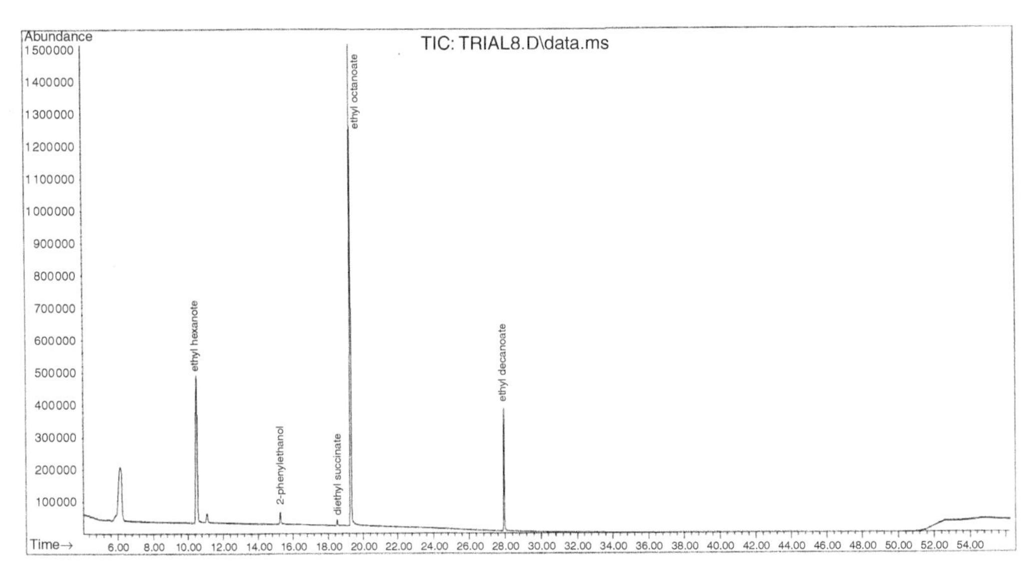

Figure 16.25 The chromatogram for Buckley's Cove 2009 Chardonnay from South Eastern Australia differs slightly from Buckley's Cove Shiraz (Figure 16.4). 2-Phenylethanol and diethyl succinate which were found in the Shiraz were not observed in this sample.

Although this simplified method does not engage the full potential of HS-SPME techniques, it is sufficient to demonstrate the theory and application of such techniques in a college/university laboratory. Additional streamlined methods may be developed for other consumable complex matrices such as whiskey or vinegar (Figure 16.28).

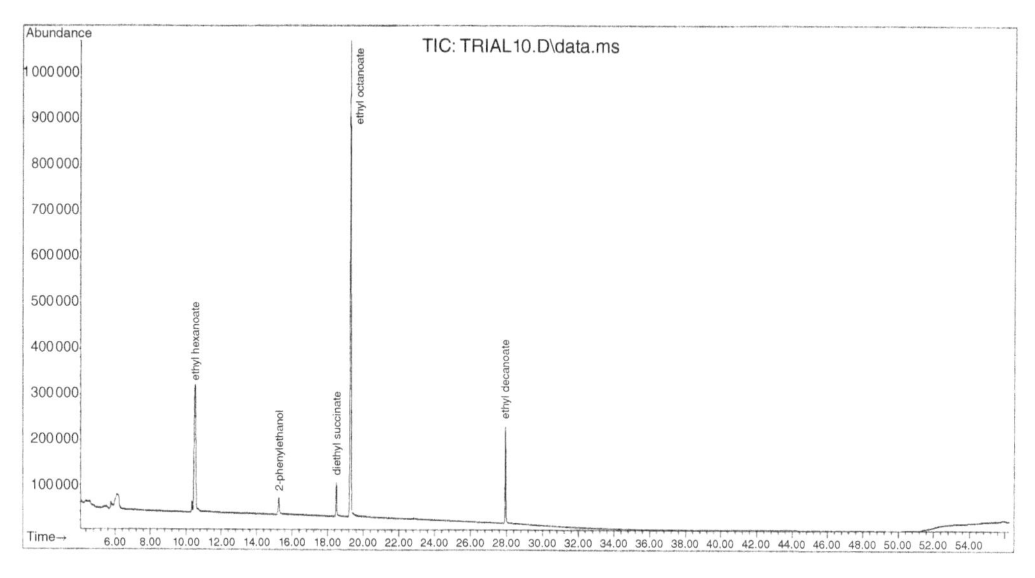

Figure 16.26 The chromatogram for Walla Walla Village Winery's Red Table Blend (82% Cabernet Sauvignon, 9% Merlot, and 9% Cabernet Franc). Contrary to our prediction, no additional compounds were detected in the higher-quality wine. Changing the sample parameters could aid the detection of additional volatiles.

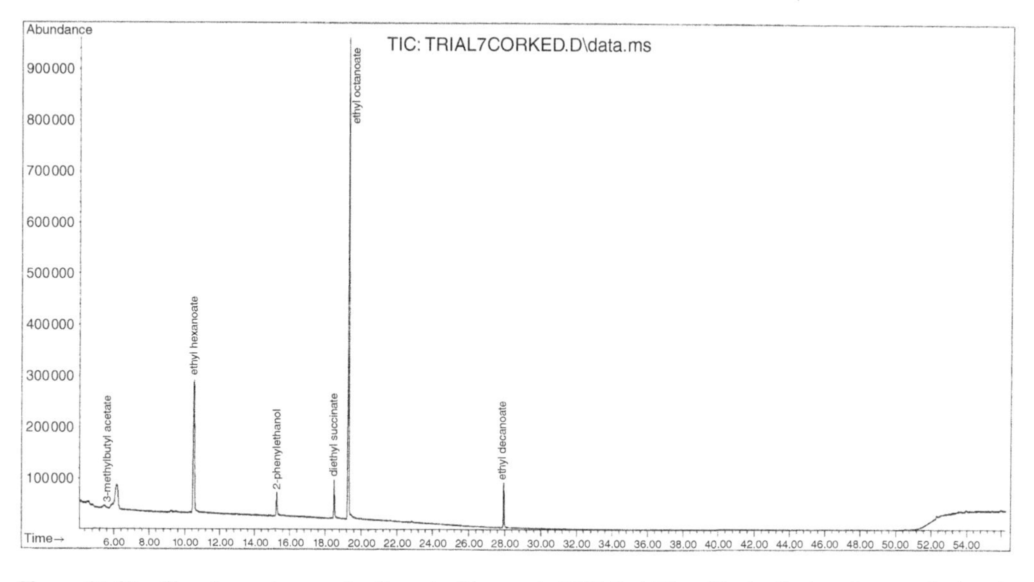

Figure 16.27 The chromatogram for Foundry Vineyards 2005 Red Wine. "Corked" wines have a displeasing aroma caused by trichloroanisole or tribromoanisole. The streamlined method could not detect the presence of these compounds. However, changing the sample parameters such as temperature or pH could improve detection of these contaminants (Source: Adapted from Boutou and Chatonnet, 2007) .

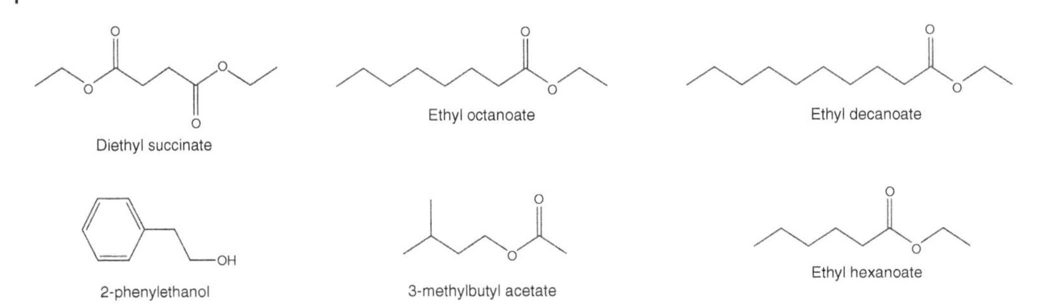

Figure 16.28 The compounds repeatedly found in all red wine samples by the proposed method.

16.10 Quantitative Determination of the Presence of Captan on Organic and Nonorganic Strawberries by Peter Mullin, Eric Ying, Jon Na, Sharon Ndayambaje, and Soren Sandeno

16.10.1 Introduction

With a growing population, farmers have been put under pressure to artificially increase crop yield by spraying chemicals that can hinder pest infestations, fungus growth, and unwanted weed species. However, these chemicals can be toxic to humans and thus their presence must be monitored on consumables such as fruit. The fungicide captan is one such toxic chemical. Its use in farming is primarily to protect against fungal diseases but it is also an additive that improves the external appearance of many red fruits such as apples, strawberries and tomatoes (USDA PDP). While it is practically nontoxic to most wildlife, the EPA has classified captan as a potential human carcinogen in high doses, far beyond levels encountered in an average diet (EPA, 2013).

Furthermore, captan has been shown to be particularly harmful to the eyes resulting in permanent damage (EPA, 2004).

Because captan's ties to efficient farming conflict with its known toxicity, quantitative methods of determining the fungicide's presence on fruit are of utmost importance. However, there is no confident consensus surrounding the best quantitative method for captan determination from a solid sample. EPA Method 1699 outlines a general procedure for determination of organochlorine and organophosphorus pesticides in environmental samples that can be applied to captan (United States, EPA 2016). This procedure entails solid-phase extraction using methylene chloride, methylene chloride:hexane (1:1), or acetone:hexane (1:1) in a Soxhlet extractor. While this is a functional procedure, it is by no means tailored to captan and instead generalized to encompass the extraction of many pesticides. Many procedures, in order to avoid the emulsion created by blending the sample, involve maceration to achieve higher extraction percentages (USEPA, 2024a).

The most common solvents involved in captan extraction are polar organics including acetonitrile, methanol, and methylene chloride but many extraction procedures also involve acidic aqueous washes and benzene (USEPA, 2024b). Once the sample has been extracted, analysis is most often performed using GC–MS which can provide both analyte separation and structural information' (ACS, 1967).

The purpose of this lab experiment was to learn, evaluate, and develop analytical procedures for determining the captan content on the surface of strawberries and raspberries. The process involved testing the viability of an extraction method using consecutive solvent washes, analyzing standards and extracted samples on GC–MS and GC–FID, and programming both sequence and method files to target optimal oven temperatures and take advantage of autosampling.

16.10.2 Materials

16.10.2.1 Captan (CAS 133-06-2) (Figure 16.29)

Methylene chloride (CAS 75-09-2)
Commercially available organic strawberries and nonorganic strawberries.

16.10.3 Methodology

Standard solution preparation. Captan (0.100 g) was added to a 100.0 mL Class A volumetric flask, diluted to volume using the dichloromethane and sonicated to create a ~1000 mg/L solution of DCM. A calibration curve is then constructed using the DCM solution from 1 to 200 ppm.

Strawberry sample extraction. Five organic strawberries were randomly selected and weighed. Each berry is individually rinsed with 25 mL of DCM a total of 45 seconds a total of 3 times. The three rinses of each five berries were collected in an Erlenmeyer flask. The flasks were then left in a fume hood for 1 week to allow for complete solvent evaporation. The remaining film in each flask was redissolved in three small portions (less than 2 mL) of DCM solution, swirled for 30 seconds and filtered via glass wool in a pasture pipet to remove solids. The final volume was adjusted in a three or five mL volumetric flask. The same procedure was repeated for five nonorganic strawberries.

16.10.4 Results and Discussion

Masses of strawberries extracted in this experiment are given in Tables 16.4 and 16.5.

To construct a calibration curve, peaks in the chromatogram were integrated and the area of the peak corresponding to captan in each standard and sample was recorded. While this method does not take the internal standard into account, it can still provide sufficient data to construct a calibration curve (shown in Figure 16.30).

Figure 16.29 Structure, name, and relevant physical properties of captan, the investigated fungicide.

Captan
1,2,3,6-tetrahydro-N-(trichloromethylthio)phthalimide

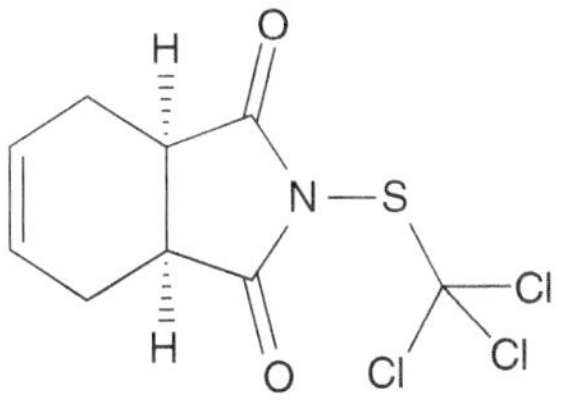

Molecular weight: 300.6 g/mol
Melting point: 178 °C
Boiling point: 314 °C
Aqueous solubility: 5.1 mg/L

Table 16.4 Organic strawberry mass (in grams).

	Beaker 1	Beaker 2	Beaker 3	Beaker 4	Beaker 5
Strawberry 1	20.025	17.158	16.624	12.224	19.073
Strawberry 2	24.717	18.369	15.788	14.448	10.358
Strawberry 3	22.202	16.899	14.914	16.563	13.815
Strawberry 4	16.368	34.848	19.892	14.369	16.343
Strawberry 5	21.494	18.827	12.257	14.903	9.797
Average	20.9612	21.2202	15.895	14.5014	13.8772

Total mass = 432.275 g.

Table 16.5 Nonorganic strawberry mass (in grams).

	Beaker 1	Beaker 2	Beaker 3
Strawberry 1	47.084	31.475	34.021
Strawberry 2	37.146	39.834	25.261
Strawberry 3	30.291	30.841	28.478
Strawberry 4	20.141	24.875	20.57
Strawberry 5	19.746	38.188	31.966
Average	30.8816	33.0426	28.0592

Total mass = 459.917 g.

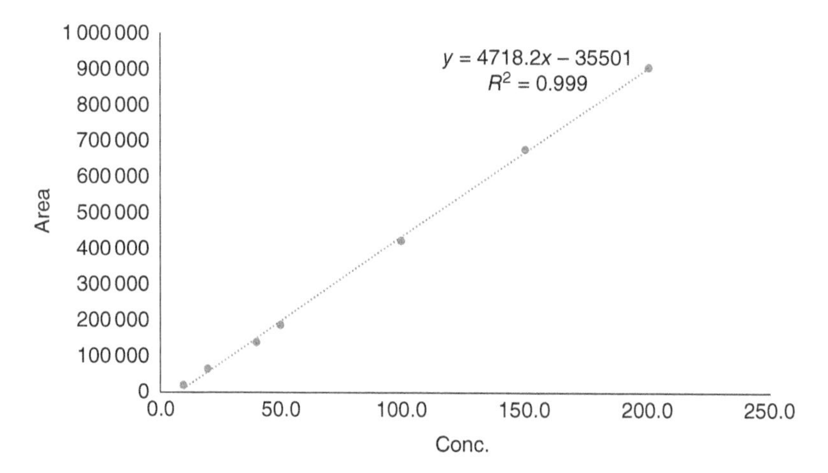

Figure 16.30 A calibration curve of captan standard concentrations versus the area of their peaks at a retention time of 16.0. A linear least-squares trendline (shown in blue) was fit to the standard concentration.

An important feature of this calibration curve is that its *y*- and *x*-intercepts do not come close to the (0,0) mark, in fact, the *x*-intercept is at 7.52 mg/L, which brings into question the accuracy of the calibration. The most probable reason behind this inaccuracy is likely incorrect concentrations from transfer error, and the lower concentrations being thusly below the detection limit. For the strawberries, only the nonorganic strawberries showed detectable amounts of captan. Plotting the area of these peaks on the calibration curve resulted in a concentration of 10.5 μg/L of captan on a berry with a mass of 30.7 g. Organic strawberries were below the detection limit and no conclusions can be made about their concentrations of captan (Table 16.6).

16.10.5 Conclusion

We can conclude that there was captan, a fungicide and reddening agent for organic produce, present on the nonorganic strawberries. We cannot, however, make any conclusions for the trace presence of captan on organic strawberries, as the data were below the detection limit.

Table 16.6 Captan concentration results.

	Nonorganic	Organic
ppm (calculated from LLS) × 0.003 L DCM per beaker = mg captan per beaker (mg)	0.0526	Data below detection limit
mg captan total (mg)	0.158	Data below detection limit
Total mass captan/total weight of strawberries (μg/g)	0.343	Data below detection limit
Average weight of single strawberry (g)	30.7	Data below detection limit
Average mass captan per 1 strawberry (μg)	10.5	Data below detection limit

16.11 Determination of Nicotine in Human Urine Using HPLC–MS by Ashley Nguyen, Lea Molacek, Maxwell Brown

16.11.1 Introduction

Nicotine is an alkaloid in tobacco that is highly addictive in nature. Its addictiveness comes from causing a release of dopamine, glutamate and gamma-aminobutyric acid in the brain (Benowitz, 2009). This leads to the calm, euphoria or "high" from ingesting or smoking a tobacco product. It can be absorbed through a variety of ways, most importantly from direct inhalation of smoke via cigarettes or vapes. In the body, about two-thirds of nicotine is metabolized to cotinine and at least 16 other metabolites (Hukkanen et al., 2005). In urine, about 8–10% of the nicotine initially introduced to the body is excreted as nicotine, while 10–15% has been metabolized to cotinine, and 33–40% has been metabolized to hydroxycotinine. The concentrations of nicotine and cotinine in human urine, semen, and plasma have been determined for forensic toxicology and as a biomarker for nicotine consumption by both liquid chromatography and gas chromatography (Abu-Awwad et al., 2016; Brajenović et al. 2015; Abdallah et al., 2016).

Being able to measure trace levels of chemicals in biological matrices such as urine and to provide exposure to such chemicals is important for assessing the health of adults and children. In the case of nicotine, it is important to examine the health impacts of tobacco use from different sources. However, biological matrices like urine are complicated to analyze due to many other chemicals and metabolites being present (Wei et al., 2014). To overcome this problem, these types of samples require pre-treatment to minimize interference and matrix effects and to more accurately measure the target analyte. These treatments could include liquid-liquid extraction, SPE, supercritical fluid extraction, microwave-assisted extraction, and more (Wei et al., 2014). SPE allows separation of the target analyte from contaminants which remain in the sorbent and can be washed out using different affinity liquids to isolate the target.

The purpose of this experiment was to learn the SPE and chromatography instrumentation for analyzing nicotine in urine. Urine samples were tested from college students who consume nicotine through a variety of methods including vaping (electronic vaporizers), tobacco consumption, Zyn nicotine pouches, and cigarettes. The samples were pretreated using SPE and the target analyte (nicotine) was chromatographically resolved using high-performance liquid chromatography

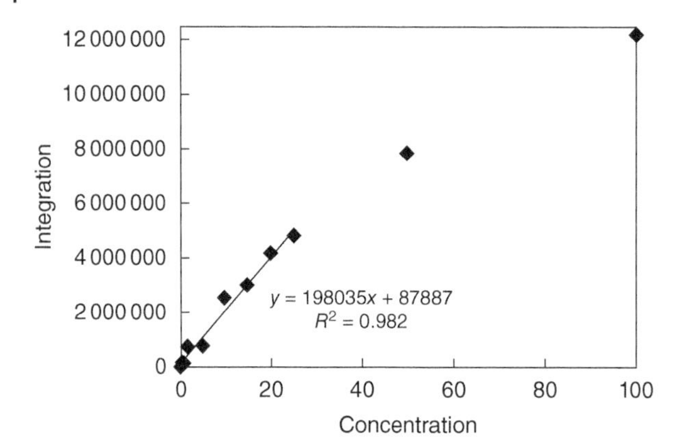

Figure 16.31 Structure and chemical data for nicotine.

coupled with mass spectrometry (HPLC–MS) in order to determine the concentration of nicotine that remained in the urine. The structure and data for Nicotine are should in Figure 16.31.

Chemicals

- 0.1% formic acid (aqueous, 3 L)
- ammonium hydroxide (50 mL)
- nicotine (0.2545 g)
- methanol (1 L)
- acetonitrile (1 L)

16.11.2 Methodology

Two 10-mL samples of urine were gathered each from a lifetime nonsmoker and a vaper and their nicotine concentrations determined against a calibration curve of nicotine external standards in simulated human urine matrix.

16.11.2.1 Preparation of Nicotine External Standards

- Simulated human urine matrix (1 L), or 5% ammonium hydroxide in methanol, was prepared by dilution of ammonium hydroxide (50 mL) in methanol (950 mL).
- Stock nicotine solution (1018 mg/L) was prepared by dilution of nicotine (0.2545 g) in simulated human urine matrix (250 mL) in a class A volumetric flask.
- Nicotine external standards (100 mL each) were created at concentrations ranging from 0.00509 to 101.8 mg/L by performing serial dilution of stock nicotine solution in simulated human urine matrix in class A volumetric flasks.

16.11.2.2 Preparation and Filtration of Human Urine Samples

- Packed beds from the Discovery DSC-MCAX mixed-mode cation exchange SPE system were attached to a vacuum and conditioned first with methanol (1 SPE column volume) and sodium phosphate (pH 6, 2 SPE volumes).
- Human urine sample (10 mL portion) was loaded and allowed to filter.

- Hydrophilic interfering compounds were subsequently washed off with acetic acid (1 M, 1 SPE column volume).
- Hydrophobic interfering compounds were washed off with methanol (1 SPE column volume).
- Simulated human urine matrix (2 mL) was pushed through the filter to extract the nicotine and then collected and transferred to an HPLC–MS sample vial.

16.11.2.3 Determination of Nicotine Concentration Using HPLC–MS

- Standards and samples were run on the HPLC–MS by an autosampler. The solvent gradient was programmed to transition the mobile phase from 90% deionized water (0.1% formic acid) and 10% acetonitrile) to 10% deionized water with 0.1% formic acid and 90% acetonitrile) over a period of 20 minutes.

16.11.3 Results and Discussion

Based on the HPLC–MS integrations and known concentrations of our nicotine external standards, a calibration curve was prepared for nicotine and fit over the linear range ($R^2 = 0.982$ up to 25 mg/L nicotine; see Figure 16.31). Peak integrations corresponding to retention times around 1.9 seconds all showed the characteristic mass-to-charge ratio of nicotine (163.2 g/mol). All urine samples and nicotine standards showed peaks around 2.6 minutes from an unknown matrix element, and the samples of human urine for both the nonsmoker and vaper showed an additional contaminant at around 1.7 minutes. Nicotine was not detectable in external standards prepared to concentrations below 0.509 mg/L, so these standards were omitted from the calibration curve. HPLC data from remaining standards exhibited nicotine peaks consistently at around 1.9 minutes, with linearly increasing integrations up to the 25 mg/L standard as shown in Figure 16.32 and Table 16.7.

Integration of the nicotine peaks from the urine of the nonsmoker and the vaper showed a significant difference in nicotine concentration. HPLC data from the nonsmoker showed no

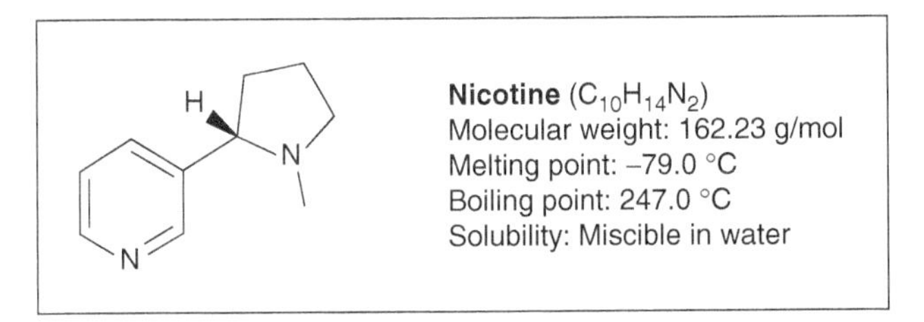

Figure 16.32 Calibration curve of nicotine external standards ranging from 0 mg/L (blank) to 101.8 mg/L nicotine in simulated human urine matrix (5% ammonia hydroxide in methanol).

Table 16.7 Integration from HPLC of urine samples from a nonsmoker and a vaper.

	Sample	Integration	RT	Nicotine Concentration (mg/L)
Nonsmoker	1	0	—	Not detectable
	2	0	—	Not detectable
Vaper	1	$8.41\text{E} \times 10^5$	1.893	3.80_3
	2	8.37×10^5	1.888	3.78_3

detectable peaks around the retention time of 1.9 minutes which would otherwise indicate nicotine present in those urine samples, meaning <0.509 mg/L is present in accordance with the detectable range of our external standards (though the actual value for the lifetime nonsmoker is predicted to be much closer to 0 mg/L). HPLC data for the vaper, on the other hand, displayed a sharp peak at 1.888 and 1.893 minutes for the respective urine samples; from the integration of these peaks relative to our calibration curve, the nicotine concentrations in the two samples were determined to be 3.80 and 3.78 mg/L, respectively. Taking the average of these two values, the concentration of nicotine in the vaper's urine was therefore determined to be 3.79 mg/L.

16.12 Analysis of Caffeine in Urine Samples Using GC–FID by Elsa Nader, Ralph Huang, Aaron Lieberman, Jane Duncan, and Matt Sousa

16.12.1 Introduction

Drinking coffees and teas is responsible for major caffeine injections to human bodies. Studies have shown that over 49% of US adults drink coffee on daily basis (Loftfield et al., 2016). The catabolism for caffeine in human bodies may vary according to the biological sex and individual metabolism cycles. In this research, we studied the kinetics of caffeine in human bodies by examining concentrations of caffeine in urine during a day using gas chromatography flame ionization detector technique (GC–FID) and external standards method. The GC–FID is the correct instrument for this procedure because caffeine is volatile and thermally stable, therefore ideal for a gas chromatograph. We chose not to use the GC–MS even though the results are confirmatory, because the better detection limits are unnecessary for the purposes of this experiment. We only have one peak to analyze and we are only analyzing caffeine, so the cheaper instrument is accurate enough. The GC–FID is the more cost-effective instrument for this analysis.

In this experiment, we collected samples from two students from Whitman College, one male and one female. They were asked to drink coffee immediately after waking up. Urine samples were collected roughly every hour during the day and times were recorded on sample vials. Caffeine in urine is relatively stable and nonvolatile in urine, therefore can be stored for over weeks and brought to GC analysis later. External standards were prepared at concentrations at 2, 4, 6, 8, 10, 12, 20, 50, 100, and 200 ppm using analytical anhydrous caffeine.

Caffeine from urine samples was extracted using previous procedure indicated above. 25 mL of samples were pipetted and added to 5 mL of 9:1 dichloromethane (DCM) and methanol mixture (measured in volume by volumetric cylinder). 1 ml concentrated sodium chloride solution and one scoop of solid sodium chloride was also added to the sample solution to separate water. Sample solutions were then shaken vigorously and brought to centrifuge for 20 minutes for separation purposes.

Note: Urine extraction samples are notorious for forming an emulsion. We avoided the emulsion by gentle adding the 9:1 DCM:methanol extraction solvent and placing the sealed vial on a rocker mixer, on the most gentle of settings for 30 minutes.

The GC–FID calibration curve (line) is shown in Figure 16.33. Caffeine concentrations in each student are show in Table 16.8 with data plotted in Figure 16.34.

16.12.2 Data Analysis

Two sample sets showed a similar trend according to our data. In the morning, immediately after consumption, caffeine concentrations in urine steadily increased, indicating the caffeine catabolism. Approximately two or three hours later, a peak was shown indicating the highest

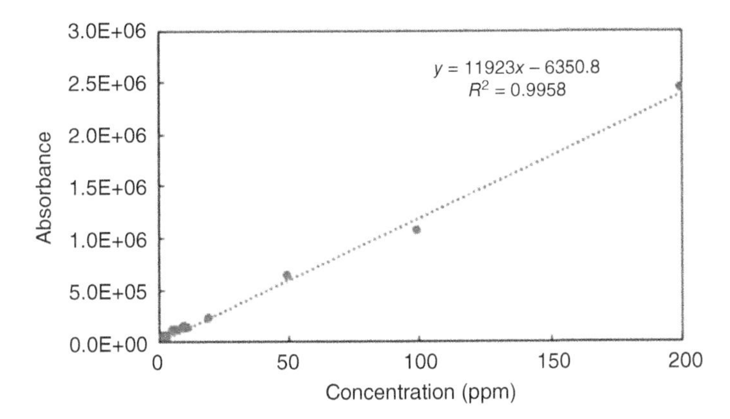

Figure 16.33 Calibration curve using external standards showing R^2 value at 0.9958 Linear squares equation: $y = 11923x - 6350.8R^2 = 0.9958$

.

Table 16.8 Nicotine concentrations in student urine as a function of time.

Time (minutes)	Concentration (ppb)
Gabe	
100	5.3
162	10.8
186	8.6
242	8.0
333	6.6
447	4.6
515	3.0
588	2.4
Hannah	
60	27.8
120	30.9
180	36.5
240	29.6
300	24.9
360	24.1

concentration for both samples. The concentration of caffeine in urine then began to drop readily. From the data, we believe that Gabe's data shows a lower caffeine concentration in his body while Hannah had a much higher concentration of caffeine.

16.12.3 Future Improvement

Most of our data for Gabe was at a lower range of the calibration curve. Therefore, future research may be required to either increase the concentration of samples or decrease the concentration of

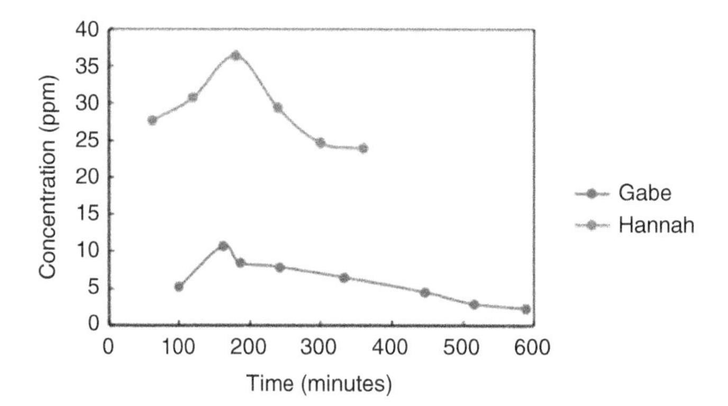

Figure 16.34 Data from two sample sets showing the kinetic curves of concentrations in urine over time.

standards prepared. Further statistical tests could be conducted to make sure our data is within a statistically significant range.

Adding more controls and an increase in sample size would be useful. We did not control diet which could influence caffeine concentration. It would also be interesting to see if weight or sex is more of an influence in caffeine concentration in the body, since either of those variables could currently be at play. A more regulated time sheet for urination may also be necessary for more precise kinetics data, although this is a difficult factor to control.

16.13 Caffeine in the Walla Walla (WA) Wastewater Effluent by MacKenzie Cummings, Mia Groff, Roya Nasseri, Noah Willis, and Clara Wheeler

16.13.1 Introduction

The world population continues to increase, and as this happens, the demand for clean and safe drinking water also increases. Pharmaceuticals and other organic wastewater contaminants (OWCs) can be measured in terms of concentration to determine the severity and quantity of contamination in water (Kolpin et al., 2002). Caffeine is a common OWC found in wastewater which originates from residential sources. Modeling and measurement of this pharmaceutical pollutants in wastewater samples allows for general analysis of residential effects on water quality within a specific region. Because caffeine is commonly consumed in the US through the consumption of tea, coffee, and pharmaceuticals such as aspirin, it is a great marker for assessment of wastewater contamination (Bruton et al., 2010).

While caffeine may not be largely considered a dangerous pollutant, traces in wastewater are released directly into the environment, as treatment plants are not entirely effective at removing pharmaceuticals from the effluent (Kolpin et al., 2002). Because pharmaceuticals such as caffeine are designed to initiate certain biological effects in humans, they are a potential biological threat to plants and animals (Li et al., 2020). Through the continuous exposure of caffeine via the environment, ecosystems can become negatively affected and dysregulated. Despite the fact that most wastewater treatment plants account for caffeine removal due to its high consumption and abundance, the amount that enters the effluent is much greater than the amount which is degraded (Li et al., 2020).

This experiment was conducted by the Environmental Chemistry Department at Whitman College to assess the presence of pharmaceuticals in the Walla Walla region. The experiment was modeled after a previous study conducted in the same region, by previous Whitman College chemists.

16.13.2 Methods

To begin the process, we made a buffer solution of $C_2H_3N:H_2O$ (acetonitrile; ACN) (1:9) buffered with 10 mM $NH_4CH_2O_2$/10 mM HCH_2O_2 to a pH of below 4.00. This was checked with a pH meter. 0.1, 0.5, 1, 5, 10, 50, and 100 ppm standards were made with the appropriate mass of caffeine and buffer solution. Our gradient was made with ACN and the buffer solution. Solution A was 9.5–0.5 parts buffer to ACN, while solution B was 1–9 parts buffer to ACN.

The SPE cartridges (Supleco, Inc. Discovery DSC-MCAX 3-mL, 100-mg tube) were first conditioned with acetone. Then, one liter of each of our four samples was drawn through the extraction cartridge with a vacuum. Once they were dry, they were rinsed with 25 mL of methanol, using a pipette bulb to force the methanol through the cartridges into 100 mL round bottom flasks. These were then rotovapped until there was no liquid left, only a dry tar-like substance coating the flask.

2.00 mL of the 1:9 ACN:aqueous buffer solution was transferred to the flasks with the dry caffeine and sonicated for 2 minutes. The liquid was then filtered into HPLC vials with a luer-lock syringe and a 22 mm PTFE filter.

The calibration curve (Figure 16.35) and samples were then run through the HPLC. We first ran one black twice, followed by the complete calibration curve, another blank, and then the samples. The calibration standards had a caffeine peak at 1.877 and the samples had peaks at 1.791, perhaps owing to the different solvents each were in Peaks were integrated manually.

16.13.3 Results

All samples contained over 200 ppb of caffeine. We detected 233, 239, 226, and 287 ppb in each of our four samples. All of our samples were of Walla Walla wastewater effluent.

The 0.1, 0.5, and 1 ppm standards did not create detectable peaks, and the concentrated standards had concentrations of over 100 ppm, according to our calibration curve. If this experiment is repeated, it would be useful to expand the calibration curve, perhaps adding a 110 and 125 ppm standard.

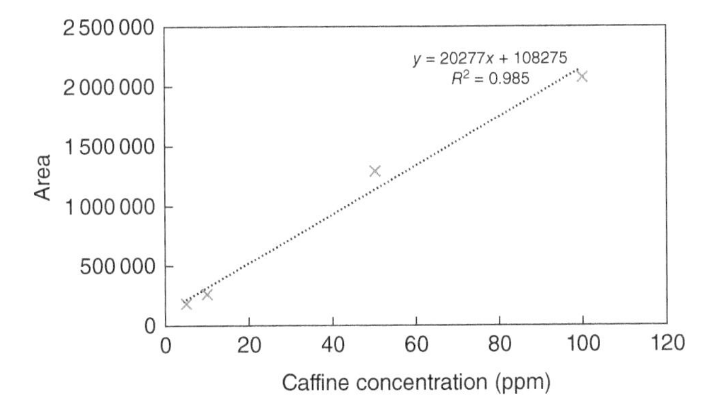

Figure 16.35 Calibration curve (line) for caffeine.

16.13.4 Conclusion

The results of this experiment depict high concentrations of caffeine in the Walla Walla wastewater effluent. However, the concentrated samples were expressed as values above the calibration curve. Nonetheless, this data reflects the high levels of caffeine which we infer to be connected by correlation with high levels of other OWCs. Extremely high levels of caffeine in wastewater are considered to be around 192 ppb while wastewater effluent concentrations generally range from <0.1 to 126 ppb. The effluent at the Walla Walla wastewater treatment plant effluent either contains huge amount of caffeine due to a reflection of the community-wide consumption of the stimulant and/or the plant has no good way to remove the OWC from the influent. Either way, this effluent has the potential to have huge effects on the downstream ecosystem because of the large concentrations of caffeine and potentially large amounts of other OWCs as well.

16.14 Gasoline Analysis by GC–FID and GC–MS by Theodore Pierce and Austin Shaff

16.14.1 Introduction

This research examines the concentrations of various potentially hazardous compounds present in gasoline. Gasoline is a generic name for volatile and flammable petroleum fuels (IARC, 2016). The name itself therefore does not tell much about what is actually present in this incredibly complicated solution. What is sold as gasoline can vary significantly in composition and must be heavily regulated. There are over 500 hydrocarbons present in standard gasoline plus a variety of artificial additives (Benbrahim-Tallaa et al., 2012). These hydrocarbons include varying groups such as alkanes, alkenes, and aromatics. Of these, some immediately fall under closer scrutiny as they have previously been identified as carcinogenic in humans (Benbrahim-Tallaa et al., 2012). Several of the most concentrated of these potentially hazardous compounds are benzene, cyclohexane, *n*-heptane, hexane, pentane, toluene, and xylenes (Benbrahim-Tallaa et al., 2012). A brief exposure to low concentrations of these compounds has not been found to be hazardous, but long-term exposure to the carcinogens could be cause for concern.

Take for example the removal of lead from gasoline. While lead levels in gasoline were never extremely high, continuous exposure, especially in young children, led to elevated levels of lead in blood (Wilson and Horrocks, 2008). The removal of lead from gasoline resulted in a decrease of 95% in blood lead levels across the US population (Dunnivant, 2017). This demonstrates the danger of even low-level toxic components of gasoline given its ubiquity and volatility. Many individuals spend several hours a day around automobiles or other gasoline-powered machinery. Gas attendants in particular have a high potential exposure to these chemicals. Clinical trials involving exposure of animals to high concentrations of gasoline vapors have been shown to can cause dizziness, muscle weakness, kidney damage, and even liver cancer (Healthline: Gasoline and Health, n.d.).

Toluene is of particular interest in gasoline. Toluene has a relatively high mass fraction in gasoline of approximately 0.055 (Approximate Mass Fraction of Gasoline, n.d.). Toluene has been found to have detrimental effects at only 500 ppm and the eight-hour exposure limit as determined by OSHA is 100 ppm (USDL, n.d.). Benzene is another hazardous chemical to human health and it has an exposure limit of only 1 ppm for an 8-hour period (Approximate Mass Fraction of Gasoline, n.d.; OSHA, n.d.). Even at concentrations of less than 1% by volume in gasoline, benzene levels need to be strictly monitored. Hexane and cyclohexane have been found to have narcotic effects at high, 1000 ppm, concentrations (Approximate Mass Fraction of Gasoline, n.d.; OSHA, n.d.).

By formulating a methodology for the analysis of components in gasoline, it is possible to more closely monitor some of these compounds of interest. An effective methodology will allow for time and cost-effective analysis of various sources of gasoline to ensure acceptable levels of hazardous compounds are being maintained at all times. The purpose of this study was to develop a procedure for use in an Instrumental Methods course to monitor vapors from a normal automobile gasoline-filling operation. Issues with the collection system led to the decision to directly analyze gasoline in its liquid form.

16.14.2 Materials and Methods

16.14.2.1 Chemicals

The components of gasoline to be analyzed were selected based on their relatively high concentration in gasoline. The components selected were heptane, benzene, toluene, *m*-xylene, *p*-xylene, cyclohexane, and ethylbenzene. An internal standard of decane at 50 ppm was also selected.

16.14.2.2 Calibration Curve

In order to create a calibration curve, a 1 L stock solution of 50 ppm decane (as internal standard) in pentane was prepared for dilution of the standards. A 1 L 1000 ppm solution of all target compounds was prepared in order to dilute into calibration curve standards. From these two stock solutions, 25 mL standards of 0.1, 0.5, 1.0, 5.0, 10, 20, 40, 60, 80, 100, 120, and 160 ppm of target compounds were prepared, each of which contained the 50 ppm internal standard of decane. External standards and diluted samples were analyzed by GC–FID.

16.14.2.3 Retention Times

After the calibration standards had been taken, a small, but detectable amount of each target component was prepared in pentane in order to measure the retention times.

16.14.2.4 Gasoline Sample

Regular-grade gasoline was collected for analysis from a local gas station. It was diluted to 18 μL in a 25 mL volumetric flask with 50 ppm solution of decane in pentane.

16.14.2.5 Instrumentation

An Agilent Gas Chromatograph/6890N Flame Ionization Detector with a 7683 series autoinjector was used to analyze all samples. An Agilent Gas Chromatograph/7820A Mass Spectrometer/5977E was used to analyze the gas sample in order to confirm the presence of the target compounds in gasoline.

16.14.2.6 GC–FID Conditions

The GC–FID was run splitless and utilized a front injection DB-5 column with a 1 μL injection volume. The inlet temperature was 230 °C and the detector temperature was 250 °C. The starting oven temperature was 40 °C held for four minutes, then ramped up at 1 °C a minute until 60 °C. The temperature was held at 60 °C for 10 minutes at which point it ramped at 15 °C a minute up to 240 °C. A blank of 50 ppm decane solution was run before and in between each standard, and a rinse of pentane was used twice before and twice after each injection. This method was used for all standard samples, retention times, and for the gasoline sample.

16.14.2.7 GC–MS Conditions

The GS–MS was run with a front injection Agilent HP-5ms Ultra Inert column with a 1.0 μL injection volume. The inlet temperature was 250 °C. The oven was ramped at the same rate as the GC–FID, starting at 40 °C and ending at 240 °C. A blank was run in between and before each sample, and the syringe was rinsed with pentane 3 times before each injection. This method was used for analysis of the gasoline sample, and one standard of 20 ppm.

16.14.3 Results

Analysis of the GC–FID results could only be performed on four of the seven target compounds due to issues with the solvent peak obscuring some of the more volatile compounds, and some compounds being indistinguishable from each other. The GC–FID was found to have a detection limit of roughly 1 ppm. So, the standards below that were discarded. The compounds that were analyzed were toluene, *m*-xylene, *p*-xylene, and heptane. The GC–FID, Figure 16.36, response showed the percent composition of gasoline as 1.52% *n*-heptane, 4.50% toluene, 1.53% *p*-xylene and

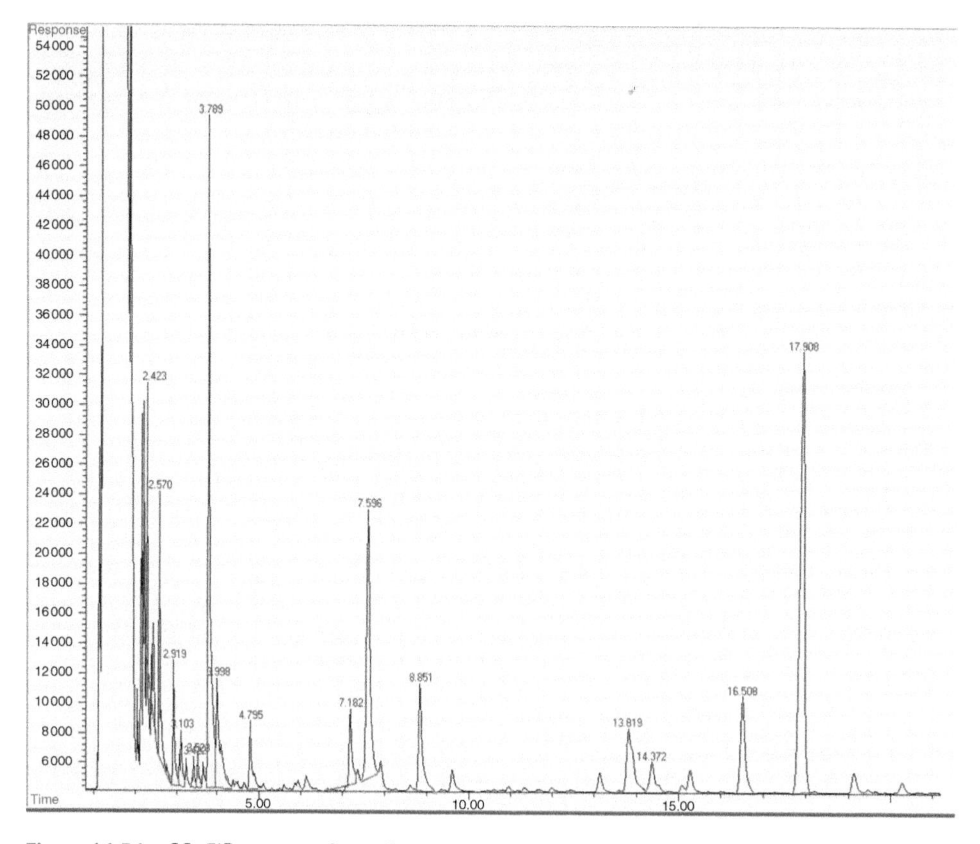

Figure 16.36 GC–FID spectra of gasoline sample with labeled peaks.

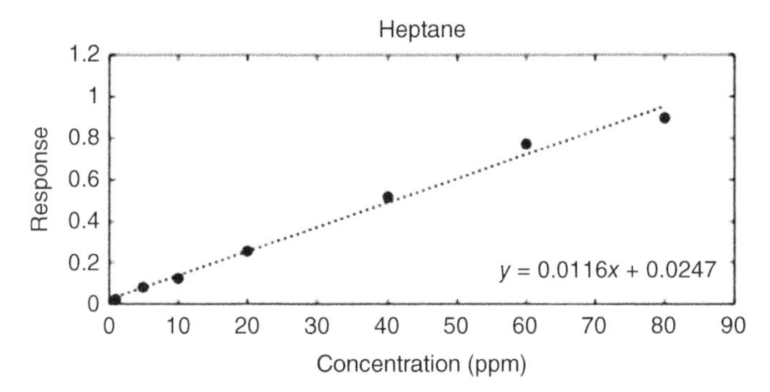

Figure 16.37 Example GC–FID calibration curve (line) for heptane.

0.34% *m*-xylene. An example calibration curve (line) for heptane is shown in Figure 16.37. These results are similar to the literature value for the percent composition of these compounds in gasoline (Kreamer and Stetzenbach, 1990). Analysis of the gasoline sample by GC–MS confirmed the presence of these compounds and shed light on the variety of other compounds present in relatively high concentrations.

16.14.4 Lesson Learned

Although the analysis of gasoline by GC–FID is possible, it is difficult due to the complex composition of gasoline. The retention times are too similar to effectively resolve some of the peaks. Furthermore, the inconsistency between the retention times in single compound runs versus the calibration curve made it difficult to assign peak identities, particularly for *m*- and *p*-xylene, and made it impossible for benzene and cyclohexane. It was also surprising to see the lack of confidence by GC–MS in the fragmentation pattern of the target compounds. The library had a difficult time determining between similar compounds, such as *o*- and *p*-xylene. Even some of the compounds identified were only with a confidence of ~70%.

Most notably, if this experiment is repeated, we recommend using carbon disulfide as the solvent. While carbon disulfide is a carcinogen, it produces a very small, if any, solvent peak in the GC–FID and as a result, will allow separation of most components found in gasoline, especially the early eluting analytes.

16.15 GC–MS as an Effective Instrument for Detecting Cocaine on US Currency by Jessica Boyers and Kacey Godwin

16.15.1 Summary

Traces of illicit recreational drugs such as cocaine have been found in US currency for several decades now. Using a relatively fast, efficient, and nondestructive method for analysis on paper bills could prove to be useful information for areas of study such as toxicology. GC–MS is a confirmatory technique that can perform analyte identification and their relative concentrations when calibrated. Our lab was able to detect cocaine on the majority of dollar bills donated by the Walla Walla community.

16.15.2 Introduction

There is a common assumption that a large percentage of currency contains some traces of the illicit drug known as cocaine. Cocaine is a stimulant that is known to boost dopamine and serotonin levels (National Institute on Drug Abuse, 2019). It is naturally occurring from an indigenous plant in South America, but it can be found in the form of a white crystalline powder sold illegally around the world. This substance is highly addictive and can be used through snorting, consumption, or direct bloodstream injection.

Studies have shown that approximately 90% of US dollar bills have some trace amount of cocaine (Dell'Amore, 2009). This would be attributed to the use of paper money to snort this drug. Any remaining powder is later transferred through close contact within wallets and spread through general spending and exchange practices.

This lab utilized GC–MS for analysis. GC–MS allows us to produce accurate and precise data. Mass spectrometry is considered the only confirmatory technique in instrumental chemistry. Another current application of GC–MS is testing for illicit drugs in hair samples. This technique can be helpful in solving crimes and in validating someone's addiction or chronic intoxication (Uhl, 1997).

16.15.3 Materials and Methods

Our basic procedure was adapted from Whitman College graduates Noah Porter and Rintaro Moriyasu who conducted a similar experiment in 2016.

16.15.4 Sample Collection and Preparation

Eleven different US $1 bills and a single $5 bill were donated by students at Whitman College for this project. All of the bills were prepared by adding 10 mL of ultra-pure methanol to a capped tube and subsequently mixing for over 24 hours. Three clean dollar bills, courtesy of Chase Bank, were also processed the same way. These samples were intended to act as a baseline comparison for the used money that we collected. All of the samples were filtered before analysis and transferred to a GC glass vial by a 3 mL syringe.

16.15.5 External Standards Instrumental Calibration

An 888 ppm cocaine stock solution was prepared by adding cocaine hydrochloride (0.0222 g) to a 25 mL class A volumetric flask and filling it to the mark with high-grade methanol. Standards at concentrations of 2.17, 4.53, 9.43, 19.64, 40.92, 85.25, and 177.6 ppm were prepared by serial dilution and used for external calibration of the instrument.

16.15.6 GC–MS Conditions

Samples were analyzed on an Agilent 6890 capillary column GC interfaced with the 5973N mass spectrometer. Each run was approximately 46 minutes long. An autosampler was used to carry out the sequence overnight.

16.15.7 Data

Figure 16.38 shows a GC–MS calibration curve (line) for the cocaine external standards ranging from 19.64 to 177.6 ppm ($R^2 = 0.996$, $y = 627476x - 1.21E7$).

Figure 16.39 gives visual comparison of cocaine concentrations found on the US dollar bill samples (Table 16.9).

16.15.8 Discussion

We were able to find traces of cocaine on 11 out of the 12 donated US dollar bills. Out of those 11 cocaine-positive samples, 8 of them had quantifiable concentrations. This matches the statistics from other studies published in *National Geographic* (2009). The clean money from Chase Bank was

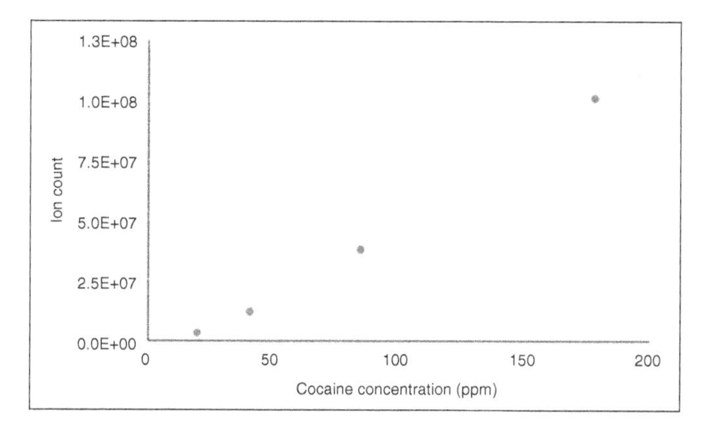

Figure 16.38 GC–MS calibration curve for the cocaine external standards ranging from 19.64 to 177.6 ppm ($R^2 = 0.996$, $y = 627476x - 1.21E7$).

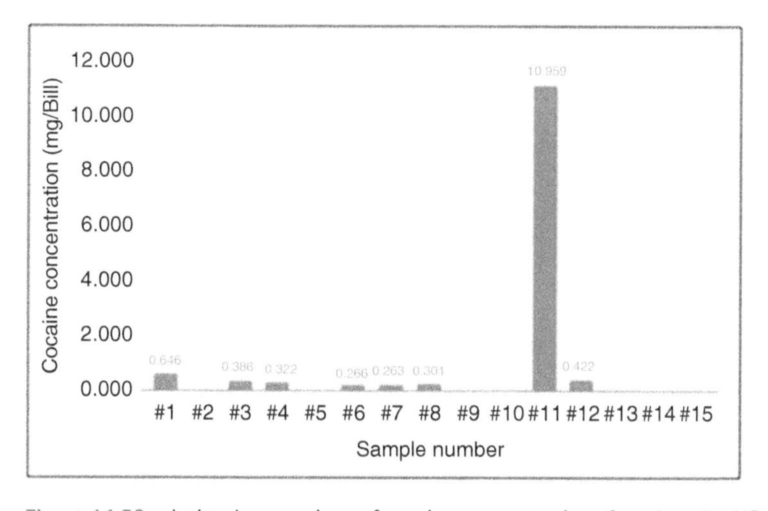

Figure 16.39 A visual comparison of cocaine concentrations found on the US dollar bill samples.

Table 16.9 Cocaine concentrations on 15 different US dollar bills determined by GC–MS.

Sample Number	Category	Ion Count	Retention Time	mg Cocaine per Bill
1	Donated	28 404 005	35.73	0.646
2	Donated	None	N/A	0
3	Donated	12 145 609	35.727	0.386
4	Donated	8 127 842	35.722	0.322
5	Donated	Trace amounts	35.71	Below d.l.
6	Donated	4 567 706	35.726	0.266
7	Donated	4 419 032	35.726	0.263
8	Donated	6 806 001	35.727	0.301
9	Donated	Trace amounts	35.72	Below d.l.
10	Donated	Trace amounts	35.72	Below d.l.
11	Donated	675 567 744	35.734	11.0
12	Donated	14 350 181	35.728	0.422
13	"Clean"	None	N/A	0
14	"Clean"	None	N/A	0
15	"Clean"	None	N/A	0

free of cocaine with no peak in the 35–36 minute range on the chromatogram. Our detection limit was ~0.2 (0.1964) mg/bill. This was the lowest of our standards that received a discernible peak and was able to have the ion count integrated by the GC–MS. Our lower concentration standards—2, 4, and 8 ppm—resolved no visible peak and only displayed noise. The final calibration curve ranged from approximately 19–177 ppm, providing an adequate concentration range for our samples with detectable ion counts of cocaine. Furthermore, there were a couple of bills that had the right retention time and relative fragmentation pattern, but we were unable to integrate them. Samples of this kind were identified as having trace amounts of cocaine. The 11th sample had an 11.0 mg which is more than the maximum pharmaceutical dosage that they give to relieve severe pain. This particular sample extended beyond our calibration range.

16.15.9 Conclusion

A future possibility would be testing out different extraction methods to see which would be more cost effective. We really wanted to try and get a regional survey of the money, but the collection process was more difficult than we anticipated. These Gen-Z kids don't carry around physical cash anymore. Although we had willing participants/donors, their go to means of payment was Venmo which is not useful for the purposes of this experiment. Additionally, it would be interesting to use GC–MS to identify other illicit drugs or substances on money. Overall, we are very grateful for this opportunity to work on an independent project and familiarize ourselves with the intricacies of the instrument.

16.16 Analytical Quantification of CBD-A Content in *Cannabis sativa* (Hemp) Using GC–FID and GC–MS by Lauren Yumibe, Sam Weiss, Maddie Bowers, Tori Li, Asher Bachtold, and Brandon Neifert

16.16.1 Introduction

Industrial hemp (*Cannabis sativa* L.) has grown in interest in the scientific community in recent years due to increasing decriminalization efforts and the plethora of therapeutic, agricultural, and economic benefits (Andre et al., 2016). Of the 750 compounds identified in *C. sativa*, cannabinoids make up the majority of bioactive compounds (Liu et al., 2022). The two dominating phytocannabinoids of interest in *C. sativa* are (1) Δ9-tetrahydrocannabinolic acid (THCA), which is the precursor to THC which provides the psychoactive characteristic of the drug-type *C. sativa* and (2) cannabidiolic acid (CBDA), which predominates in hemp varieties. The decarboxylated form of CBDA is CBD (Citti et al., 2018) and is able to mediate the psychoactive effects of THC, has analgesic properties, can inhibit anandamide absorption, and is cytotoxic to breast cancer cells (Formato et al., 2020). Of recent interest cannabinoids have also been shown to prevent and treat infection from SARS-CoV-2 (Van Breemen et al., 2022).

Extraction of the CBD cannabinoid using organic solvents, especially from inflorescences of *C. sativa*, is the most convenient method although there are several more advanced and greener extraction technologies that have been implemented in cannabinoid analyses (Liu et al., 2022), including supercritical fluid, microwave-assisted, and ultrasound-assisted extraction processes (Lazarjani et al., 2021). Following extraction using organic solvents, GC–MS is often used for identification of cannabinoids (Ciolino et al., 2018) as well as GC–FID to determine the main compounds in *C. sativa* plant material (Baranauskaite et al., 2020). When using GC techniques, there should be no detection of CBDA due to the high temperatures, which causes heat-induced decarboxylation of CBDA into its CBD derivative (Baranauskaite et al., 2020; Formato et al., 2020). Therefore, these analytical methods should produce CBD yield only from hemp flowers in *C. sativa*, which lack THC.

As *C. sativa* becomes increasingly legalized, researched, and industrialized for its recreational and medicinal benefits, there is a growing need to standardize a method for determining the diverse bioactive compounds of *C. sativa*. Furthermore, the percent content of CBD displayed on product labels has been shown to not reflect the actual content of the product itself (Gurley et al., 2020), which may have negative consequences for individuals with particular tolerance to the effects of CBD. This study aims to generate a reproducible analytical procedure using GC–FID to obtain the yield of CBD from dried *C. sativa* flowers (Brand: Plain Jane), along with GC–MS for confirmatory detection. Additionally, in this experiment, we will provide a procedure for quantitative analysis from CBD oil extraction. We will then compare our quantitative results from each sample bud with the labeled CBD content to assess the legitimacy of the claimed value.

16.16.2 Experimental

Our group was asked to confirm whether a sample of commercial hemp buds contained 20% CBD-A by mass per bud as advertised. A sample size of 10 hemp buds was extracted in organic solvent to prepare for analytical and quantitative analysis. Five buds were analyzed using analytical methods and the remaining five were analyzed with quantitative methods. Analytical analysis was performed with a combination of GC–FID and GC–MS instruments for quantification of mass by calibration curve and confirmatory analysis respectively. Quantitative analysis was performed by using the roto-vap to isolate CBD from other effluent in the extractant and weighing the remaining

CBD oil. The data collected for each extractant was then compared to its original bud to confirm whether the calculated mass of CBD agreed with the expected mass. Example procedures are provided below, as applied by our laboratory.

16.16.3 Materials

Hemp buds (obtained from Plain Jane) were extracted with high-purity Petroleum Ether. For the analytical procedure stock and sample solutions were prepared with Petroleum Ether as solvent. High-purity CBD-A standard (1000 ppm, 1.000 mL) was used to prepare stock solutions. To ensure no water contamination $MgSO_4$ was used to dehydrate the extractant.

16.16.4 Extraction Preparation

Ten 20-mL centrifuge vials were massed and numbered 1–10 to keep track of individual buds throughout the experiment. One hemp bud was added to each centrifuge vial, and each vial was massed again to determine the bud mass. Ten mL of Petroleum Ether was added to each vial to extract CBD from the buds. The vials were capped and left to rotate at 25 rpm for approximately 72 hours to ensure all CBD was extracted. Following extraction, extractant fluid was pipetted through a column with glass wool and $MgSO_4$ (to remove any water) into a 50 mL volumetric flask (also numbered 1–10 corresponding to the respective extractant). Once extractant fluid was transferred, the buds were rinsed twice more with 10 mL Petroleum Ether per rinse, with each rinse pipetted through the column. Once filtration was completed, each flask was diluted to volume with Petroleum Ether.

16.16.5 Calibration Stock Preparation

From the provided CBD-A standards, calibration stock solutions of 1, 5, 10, 20, 50, 100, and 150 ppm were prepared. 2 mL volumetric flasks were used for each stock solution, dilutions were performed using either a 100- or 10-microliter syringe. Each stock solution was then brought to volume with Petroleum Ether using a glass pipette.

16.16.6 Analytical Procedure

Volumetric flasks numbered 1–5 were chosen for the analytical procedure. Each flask was diluted by 50×, where 2 mL was transferred from the 50 mL flask into a 100 mL volumetric flask, via a volumetric pipette. An aliquot was then filtered through a Pasture pipette containing a small amount of glass wool to remove particles. Approximately, 2 mL of each sample filtrate was then transferred to labeled vials compatible with the GC–FID autosampler. Each of the calibration stock solutions was similarly transferred to labeled vials, along with a blank of Petroleum Ether. Vials were arranged on the sample trays with the blank first, followed by the calibration stock vials from lowest to highest concentration, followed by sample vials 1–5. Each vial was then analyzed using the GC–FID. Once the calibration curve generated by the GC–FID was validated, each vial was then analyzed with GC–MS to confirm that identified peaks from GC–FID were CBD.

16.16.7 GC–FID Analysis

Our sequence analyzed the Petroleum Ether blank twice, followed by each calibration standard from lowest to highest concentration, and then two more blanks were run. Following blank readings analysis of sample vials then proceeded. Temperature programming settings began at 60 °C incrementing by 10 degrees per minute until reaching a final reading of 250 °C. Following the GC–FID run, each data file was integrated, and the peak area and retention time were recorded from the integration results of the most dominant peak.

16.16.8 GC–MS Analysis

The sequence proceeded identically to that of the GC-–FID sequence. Temperature programming was similarly consistent. Following the GC–MS run, each data file's spectra were integrated, the corresponding MS of the most dominant peak was analyzed, and the corresponding peak area and retention time were also recorded.

16.16.9 Quantitative Procedure

The contents of volumetric flasks 6–10 were quantitatively transferred into individual preweighed round-bottom flasks. Excess petroleum ether was removed using rotary evaporation, and the flasks (now with a green residue) were reweighed. The weight of the empty flasks was subtracted from the weight of the flasks with residue to determine the mass of the residue (hypothetically mostly CBDA).

16.17 Results and Discussion

16.17.1 GC–FID Analysis

Each of our calibration standards generated prominent peaks with retention times consistently within 25.1–25.8. The peak areas of our calibration standards produced a calibration curve with an R^2 value of 0.95. However, the peak area from the 150 ppm standard was excluded from the calibration curve as it drastically exceeded the linear range, suggesting a dilution error occurred during standard preparation. This resulted in a calibration curve that did not encompass the concentration range of our samples. As such, while our calibration curve (Figure 16.40) is robust, there is significant uncertainty regarding the validity of our results with the analytical method.

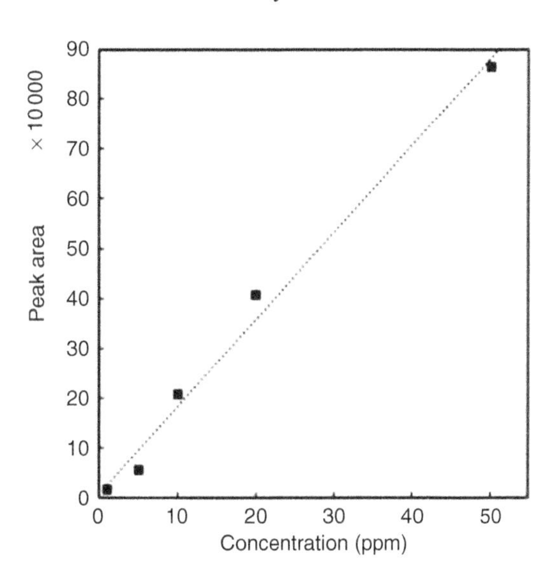

Figure 16.40 Integrated area of each peak at each concentration of GC–FID standard solution.

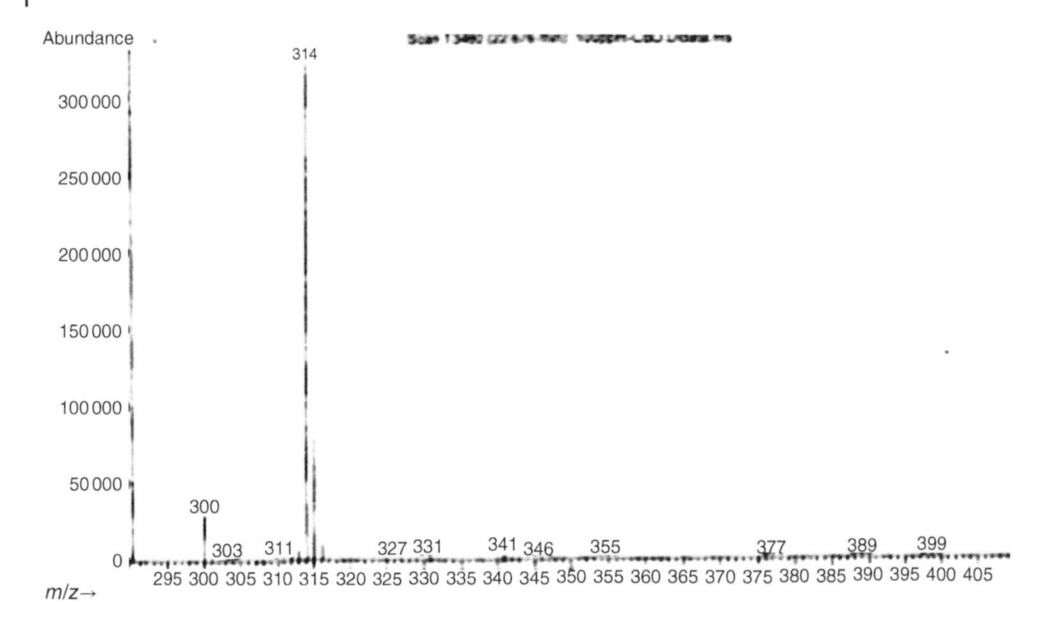

Figure 16.41 MS results from our 100 ppm standard.

16.17.2 GC–MS Analysis

The presence of CBD was confirmed in our calibration standards and hemp samples through presence of a prominent 314 *m/z* peak that decreased in abundance with decreasing ppm CBD in our standards (Figure 16.41). The 314 peak was not observed in our Petroleum Ether blanks which confirms both the peaks identified in GC–MS and GC–FID as CBD, as there were no other chemicals added during sample preparation. Predictably, other peaks grow more prominent relative to the 314 *m/z* peak as the concentration of CBD decreased in our standards. The peaks rendered visible in lower ppm concentrations were still identified in the higher ppm standards, albeit with negligible abundance, suggesting none of our prepared standards were affected by external contamination. The MS generated from our samples was very consistent with the MS generated from our higher-concentration calibration standards, producing a prominent 314 *m/z* peak, a smaller 310 *m/z* peak, and no other visible peaks. This suggests our extraction procedure succeeded in isolating CBD within our hemp buds.

Though initially planned, a calibration curve using the integration results from the GC–MS proved unviable. The peak area for the 150 ppm standard drastically exceeded the linear range by an even larger margin than the GC–FID. The 5 ppm standard also possessed a lower peak area than the 1 ppm standard, which violated what we would expect from both our prepared standards and the GC–FID results. Due to these inconsistencies and the presence of a linear calibration curve from GC–FID data, no calibration curve was generated from the GC–MS data.

16.18 Conclusions

While not every bud corresponded to the same expected ratio of CBD to bud mass, the results were on average consistent with the 20% content, but with considerable variability (plus or minus over 20% differences). This is significant because understanding the amount (dose) of CBD in a hemp bud is important in determining its effects in a medicinal or practical setting. CBD has many anti-inflammatory, pain-relieving, and other medicinal properties. Different hemp strains contain different amounts of CBD, which is important to consider in order to accurately dose it for therapeutic purposes.

References

Abdallah, I., Hammel, D., Stinchcomb, A., and Hassan, H. (2016). A fully validated LC-MS/MS method for simultaneous determinations of nicotine and its metabolite cotinine in human serum and its application to a pharmacokinetic study after using nicotine transdermal delivery systems with standard heat application in adult smokers. *Journal of Chromatography B* 1020: 67–77.

Abu-Awwad, A., Arafat, T., and Schmitz, O.J. (2016). Simultaneous determination of nicotine, cotinine, and nicotine N-oxide in human plasma, semen, and sperm by LC-Orbitrap MS. *Analytical and Bioanalytical Chemistry* 408: 6473–6481.

ACS (1967). https://pubs.acs.org/doi/pdf/10.1021/jf60070a007?casa_token=Og7Nn6Rl8eAAAAAA%3A_2ltSboSXxmbcwS4ZXZeIsHwou4wUQTDDUEzLF_RVrof50tffjHDU_eSKszChvfWPuOVq8E9NJ-IjuSq& (accessed 20 January 2024).

Andre, C.M., Hausman, J.F., and Guerriero, G. (2016). *Cannabis sativa*: the plant of the thousand and one molecules. *Frontiers in Plant Science* https://doi.org/10.3389/fpls.2016.00019.

Andujar-Ortiz, I., Moreno-Arribas, M.V., Martín-Álvarez, P.J., and Pozo-Bayón, M.A. (2009). Analytical performance of three commonly used extraction methods for the gas chromatography-mass spectrometry analysis of wine volatile compounds. *Journal of Chromatography. A* 1216(43): 7351–7357.

Approximate Mass Fraction of Gasoline. (n.d.). *Groundwater Management Review*. http://bcn.boulder.co.us/basin/waterworks/gasolinecomp.pdf (accessed 14 December 2016).

Baranauskaite, J., Marksa, M., Ivanauskas, L, et al. (2020). Development of extraction technique and GC/FID method for the analysis of cannabinoids in *Cannabis sativa* L. spp. *Santicha* (Hemp). *Phytochemical Analysis* 31: 516–521.

Benbrahim-Tallaa, L., Baan, R.A., Grosse, Y., et al. (2012). Carcinogenicity of diesel-engine and gasoline-engine exhausts and some nitroarenes. *The Lancet Oncology* 13(7): 663–664

Benowitz, N. (2009). Pharmacology of: addiction, smoking-induced disease, and therapeutics. *Annual Review of Pharmacology and Toxicology* 49: 57–71.

Boutou, S. and Chatonnet, P. (2007). Rapid headspace solid-phase microextraction/gas chromatographic/mass spectrometric assay for the quantitative determination of some of the main odorants causing off-flavours in wine. *Journal of Chromatography A* 1141(1): 1–9.

Brajenović, N., Karačonji, I.B., and Bulog, A. (2015). Evaluation of urinary Btex, Nicotine, and Cotinine as biomarkers of airborne pollutants in nonsmokers and smokers. *Journal of Toxicology and Environmental Health. Part A* 78(17): 1133–1136. https://doi.org/10.1080/15287394.2015.1066286

Bruton, T., Alboloushi, A., de la Garza, B., et al. (2010). *Contaminants of Emerging Concern in the Environment: Ecological and Human Health Considerations* 1048, 257–273. American Chemical Society.

Burger, P., Plainfossé, H., Brochet, X., et al. (2019). Extraction of natural fragrance ingredients: history overview and future trends. *Chemistry & Biodiversity* 16(10): e1900424. https://doi.org/10.1002/cbdv.201900424.

Ciolino, L.A., Ranieri, T.L., and Allison, M.T. (2018). Commercial cannabis consumer products part 1: GC–MS qualitative analysis of cannabis cannabinoids. *Forensic Science International* 289: 429–437.

Citti, C., Pacchetti, B., Vandelli, M.A., et al. (2018). Analysis of cannabinoids in commercial hemp seed oil and decarboxylation kinetics studies of cannabidiolic acid (CBDA). *Journal of Pharmaceutical and Biomedical Analysis* 149: 532–540.

Cleveland Clinic (2022). *Caffeine & Headaches: Treatment & Sources*. Available at: http://my.clevelandclinic.org/health/diseases_conditions/hic_Overview_of_Headaches_in_Adults/hic_Caffeine_and_Headache (accessed 04 May 2022).

Dell'Amore, C. (2009). *Cocaine on Money: Drug Found on 90% of U.S. Bills*. National Geographic. August 16.

Diesel and Gasoline Engine Exhausts and Some Nitroarenes. http://monographs.iarc.fr/ENG/Monographs/vol105/mono105.pdf (accessed 14 December 2016).

Dominguez, C., Guillen, D.A., and Barroso, C.G. (2002). Determination of volatile phenols in Fino sherry wines. *Analytica Chimica Acta* 458(1): 95–102.

Dunnivant, F. (2017). *Environmental Success Stories: Solving Major Ecological Problems and Confronting Climate Change*. Columbia University Press.

EPA (2004). *Amendment to the 1999 Captan Reregistration Eligibility Decision (RED)*. Washington, DC: U.S. Environmental Protection Agency, Office of Chemical Safety and Pollution Prevention, U.S. Government Printing Office.

EPA (2013). *Captan: Human Health Risk Scoping Document in Support of Registration Review*. Washington, DC: U.S. Environmental Protection Agency, Office of Chemical Safety and Pollution Prevention, U.S. Government Printing Office.

Formato, M., Crescente, G., Scognamiglio, M., et al. (2020). (–)-Cannabidiolic acid, a still overlooked bioactive compound: an introductory review and preliminary research. *Molecules* 25: 2638.

de Groot, A.C. and Schmidt, E. (2016). Essential oils, Part I: introduction. *Dermatitis* 27(2): 39–42. https://doi.org/10.1097/DER.0000000000000175.

Gurley, B.J., Murphy, T.P., Gul, W., et al. (2020). Content versus label claims in cannabidiol (CBD)-containing products obtained from commercial outlets in the state of Mississippi. *Journal of Dietary Supplements* 17(5): 599–607. https://doi.org/10.1080/19390211.2020.1766634.

Healthline: Gasoline and Health. (n.d.). http://www.healthline.com/health/gasoline#Overview1 (accessed 14 December 2016)

Hukkanen, J., Pleyton Jacob III, and Benowitz, N.L. (2005). Metabolism and disposition kinetics of nicotine, *Pharmacological Reviews* 57(1): 79–115. https://doi.org/10.1124/pr.57.1.3

IARC (2016). *Gasoline*. http://monographs.iarc.fr/ENG/Monographs/vol45/mono45-8.pdf (accessed 14 December 2016).

Jjunju, F.P.M., Giannoukos, S., Marshall, A., and Taylor, S. (2019). In-situ analysis of essential fragrant oils using a portable mass spectrometer. *International Journal of Analytical Chemistry* 2019: e1780190. https://doi.org/10.1155/2019/1780190.

Johansen, J.D. (2003). Fragrance contact allergy. *American Journal of Clinical Dermatology* 4(11): 789–798. https://doi.org/10.2165/00128071-200304110-00006.

Kolpin, D.W., Furlong, E.T., Meyer, M.T., et al. (2002). Pharmaceuticals, hormones, and other organic wastewater contaminants in U.S. streams, 1999–2000: a national reconnaissance. *Environmental Science & Technology* 36: 1202–1211.

Kreamer, D. and Stetzenbach, K. (1990). Development of a standard, pure-compound base gasoline mixture for use as a reference in field and laboratory experiments. *GWMR* 135–145. https://info.ngwa.org/GWOL/pdf/901155372.PDF (accessed 14 December 2016).

Kumar, R. and Tripathi, Y.C. (November 2011). file:///Users/auroraanderson/Downloads/Chapter-TrainingManual-GettingFragrancefromPlants.pdf.

Lapczynski, A., Bhatia, S.P., Foxenberg, R.J., et al. (2008). Fragrance material review on geraniol. *Food and Chemical Toxicology* 46(11 Supplement): S160–S170. https://doi.org/10.1016/j.fct.2008.06.048.

Lazarjani, M.P., Young, O., Kebede, L., and Seyfoddin, A. (2021). Processing and extraction methods of medicinal cannabis: a narrative review. *Journal of Cannabis Research* 3: 32.

LeGrega, M.D., Buckingham, P.L., and Evans, J.C. (1994). *Hazardous Waste Management*. NY: McGraw-Hill, Inc.

Letizia, C.S., Cocchiara, J., Lalko, J., and Api, A.M. (2003). Fragrance material review on linalool. *Food and Chemical Toxicology* 41(7): 943–964. https://doi.org/10.1016/S0278-6915(03)00015-2.

Li, S., Wen, J., He, B., Wang, J., et al. (2020 March). Occurrence of caffeine in the freshwater environment: implications for ecopharmacovigilance. *Environmental Pollution*. https://www.sciencedirect.com/science/article/pii/S0269749119365108 (accessed 04 May 2022).

Liu, Y., Liu H.Y., Li S.H., et al. (2022). *Cannabis sativa* bioactive compounds and their extraction, separation, purification, and identification technologies: an updated review. *Trends in Analytical Chemistry* 149: 0165–9936.

Loftfield, E., Freedman, N.D., Dodd, K.W., et al. (2016). *The Journal of Nutrition* 146(9): 1762–1768.

Matulyte, I., Marksa, M., Ivanauskas, L., et al. (2019). GC–MS analysis of the composition of the extracts and essential oil from myristica fragrans seeds using magnesium aluminometasilicate as excipient. *Molecules* 24(6): 1062. https://doi.org/10.3390/molecules24061062.

National Institute on Drug Abuse. (2019). *Cocaine*. NIDA. https://www.drugabuse.gov/publications/drugfacts/cocaine (accessed 03 May 2019).

Ortega-Heras, M., Gonzalez-SanJose, M.L., and Beltran, S. (2002). Aroma composition of wine studied by different extraction methods. *Analytica Chimica Acta* 458(1): 85–93.

OSHA. (n.d.). *Hexane*. http://bcn.boulder.co.us/basin/waterworks/gasolinecomp.pdf (accessed 14 December 2016).

Polaskova, P., Herszage, J., and Ebeler, S.E. (2008). Wine flavor: chemistry in a glass. *Chemical Society Reviews* 37(11): 2478–2489.

Price, S. and Price, L. (2012). *Aromatherapy for Health Professionals*, 4e. Edinburgh, UK: Elsevier Health Sciences.

Simpson, R.F. (1978). Aroma and compositional changes in wine with oxidation, storage and aging. *Vitis* 17(3): 274–287.

Tat, L., Comuzzo, P., Stolfo, I., and Battistutta, F. (2005). Optimization of wine headspace analysis by solid-phase microextraction capillary gas chromatography with mass spectrometric and flame ionization detection. *Food Chemistry*, 93(2): 361–369.

Tisserand, R. (2017). *New Survey Reveals Dangers of Not Diluting Essential Oils*. Tisserand Institute. https://tisserandinstitute.org/new-survey-reveals-dangers-of-not-diluting-essential-oils/ (accessed 03 May 2023).

Tisserand, R. and Young, R. (2014). *Essential Oil Safety*, 2e. Edinburgh, UK: Churchill Livingstone.

Travassos, A.R., Claes, L., Boey, L., et al. (2011). Non-fragrance allergens in specific cosmetic products. *Contact Dermatitis* 65(5): 276–285. https://doi.org/10.1111/j.1600-0536.2011.01968.x.

Uhl, M. (1997). Determination of drugs in hair using GC/MS/MS. In: *Forensic Science International* 84, pp. 281–294. https://doi.org/10.1016/S0379-0738(96)02072-5

USDL. (n.d.). *Toluene*. https://www.osha.gov/SLTC/toluene/exposure_limits.html (accessed 14 December 2016)

USEPA (2016). https://www.epa.gov/sites/default/files/2016-09/documents/captan.pdf (accessed 29 April 2024)

USEPA (2024a). https://www3.epa.gov/pesticides/chem_search/cleared_reviews/csr_PC-081301_19-Feb-86_081.pdf (accessed 20 January 2024)

USEPA (2024b). https://archive.epa.gov/pesticides/chemicalsearch/chemical/foia/web/pdf/060101/060101-041.pdf (accessed 20 January 2024)

Van Breemen, R.B., Muchiri, R.N., Bates, T.A., et al. (2022). Cannabinoids block cellular entry of SARS-CoV-2 and the emerging variants. *Journal of Natural Products* 85(1): 176–184. https://doi.org/10.1021/acs.jnatprod.1c00946.

Wei, B., Feng, J., Rehmani, I.J., et al. (2014). A high-throughput robotic sample preparation system and HPLC–MS/MS for measuring urinary anatabine, anabasine, nicotine and major nicotine metabolites. *Clinica Chimica Acta* 436: 290–297. https://doi.org/10.1016/j.cca.2014.06.012.

Weldegergis, B.T., Tredoux, A.G.J., and Crouch, A.M. (2007). Application of a headspace sorptive extraction method for the analysis of volatile components in South African Wines. *Journal of Agricultural and Food Chemistry* 55(21): 8696–8702.

Wilson, N. and Horrocks, J. (2008). Lessons from the removal of lead from gasoline for controlling other environmental pollutants: a case study from New Zealand. *Environmental Health* 7: 1.

Zhang, Z. and Pawliszyn, J. (1993). Headspace solid-phase microextraction. *Analytical Chemistry* 65(14): 1843–1852.

Zhang, C.-D., Hu, X.-Y., Wang, H.-S., and Yan, F. (2021). GC–MS analysis of essential oil extracted from Acori *Tatarinowii rhizoma*: an experiment in natural product analysis. *Journal of Chemical Education* 98(9): 3004–3010. https://doi.org/10.1021/acs.jchemed.1c00451.

Further Reading

Addiction Resource. (n.d.). *Cocaine Dosage Guide for Medical and Recreational Purpose*. https://addictionresource.com/drugs/cocaine-and-crack/dosage/ (accessed 03 May 2019).

Al Bratty, M., Al-Rajab, A.J., Rehman, Z., et al. (2021). Fast and efficient removal of caffeine from water using dielectric barrier discharge. *Applied Water Science* 11: 97. https://doi.org/10.1007/s13201-021-01413-5.

Cardinali, F.L., Ashley, D.L., Wooten, J.V., et al. (2000). The use of solid-phase microextraction in conjunction with a benchtop quadrupole mass spectrometer for the analysis of volatile organic compounds in human blood at the low parts-per-trillion level. *Journal of Chromatographic Science* 38(2): 49–54.

Cleveland Clinic (2023). *11 Essential Oils: Their Benefits and How To Use Them*. https://health.clevelandclinic.org/essential-oils-101-do-they-work-how-do-you-use-them/ (accessed 03 May 2023).

Evrard, I., Legleye, S., and Cadet-Tairou, A. (2010). Composition, purity and perceived quality of street cocaine in France. *International Journal of Drug Policy* 21: 399–406.

Fan, S., Chang, J., Zong, Y., et al. (2018). GC–MS analysis of the composition of the essential oil from *Dendranthema indicum* var. aromaticum using three extraction methods and two columns. *Molecules* 23(3): 576. https://doi.org/10.3390/molecules23030576.

Felman, A. (2018). Nicotine: facts, effects, and addiction. In: Deborah Weatherspoon, *Medical News Today*, MediLexicon International, 11 Jan, www.medicalnewstoday.com/articles/240820.php.

Gholap, V., Kosmider, L., and Halquist, M. (2018). A standardized approach to quantitative analysis of nicotine in e-liquids based on peak purity criteria using high-performance liquid chromatography. *Journal of Analytical Methods in Chemistry* 2018: 1720375.

Harris, D.C. (1999). *Quantitative Chemical Analysis*, 5e. NY: Freeman and Company.

Klaschka, U. (2016). Natural personal care products—analysis of ingredient lists and legal situation. *Environmental Sciences Europe* 28(1): 8. https://doi.org/10.1186/s12302-016-0076-7.

Oyler, J., Darwin, W.D., and Cone, E.J. (1996). Cocaine contamination of United States paper currency. *Journal of Analytical Toxicology* 20(4): 213–216.

Robert, P. and Belardi, J.B.P. (1989). Application of chemically modified fused silica fibres in the extraction of organics from water matrix samples and the rapid transfer to capillary columns. *Water Pollution Research Journal of Canada* 24: 179–191.

Skoog, D.A., Holler, F.J., and Nieman, T.A. (1998). *Principles of Instrumental Analysis*, 5e. Philadelphia, PA: Harcourt Brace College Publishing.

USGS (2022). USGS-OFR-02-94 Table 4, Method 3. Available at: http://toxics.usgs.gov/pubs/OFR-02-94/table4.html (accessed 04 May 2022).

Zuo, Y., Zhang, K., Wu, J., et al. (2008). An accurate and nondestructive GC method for determination of cocaine on US paper currency. *Journal of Separation Science* 31(13): 2444–2450.

Index

Essential Methods of Instrumental Analysis, First Edition. Frank M. Dunnivant and Jake W. Ginsbach.

Companion Website: www.wiley.com/go/essentialmethodsofinstrumentalanalysis1e